Lecture Notes in Electrical Engineering

Volume 1038

The book series *Lecture Notes in Electrical Engineering* (LNEE) publishes the latest developments in Electrical Engineering—quickly, informally and in high quality. While original research reported in proceedings and monographs has traditionally formed the core of LNEE, we also encourage authors to submit books devoted to supporting student education and professional training in the various fields and applications areas of electrical engineering. The series cover classical and emerging topics concerning:

- Communication Engineering, Information Theory and Networks
- Electronics Engineering and Microelectronics
- Signal, Image and Speech Processing
- Wireless and Mobile Communication
- Circuits and Systems
- Energy Systems, Power Electronics and Electrical Machines
- Electro-optical Engineering
- Instrumentation Engineering
- Avionics Engineering
- Control Systems
- Internet-of-Things and Cybersecurity
- Biomedical Devices, MEMS and NEMS

For general information about this book series, comments or suggestions, please contact leontina.dicecco@springer.com.

To submit a proposal or request further information, please contact the Publishing Editor in your country:

China

Jasmine Dou, Editor (jasmine.dou@springer.com)

India, Japan, Rest of Asia

Swati Meherishi, Editorial Director (Swati.Meherishi@springer.com)

Southeast Asia, Australia, New Zealand

Ramesh Nath Premnath, Editor (ramesh.premnath@springernature.com)

USA, Canada

Michael Luby, Senior Editor (michael.luby@springer.com)

All other Countries

Leontina Di Cecco, Senior Editor (leontina.dicecco@springer.com)

** This series is indexed by EI Compendex and Scopus databases. **

Amit Kumar · Vinit Kumar Gunjan · Yu-Chen Hu ·
Sabrina Senatore
Editors

Proceedings of the 4th International Conference on Data Science, Machine Learning and Applications

ICDSMLA 2022, 26–27 December,
Hyderabad, India

 Springer

Editors
Amit Kumar
BioAxis DNA Research Centre Private Ltd.
Hyderabad, Telangana, India

Yu-Chen Hu
Providence University
Taichung, Taiwan

Vinit Kumar Gunjan
Department of Computer Science
and Engineering
CMR Institute of Technology
Hyderabad, Telangana, India

Sabrina Senatore
Department of Computer Engineering,
Electrical Engineering and Applied
Mathematics
University of Salerno
Fisciano, Salerno, Italy

ISSN 1876-1100   ISSN 1876-1119 (electronic)
Lecture Notes in Electrical Engineering
ISBN 978-981-99-2057-0   ISBN 978-981-99-2058-7 (eBook)
https://doi.org/10.1007/978-981-99-2058-7

This Springer imprint is published by the registered company Springer Nature Singapore Pte Ltd.
The registered company address is: 152 Beach Road, #21-01/04 Gateway East, Singapore 189721,
Singapore

Contents

Social Media + Machine Learning to Offer Clues on Suicide Ideation Concerns

Lakshmi Prayaga, Chandra Prayaga, and Amrutha Gunuru

Abstract Mental health concerns including suicide ideation are a growing concern especially among younger population. Social media also is providing a platform for people with mental health concerns to vent their frustrations and other psychological mental health concerns including suicide ideation. This research is a study on the application of topic modeling, a machine learning algorithm on big data gathered from social media to discover clues that may play a role resulting in adverse outcomes such as suicide ideation.

Keywords Social media Clues · Machine learning · Tweets · Suicide ideation · Topic modeling · Latent Dirichlet Allocation-LDA model

1 Introduction

Social media has become an integral part of our everyday lives, and its impact on our lives is extensive. For the people living in the digital world, negative criticism, mocking, or discrimination on social media tend to lead them into depression which ultimately may drive them toward drastic and unwanted outcomes including suicide ideation or suicide. Through this research, we attempt to decipher information from social media that can point to some signs associated with suicidal tendencies and can assist in detecting these early symptoms that can trigger interventions to prevent this tragic outcome. Latent Dirichlet allocation (LDA), an unsupervised machine

L. Prayaga (✉)
Department of Information Technology, University of West Florida, Pensacola, USA
e-mail: Lprayaga@uwf.edu

C. Prayaga
Physics Department, University of West Florida, Pensacola, USA

A. Gunuru
Information Technology, University of West Florida, Pensacola, USA

learning algorithm, was used in this study on a set of tweets to study signs of suicide ideation. We present the findings in this paper.

2 Literature Review

WHO (2020) [1, 2] reports that there were 800,000 deaths worldwide due to suicide. Suicidal tendencies start during adolescence and continue to grow during early adulthood and old age [3–5]. Suicide ideation is the wish to die, plans actions to die, playing and encouraging thoughts about dying [6, 7]. The pandemic of COVID-2019 has also aggravated the situation. Harmer et al. [6] used surveys to observe the impact of Covid on suicide ideation. This group of researchers reported that during the pandemic suicide ideation was high among "respondents aged 18–24 years (25.5%), minority racial/ethnic groups (Hispanic respondents [18.6%], non-Hispanic black [black] respondents [15.1%]), self-reported unpaid caregivers for adults (30.7%), and essential workers (21.7%)".

Though suicide is rising among young adults it is often not easy to pinpoint the causes and offer interventions. However, social media is a new platform where consumers share the good and the bad in their lives in a free format. Data from social media is also accessible to researchers and academicians to harvest, process, and analyze this data.

Data collected from social media is often called big data due to its veracity, volume, and variety. However, this data is unstructured and is a core property of big data. Big data is not easy to organize and analyze due to the size and the variety of data. In this context, machine learning techniques offer the processing power required to automate the process of data collection, data preparation, data analysis, and data visualizations for big data.

Czeiser et al. [8] conducted a survey to observe the impact of Covid on adults. Regression analysis in R was used for this study, and they reported that 5186 people participated in the study. 33% of the population reported anxiety or depression, 29.6% reported PTSD or trauma due to COVID-19, 15.1% reported increased substance use, and 11.9% reported having symptoms of suicide ideation. These symptoms were also more prevalent in younger adults than older age groups (> 65).

After an year of the outbreak of COVID-19, [3] a study on suicidality and COVID-19 in Dec 2021, discovered an increase in the suicide ideation among patients with Covid. The report suggests that prior to the pandemic suicide was a single point of concern but the pandemic with stress due to social isolation, financial needs, depression, and limited healthcare options, made this a dual pandemic of Coronavirus and suicide since the impact of the pandemic is just not only the physical health but on mental health as well.

Suicide ideation and self-harm (SH) among combat veterans is also a concern. However, SI and SH are a result of a complex mix of variables. A recent study [9] used 738 surveys collected from combat veterans. These surveys contained 192 variables. Data collected included variables that were multifaceted and not just related to mental health or isolation which are usually flags for tendencies leading to SI or SH. In this context, the authors observed that machine learning was able to take ten of the variables that were not related to identifying SH or SI and yet detect the presence of SH or SI with a 75.3% accuracy. This study suggests that machine learning can be a good instrument to analyze large datasets and find predictors that can be used in SI SH risk assessment of patients.

Other researchers [10] observed that there was a strong correlation and confirmation on the association between depressive symptoms and social isolation with suicide ideation. Machine learning algorithms, namely random forest, K nearest neighbor, and neural networks, were used to study this relationship. The results also suggest that women had relatively higher suicidal tendencies 33.3%. It was determined that social influence has both direct and indirect relation to depression.

Another characteristic feature of suicide ideation is that suicidal thoughts may not necessarily be constant. In a study, researchers [6] observed that the characteristics of the suicide ideation fluctuate dramatically. Suicide ideation can be either active or passive. Active ideation denotes experiencing suicidal tendencies now and consciously plan in a specific way to inflict self-harm. However, passive ideation indicates the wish to die and without a specific plan to harm themselves. This study reported that 31% of the people who committed suicide are either an inpatient or outpatient who were treated within a year and about 57% had contact with the mental care professionals. They noticed the rate of clinically depressed is low when compared to the suicides related to new-onset of depression.

Research on suicide ideation using Reddit's users [11] has revealed that the suicide posts had a higher score than non-suicidal posts. Posts on authenticity, anxiety, mentality, and depression had a higher scale and perception, and attention and mind-thinking were on the lower scale. They evaluated the process by using machine learning and deep hybrid learning with an accuracy of 95%. From the results, they derive that the users committing suicide exhibit physiological or agitation.

Our contribution to this literature review is that we study data posted on social media, specifically Twitter to observe if a. findings from posts made on this platform are in line with findings from other forms of data collection such as surveys, electronic health records and b. if posts have any additions to commonly known flags for suicide ideation. The methodology used and results from our study are presented below.

3 Data and Methodology

To study the posts made on social media related to the topic of suicide, 89,519 tweets were gathered and analyzed using topic modeling. Tweets were gathered just by using the search phrase suicide. No other criteria were used for this preliminary study.

4   Data Preparation and Cleaning

The tweets collected were converted to a text document and cleaned for data processing. Data cleaning procedures used included removing retweets, stop words, punctuation marks, hyperlinks, leading blanks, and spaces.

5   Data Exploration with Word clouds

Word clouds are a good way to explore what is contained in a block of text. They are a form of visual representation that show words in various sizes based on the frequency of their occurrence in a block of text. The more the frequency of the word the larger is its size. Figure 1 is a word cloud from the 89,519 tweets that were collected. From this visualization, some of the words that stand out are end, everyone, student, count, iPhone, Android, and lost among many other words. These words can further be attributed to commonly associated sentiments related to suicide. Word clouds thus also provide the sentiments and thoughts of the audience related to a specific topic.

Fig. 1   Word cloud

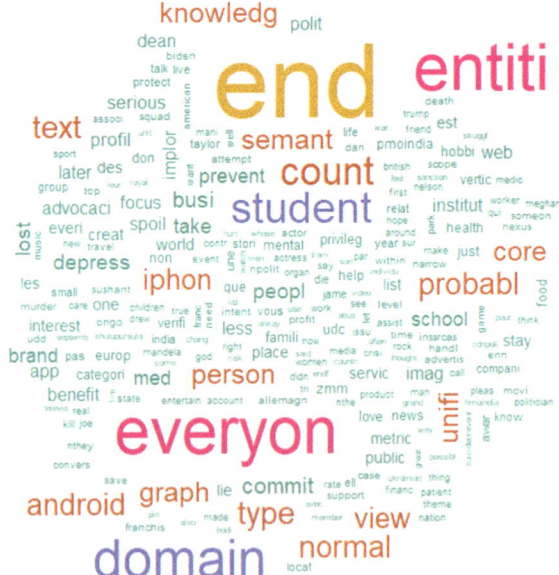

6 ML—Topic Modeling

To further analyze the sentiments of the audience and words related to a specific topic such as suicide ideation, topic modeling was used to identify major topics from the conversations of tweets. Topic modeling is an unsupervised machine learning algorithm that is used to group sets of words that make up topics from documents. Through topic modeling, we can detect the words and patterns which have similar expressions or meanings. Filtering the top 20 frequently used terms, we can derive other parameters related to the suicide ideation. Figure 2 represents the top 20 words from the body of text, and Fig. 3 shows the words that were used more than 6000 times in the document.

Figure 3 is a graphical representation of the words used more than six thousand times in the tweets collected related to suicide.

```
> freq[head(ord, n = 20)]
   suicid      end descript   entiti  everyon   domain
    69380    48704    41410    29739    29658    23995
 taxonomi  student    count     text     type   normal
    23208    18570    16644    12874    12645    12559
   probabl  android    iphon     view    graph knowledg
    12484    11855    11193    10938    10665    10638
      core   semant
    10620    10607
```

Fig. 2 Top 20 words

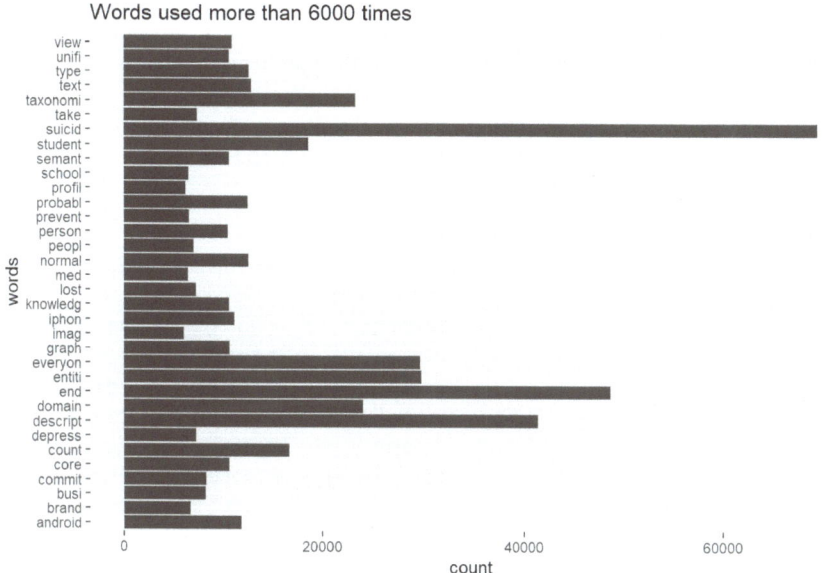

Fig. 3 Words used > 6000 instances

After ignoring some words that have a more semantic relevance than an emotional context, a set of words emerge from Fig. 3 that are of interest and candidates for further analysis. These words include student, end, depression, … At the preliminary level, these words also agree with other studies which identify "depression", "young adults/ students", thinking that the "end", is near as some common words related to suicide ideation.

7 Application of the Latent Dirichlet Allocation (LDA) Model

The LDA in topic modeling builds words per topic and topic per document. The LDA model was applied to the tweets dataset. The model identifies major topics and words that are present in each topic. Figure 4 displays the high frequency words in each topic. The number of topics (3 in this case) were chosen as a starting point number.

To determine the optimal number of topics for a given dataset, the following packages and functions were used. The Package is idatunuing, and the four functions to determine the number of topics are Griffiths2004, CaoJuan2009, Arun2010, and Deveaud2014. Results from applying these methods are presented in Figs. 5 and 6.

Figure 5 is a result of using the four methods in the ldatuning package to discover the optimal number of topics for this dataset. Figure 6 shows the two best applicable methods: CaoJuan2009 and Deveaud2014 methods to determine the optimal number of topics. These two methods were chosen as the most suitable methods are due to the fact that the two methods CaoJuan2009 and Deveaud2014 show a peak and a low at the number sixteen. The other two methods were not very informative for this dataset. Figure 7 is the results of using sixteen as the number of topics and extracting the top five terms per topic.

Figure 8 is a visual representation of the top ten terms from each topic. These words were chosen by the algorithm since they have a high beta value.

Fig. 4 Top words from three topics

```
> top10terms_3
          Topic 1      Topic 2    Topic 3
 [1,]  "descript"   "end"       "suicid"
 [2,]  "entiti"     "count"     "student"
 [3,]  "domain"     "text"      "everyon"
 [4,]  "taxonomi"   "probabl"   "iphon"
 [5,]  "view"       "everyon"   "take"
 [6,]  "graph"      "type"      "depress"
 [7,]  "knowledg"   "normal"    "lost"
 [8,]  "core"       "commit"    "school"
 [9,]  "semant"     "prevent"   "med"
[10,]  "unifi"      "profil"    "institut"
```

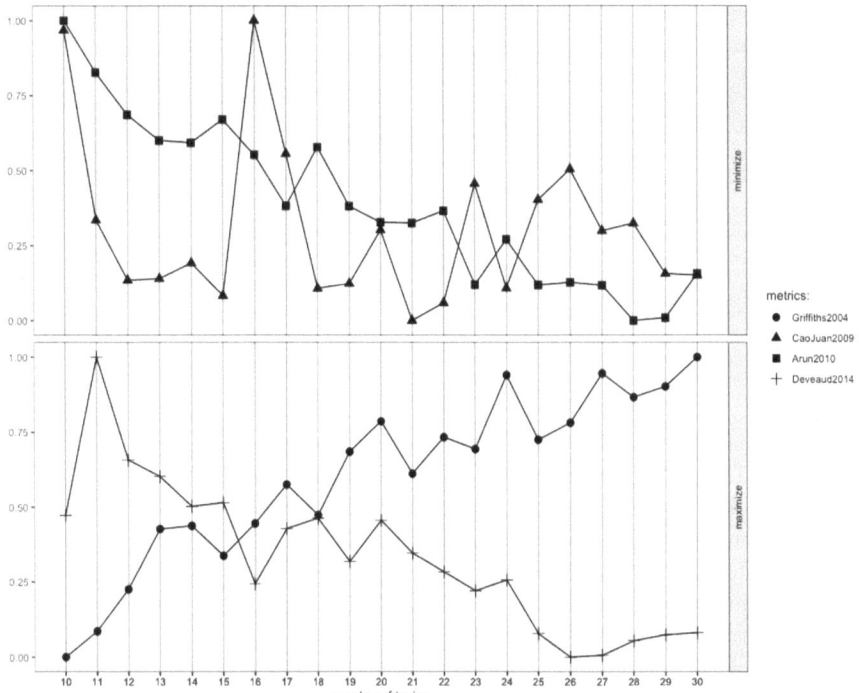

Fig. 5 Four methods to discover optimal topics

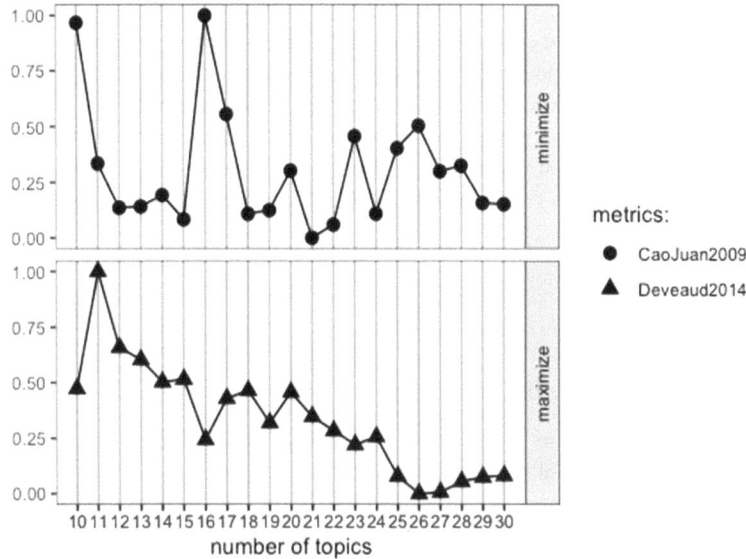

Fig. 6 CaoJuan2009 and Deveaud2014 methods to determine the optimal number of topics

	Topic 1	Topic 2	Topic 3	Topic 4	Topic 5	Topic 6	Topic 7	Topic 8
[1,]	"take"	"suicid"	"suicid"	"entiti"	"polit"	"stay"	"end"	"suicid"
[2,]	"depress"	"pmoindia"	"commit"	"descript"	"entiti"	"lie"	"everyon"	"que"
[3,]	"account"	"india"	"one"	"busi"	"world"	"suicid"	"iphon"	"someon"
[4,]	"school"	"nexus"	"die"	"domain"	"peopl"	"privileg"	"suicid"	"know"
[5,]	"suicid"	"death"	"don"	"interest"	"descript"	"zmm"	"love"	"une"
	Topic 9	Topic 10	Topic 11	Topic 12	Topic 13	Topic 14	Topic 15	Topic 16
[1,]	"end"	"les"	"student"	"app"	"count"	"android"	"suicid"	"descript"
[2,]	"text"	"est"	"suicid"	"web"	"end"	"suicid"	"prevent"	"taxonomi"
[3,]	"type"	"europ"	"med"	"end"	"profil"	"everyon"	"health"	"domain"
[4,]	"normal"	"des"	"school"	"everyon"	"imag"	"support"	"mental"	"entiti"
[5,]	"probabl"	"pas"	"lost"	"suicid"	"public"	"friend"	"help"	"view"

Fig. 7 Top five words from each topic

Beta values for each word are calculated to show the importance of each word in a topic. The higher the beta value the more important is the word in that topic. Figure 8 is a matrix with a sample of the beta value for words in each topic. The three columns, namely topic, term, and beta list the topic number to which the term belongs and its corresponding beta value. From this sample data, it can be noted that row 3 contains the words friend with a beta value of 1.40e-2 which thus is the most important word in topic number three (Fig. 9).

8 Gamma

Gama values represent the contribution or importance of a topic to the document. The higher the gama value, the more important is that topic in that document. Figure 10 is a sample of ten topics and their relative importance to the document.

As a final step of topic modeling on tweets, the top three terms from each topic were concatenated into string values to offer some descriptions for those topics. Figure 11 corresponds to these concatenated string values from each topic.

9 Discussion

The results from topic modeling provide the following observations.

1. Our research confirms prior work and observes that depression, anxiety, and loneliness are some of the main concerns in young adults that could be factors that are conducive toward suicidal ideation. Topics 1, 3 reflect these themes.
2. We also note other observations such as political and educational contexts are also playing a role in suicide ideation, possibly due to stress caused by these contexts. These would be good topics for further research. Topics 5, 11 suggest these inferences.

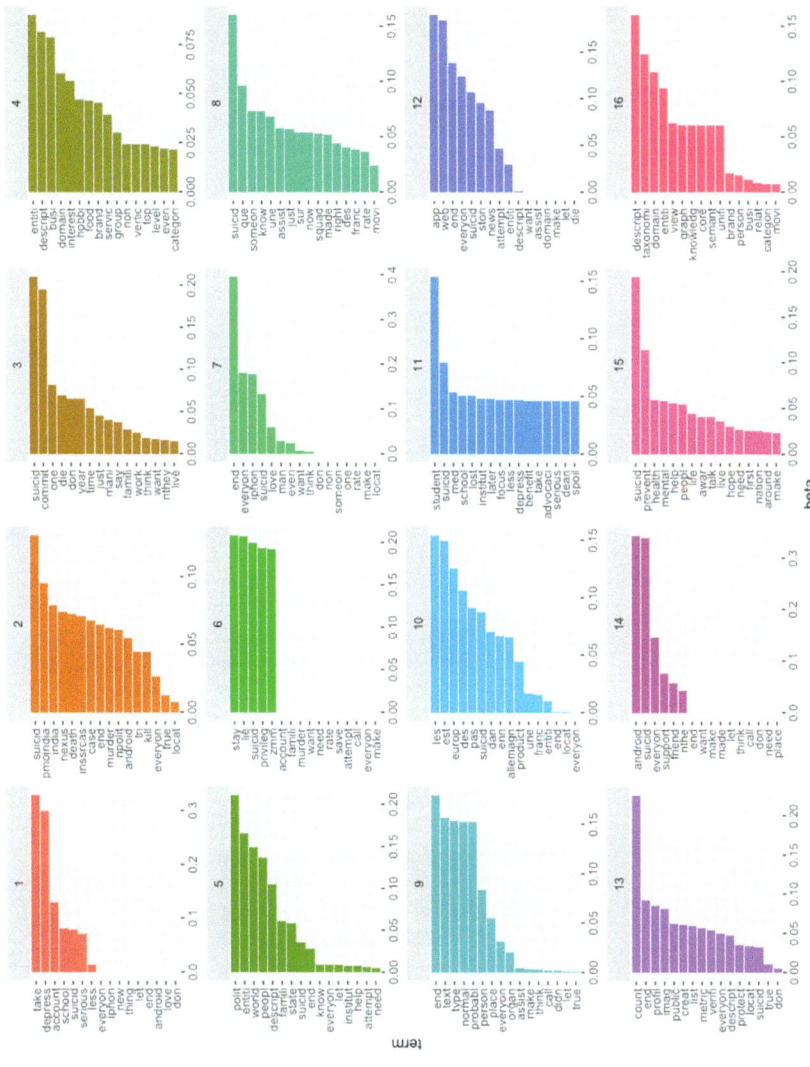

Fig. 8 Graphical representation of words from each topic

Fig. 9 Sample beta values for terms from different topics

```
> beta_topics
# A tibble: 324 × 3
   topic term        beta
   <int> <chr>      <dbl>
 1     1 friend 7.15e-14
 2     2 friend 2.89e-15
 3     3 friend 1.40e- 2
 4     4 friend 3.18e-23
 5     1 peopl  3.63e- 9
 6     2 peopl  6.72e-10
 7     3 peopl  4.90e- 2
 8     4 peopl  1.77e-35
 9     1 suicid 1.61e- 1
10     2 suicid 3.80e- 1
# … with 314 more rows
# i Use `print(n = ...)` to see more rows
> beta_top_terms <- beta_topics %>%
```

Fig. 10 Gama values per topic

```
> gamma_document
# A tibble: 232,792 × 3
   document topic   gamma
   <chr>    <int>   <dbl>
 1 2            1 0.0318
 2 3            1 0.0761
 3 4            1 0.0170
 4 5            1 0.00535
 5 6            1 0.0201
 6 7            1 0.0201
 7 8            1 0.342
 8 9            1 0.00535
 9 10           1 0.00535
10 11           1 0.0449
# … with 232,782 more rows
# i Use `print(n = ...)` to see more rows
```

Fig. 11 Concatenation of top three terms from each topic

```
"topic 1 NewTopicName: take depress account"
"topic 2 NewTopicName: suicid pmoindia india"
"topic 3 NewTopicName: suicid commit one"
"topic 4 NewTopicName: entiti descript busi"
"topic 5 NewTopicName: polit entiti world"
"topic 6 NewTopicName: stay lie suicid"
"topic 7 NewTopicName: end everyon iphon"
"topic 8 NewTopicName: suicid que someon"
"topic 9 NewTopicName: end text type"
"topic 10 NewTopicName: les est europ"
"topic 11 NewTopicName: student suicid med"
"topic 12 NewTopicName: app web end"
"topic 13 NewTopicName: count end profil"
"topic 14 NewTopicName: android suicid everyon"
"topic 15 NewTopicName: suicid prevent health"
"topic 16 NewTopicName: descript taxonomi domain"
```

3. Mobile devices such as iPhone and Android phones as noted in topics 7 and 14 are also being used a lot to suggest that people use these devices to post their feelings.
4. There is also discussion on help required for suicide ideation as shown in topic 15.

From this analysis it is can be argued that social media such as Twitter or Facebook offer a wealth of information on difficult topics such as suicide ideation and other mental health illnesses that academic researchers can glean information from. These platforms allow for free communication and some sort of anonymity that allows people to share their thoughts in an unguarded format that enriches the depth of information on the topics under discussion. It is hoped that advances in machine learning can analyze such natural language conversations making up a voluminous chunk of big data and provide informative takeaways to address challenges such as curbing suicide ideation and other mental health concerns.

10 Future Work

Limitations to our research are the number of tweets considered, since we only analyzed 89,519 tweets, the derivations or the conclusions may vary when applied to larger datasets. A future project is to compare the LDA models with other machine learning models and compare the accuracies of different models. A social research problem would also be to consider what factors in political and educational contexts are causing adverse mental health issues.

References

1. WHO. https://www.who.int/health-topics/suicide#tab=tab_1
2. WHO. https://www.who.int/news-room/fact-sheets/detail/suicide
3. Hawton K, Saunders KEA, O'Connor RC (2012) Self-harm and suicide in adolescents. Lancet 379:2373–2382. https://doi.org/10.1016/S0140-6736(12)60322-5
4. Klonsky ED, May AM, Saffer BY (2016) Suicide, suicide attempts, and suicidal ideation. Annu Rev Clin Psychol 12:307–330. https://doi.org/10.1146/annurev-clinpsy-021815-093204
5. Dendup T, Zhao Y, Dorji T, Phuntsho S (2020) Risk factors associated with suicidal ideation and suicide attempts in Bhutan: an analysis of the 2014 bhutan STEPS survey data. PLoS ONE 15:e0225888. https://doi.org/10.1371/journal.pone.0225888
6. Harmer B, Lee S, Duong TVH, Saadabadi A (2022) Suicidal ideation. In: StatPearls [Internet]. StatPearls Publishing, Treasure Island. PMID: 33351435
7. Morese R, Longobardi C (2020) Suicidal ideation in adolescence: a perspective view on the role of the ventromedial prefrontal cortex. Front Psychol 11:713. https://doi.org/10.3389/fpsyg.2020.00713
8. Czeisler MÉ, Lane RI, Wiley JF, Czeisler CA, Howard ME, Rajaratnam SMW (2021) Follow-up survey of US adult reports of mental health, substance use, and suicidal ideation during the COVID-19 pandemic, September 2020. JAMA Netw Open 4(2):e2037665. https://doi.org/10.1001/jamanetworkopen.2020.37665. https://jamanetwork.com/journals/jamanetworkopen/article-abstract/2776559

9. Colic S, He JC, Richardson JD, St. Cyr K, Reilly JP, Hasey GM (2022) A machine learning approach to identification of self-harm and suicidal ideation among military and police Veterans. J Mil Veteran Fam Health. 8:56–67. https://doi.org/10.3138/jmvfh-2021-0035

10. Kim S, Lee K (2022) The effectiveness of predicting suicidal ideation through depressive symptoms and social isolation using machine learning techniques. J Pers Med 12:516. https://doi.org/10.3390/jpm12040516

11. Yeskuatov E, Chua SL, Foo LK (2022) Leveraging reddit for suicidal ideation detection: a review of machine learning and natural language processing techniques. Int J Environ Res Public Health 19(16):10347. https://doi.org/10.3390/ijerph191610347.PMID:36011981; PMCID:PMC9407719

Opinion Mining-Based Fake Review Detection Using Deep Learning Technique

Koustav Pal, Sayan Poddar, S. L. Jayalakshmi⑩, Madhumita Choudhury, S. K. Saif Ahmed, and Soumyajit Halder

Abstract Recently, the field of text mining gained more attention due to its enormous opportunities and challenging problems in the exponential growth of unstructured textual data. People frequently express their thoughts in complicated ways, making automated labelling of textual data challenging. The usage of mislabelled data sets reduced the efficiency of automated labelling tasks. In this paper, we proposed an opinion mining-based fake review detection using deep learning technique to express word semantic sentiment. SentiWordNet and WordNet lemmas are used to retrieve word synsets and sentiment scores. The Amazon shoes review data set is used for implementation. The proposed model achieved 92% accuracy for the Amazon shoes review data set.

Keywords Opinion mining · Sentiment analysis · Attention mechanism · Amazon review analysis · Deep learning · Lemmatization · Long short-term memory (LSTM)

1 Introduction

Recent social media growth has made it possible for users to publish opinions about things, people, events, and subjects in a range of formal and informal ways. Reviews, forums, social media posts, blogs, and discussion boards are a few examples of these types of environments. The computational analytics involved with such text is referred to as the problem of opinion mining and sentiment analysis [1]. Enhancing customer happiness, boosting conversions, and improving revenue are the main goals of collecting customer reviews. Regularly updated customer reviews attract more people to the website. Online customer reviews are a crucial element of the purchasing process for e-commerce sites. A product's page will have a greater chance of getting

K. Pal · S. Poddar · S. L. Jayalakshmi (✉) · M. Choudhury · S. K. Saif Ahmed · S. Halder
Department of Computer Science, Pondicherry University, Puducherry, India
e-mail: sathishjayalakshmi02@pondiuni.ac.in

A. Kumar et al. (eds.), *Proceedings of the 4th International Conference on Data Science, Machine Learning and Applications*, Lecture Notes in Electrical Engineering 1038, https://doi.org/10.1007/978-981-99-2058-7_2

more traffic and a higher click-through rate on search engine result pages if it has an acceptable rating [2].

Many positive reviews on the platform increase the chances of encouraging clients to spend extra on services/products. Favourable reviews will increase client trust in the brand. Negative reviews may damage the company's reputation, credibility, and reliability. Numerous buyers are hesitant to buy from firms with no or many unfavourable evaluations [3]. Customers are 86% less likely to buy from companies with poor ratings. A bad review may turn off 22% of buyers, and the unfavourable reviews can turn off 59%. As a result, any poor e-commerce product reviews might harm the brand and cause a drop in sales [4]. It is time-consuming and inaccurate to classify positive and negative reviews manually; as a result, the development of automatic methods to identify the sentiment of thoughts has become essential. The majority of the opinion mining-based approach for false review detection has used conventional machine learning methods using the term frequency-inverse document frequency (TD-IDF) values, and subsequently, the discrimination of linguistic features is also estimated to enhance the accuracy of the system [5, 6]. Recently the attention-based deep model with convolutional neural network (CNN) is used for sentimental analysis [7].

This work uses both sentiment analysis and opinion mining terms interchangeably. This has been accomplished using an attention-based technique. In this work, we describe a deep learning model that automatically learns representations and features for the classification of reviews using an attention-based mining approach. We develop a pre-processing step that conducts fundamental feature extraction to decrease the number of model inputs. When compared with the conventional methods, the proposed model produced better results. This work is organized as follows: Sect. 2 reviews the related work on the classification of customer reviews. The proposed model is then described in Sect. 3. Experimental result and discussions are presented in Sect. 4. Finally, Sect. 5 presents conclusion.

2 Literature Review

Sedighi et al. [8] proposed a deep neural networks approach combined with an attention mechanism to perform classification by constructing a model to learn and capture the discriminating characteristics automatically. Their model's evaluation reveals that it performs noticeably better than conventional models and does not require any hand feature engineering. Kudakwashe et al. [9] proposed a concept called sentiment polarity, and it extracts the information about hotel services from reviews to automatically construct a sentiment data set for testing and training. A comparison analysis with Naive Bayes multinomial, minimal sequential optimisation, complement Naive Bayes, and Composite hypercubes on iterated random projections was performed to determine an effective machine learning method for the framework's classification component.

Using CNN and two different bidirectional RNN networks, the authors of [10] created an attention-based sentiment analysis. We first employ a pre-processor to enhance the data quality by removing noisy data and correcting mistakes. In order to reduce feature dimensionality and retrieve contextual information, our model employs CNN with max-pooling. Thirdly, to capture long-term dependencies, two separate bidirectional RNNs, namely long short-term memory and gated recurrent unit, are used. Sagnika et al. [11] proposed a model that cleverly combines long short-term memory (LSTM) with a CNN model. Modern deep learning models like CNN and LSTM can effectively analyse textual material and spot innate relationships and patterns with different degrees of abstraction. The suggested approach creates an ensemble model that incorporates the advantages of the two models. The addition of an attention network improves the model's effectiveness.

3 Opinion Mining-Based Fake Review Detection Using Deep Learning Technique

In this paper, we introduced an opinion mining-based fake review detection using deep learning technique that concatenates group of lexicons, a model for Word2vec, and an attention mechanism (AM) to express sentiment observation. AM is used to create sentiment data from SentiWordNet and the Liu's lexicon. However, the context of the text is first determined using the Word2vec model before analysis. These two tasks work together to handle the linguistic problems using the deep learning model. Before generating the text's sentiment features from the context, it attempts to comprehend the words in terms of their sentiment conflict. This process simulates how a person could approach this problem. As shown in Fig. 1, the system is divided into three primary components, and the following sections provide descriptions of each component.

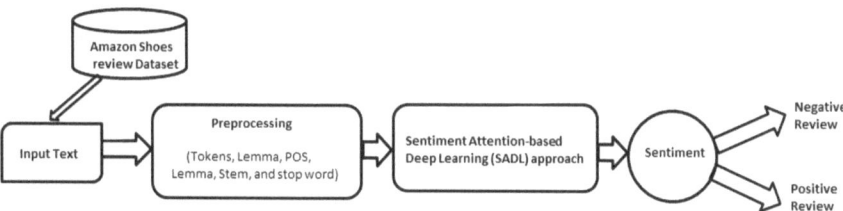

Fig. 1 Architecture of the proposed system

3.1 Text Pre-processing

In sentiment analysis, text pre-processing is one of the important step for analysing the information from social networks, this is an important task in sentiment analysis. Our pre-processing method is composed of four steps: (i) Text extraction; (ii) Text cleaning; (iii) Lemmatizing; and (iv) Negation marking. Language grammar and structure are used in this NLP-based method. By word lemmatizing with WordNet, downstream analysis tasks that involve matching terms from the text with those from lexicons are made simple.

Text extraction

In the text extraction, few sentences in the reviews are first divided into clauses based on the punctuation that is available at the clause level.

Text cleaning

This will change the capital letters to lowercase and eliminate any special characters.

Lemmatization

By using this technique, the text is changed into its stemmed form. Recent methods in the literature frequently apply an intelligent approach to collapse various word templates by seeking to remove affixes. Suffixes ending with -es (or) -s can make a noun singular or plural. For a verb to be in its present or past participle form, apply the suffixes -ing or -ed. The -est suffix allows adjectives to take comparative or superlative forms. SentiWordNet searches for synsets of a particular word are, however, impractical because of the erroneous stemmed forms produced by these algorithms. In this work, we adopted the methodology from [12], in which SentiWordNet and WordNet lemma were used to retrieve word synsets and sentiment scores.

3.2 Embedding of Attention Word

One of the most important developments in deep learning research is attention mechanism (AM). According to word sentiment scores, the proposed technique uses lexicons and AM to teach students how to concentrate on the key words in a phrase. The proposed method considers the word representations in the forms of either semantics or sentiment by using of both lexicons and Word2vec models. SentiWordNet and the Liu's lexicon are used to extract word sentiment information. To retrieve the sentiment information from the lexicons, the sentiment score calculation method is used.

3.3 Components of Opinion Mining-Based Fake Review Detection Using Deep Learning Technique Approach

Dropout

It was suggested to use dropout as a regularization strategy for neural network (NN) classifiers. When training, dropout method picks neurons at random that are not used. The activation of downstream neurons is not influenced by these neurons during the forward phase, and no updation of weight is made to the neurons during the back propagation step. The most common method of resolving overfitting is dropout. The proposed approach is also useful for handling low priority levels.

Bi-directional LSTM

The LSTM, a particular gated RNN, was introduced to handle the issue of handling long-term reliance. Fundamental building blocks of an LSTM network architecture are input gates, output gates, forget gates, and memory cells. The memory cell travels immediately through the entire chain to store the information for either long or short periods of time. What data should be removed from the cell is determined by the forget gate. Which fresh information will be added to the cell's memory is decided by the input gate. The output gate controls how much data the LSTM produced. The proposed system includes a BiLSTM to combine more data in the form of sentiment and semantics.

Dense layer

In the proposed system, the output is categorized into positive and negative opinions using a fully connected dense layer with a SoftMax function.

4 Experimental Studies

4.1 Data set Used

An Amazon shoe review [13] data set served as the data source for our project, and it is shown in Fig. 2. It reflects representative collections of consumer assessments, opinions, and variations for how a product is viewed in various geographic regions and the presence of promotional bias or intent in reviews. It contains the following columns: marketplace, which is the Web site's two-letter ISO 3166 country code. More than 100,000 consumer reviews are available as tab-separated values (TSV) files in the amazonreviews-PDS S3 bucket in the AWS US East Region. A specific review is indicated in the data file using with tab-delimited, with no quotes, and with escape characters.

A seemingly random number is allocated to a single author is called a customer ID. Each review is identified by using an unique review ID. The product ID identifies

Unnamed: 0	customerid	reviewid	productid	productparent	producttitle	rating	verifiedpurchase	reviewtitle	Review body
804898	2361047	25929976	R3TFM0HVOOS20D	B000L0O09G	281605411	Merrell Men's Moab Ventilator Hiking Shoe	5	Y	Great Hiking Shoe	I bought these shoes for a trailing hiking vac...
765165	1825603	45076746	R213KXZ4P5TE3	B003UHU5T6	866537507	New Balance Men's MR860 Running Shoe	1	Y	Returned Shoe	New Balance shoe was advertised as 13B Shoe ...
743357	1170569	6639856	R2XBYVDERMEO0D	B00FDXRMRI	245930226	Jessica Simpson Women's Wintee Dress Pump	4	Y	Four Stars	Great shoe 2014-12-28
105749	898438	18849452	R1JBBHTVGQ9YOG	B008G3BC7I	398112824	RYKA Women's Influence Cross Training Shoe	5	Y	Five Stars	Perfect fit!! Getting much use in Zumba class!
610266	1481949	43263299	R1C50G9Q895L6W	B000G615UC	234909654	Bates Men's Ulta-lites 8 Inches Tactical Sport..	5	Y	Five Stars	Great warm comfortable boots

Fig. 2 Description of Amazon shoe review data set

the review given for the particular product. Similarly, reviews for the same product written in various languages can be grouped using the same product ID in multilingual data set. The name of the item is the product title. Product category is a big product category used to group reviews. Rating is having the values from 1 (lowest) to 5 (highest). The number of constructive votes indicates the total number of votes the review has gotten. This review was inspired by the Vine programme. Next, verified purchase tells that a review is given after a real purchase. Review heading contains the title of the review, review body shows the review's text, and the review date records the date the review was written.

4.2 Performance Analysis

Figure 3 shows the various performance measures of the proposed system. The '0' represents the negative reviews, and '1' illustrates the positive reviews. It can be observed from Fig. 4 that positive reviews are clearly identified from negative (fake) reviews with the help of the proposed deep learning-based approach. The proposed model achieved the $F1$-score as 95% and accuracy as 92%. The graph showed increased accuracy for the epoch when the model was run on training and validation data sets. The accuracy on the Y-axis is given on a scale of 0–1. Figure 5 shows the accuracy with which the model classified the reviews positively and negatively on the training and validation data set. The accuracy also increased with the increasing number of the epoch. It can be observed from Fig. 5 that the presence of some anomalies in the validation data set, increased the value of loss function. In future work, the data set can be further refined, and some useless attributes from the data set can be eliminated.

	precision	recall	f1-score	support
0	0.63	0.43	0.51	4096
1	0.94	0.97	0.95	36534
accuracy			0.92	40630
macro avg	0.79	0.70	0.73	40630
weighted avg	0.91	0.92	0.91	40630

Fig. 3 Performance measure

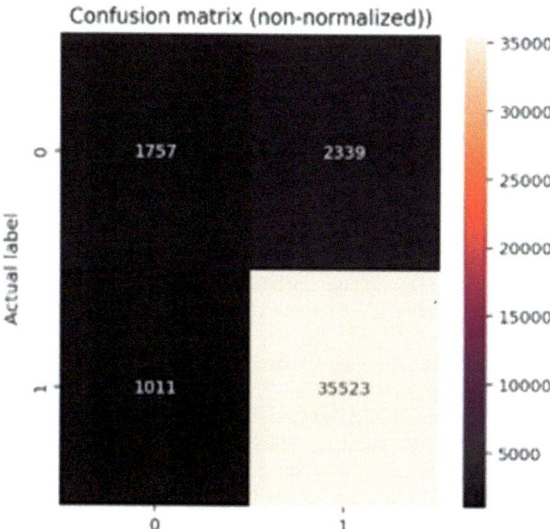

Fig. 4 Confusion matrix

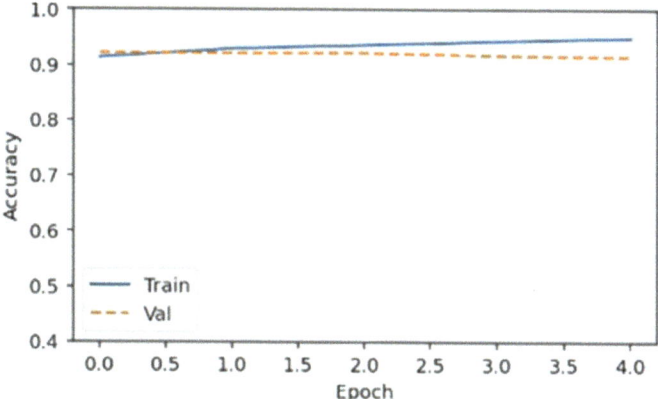

Fig. 5 Accuracy Vs Epoch

5 Conclusion

Recently, opinion mining-based approach gained more attention in the literature. In this paper, we proposed an opinion mining-based fake review detection using deep learning technique approach for detecting the fake customer reviews in an online Web site. For the same Amazon shoe review data set was used. The proposed approach used SentiWordNet and WordNet lemma to retrieve word synsets and sentiment scores. The implementation result shows that the proposed model achieved 92% accuracy. In future, the presence of anomalies in the validation data set can be further refined to improve the accuracy.

References

1. Aggarwal CC (2022) Opinion mining and sentiment analysis. Machine learning for text. Springer, Cham, pp 491–514
2. Anas SM, Kumari S (2021) Opinion mining based fake product review monitoring and removal system. In 2021 6th international conference on inventive computation technologies (ICICT-IEEE), pp 985–988
3. Sun S, Luo C, Chen J (2017) A review of natural language processing techniques for opinion mining systems. Inf Fusion 36:10–25
4. https://www.powerreviews.com/research/how-fake-reviews-destroy-consumer-trust/
5. Wankhade M, Rao ACS, Kulkarni C (2022) A survey on sentiment analysis methods, applications, and challenges. Artif Intell Rev 55:5731–5780
6. Arif SA, Hossain TB (2021) Opinion mining of customer reviews using supervised learning algorithms. In: 2021 5th international conference on electrical information and communication technology (EICT), pp 1–6
7. Kamyab M, Liu G, Rasool A, Adjeisah M (2022) ACR-SA: attention-based deep model through two-channel CNN and Bi-RNN for sentiment analysis. Peer J Comput Sci
8. Sedighi Z, Ebrahimpoor-Komleh H, Bagheri A, Kosseim L (2019) Opinion spam detection with attention-based neural networks. In: The thirty-second international flairs conference, 2019
9. Zvarevashe K, Olugbara O (2018) A framework for sentiment analysis with opinion mining of hotel reviews. In: 2018 conference on information communications technology and society (ICTAS)
10. Kamyab M, Liu G, Adjeisah M (2021) Attention-based CNN and Bi-LSTM model based on TF-IDF and glove word embedding for sentiment analysis. Appl Sci 11(23):11255
11. Sagnika S, Mishra BSP, Meher SK (2021) An attention-based CNN-LSTM model for subjectivity detection in opinion-mining. Neural Comput Appl 33(24):17425–17438
12. Husnain M, Missen MMS, Akhtar N, Coustaty M, Mumtaz S, Prasath VB (2021) A systematic study on the role of SentiWordNet in opinion mining. Front Comput Sci 15(4):1–19
13. https://s3.amazonaws.com/amazon-reviews-pds/tsv/amazonreviewsusShoesv100.tsv.gz

Analysis of the SEER Data set for Lung Cancer Diagnosis for Stage Classification and Survival Analysis

V. Deepa and S. K. B. Sangeetha

Abstract The basic usage of survival analysis is to perform a medical observation and the amount of time needed for the analysis. Survival analysis is analysed mainly in the field of biological engineering. The tumours are diagnosed using a density function. Our study involves the evaluation of the parametric analysis of the probability survival models. Our proposed work is to develop models for the stage classification of lung cancer using the region-based SEER data set. The mortality rate is calculated at the end of the cancer's progression. The classification challenge can be seen via the viewpoint of survival analysis. The survival statistics helps in analysing the cancer stages and the mortality rate.

Keywords Mortality rate · Survival analysis · Cancer stages

1 Introduction

When a patient is diagnosed with lung cancer, clinicians attempt to determine the spread of the disease which is termed as staging. The stages of cancer is required for radiologist to determine the spread of the disease throughout the body. The survival statistics is generated using the stages of cancer. The term "survival time" refers to the time between the diagnosis of an illness and the patient's death. At the end of the cancer's progression, when the disease is most aggressive, a survival analysis is projected.

V. Deepa (✉) · S. K. B. Sangeetha
Department of Computer Science and Engineering, SRM Institute of Science and Technology, Vadapalani Campus, Chennai 600026, India
e-mail: dv1019@srmist.edu.in

S. K. B. Sangeetha
e-mail: sangeets8@srmist.edu.in

1.1 Patients with Stage I Cancer

When a patient of over 50 years is diagnosed with stage I cancer, chemotherapy is done to reduce the survival rate of the patients. The treatment is mandatory to reduce the hazard. The hazard rate can be reduced with stage I cancer.

1.2 Patients with Stage II Cancer

The patient is given surgery and radiation when they are diagnosed with stage II cancer. When the age of the patient is more, the surgery and radiation is required at higher level to reduce the risk factor of the survival rate of the patient.

1.3 Patients with Stage III Cancer

When a patient below fifty age is diagnosed with stage III cancer, radiation is required. The radiation helps the patient to reduce the hazard. The survival rate reduces with the increase in age of the patients. The risk factor increases with age of the patient. The tumours spreads gradually if proper treatments are not given.

1.4 Patients with Stage IV Cancer

The patient is given cancer therapy and radiation when they are diagnosed with stage IV cancer. When the patient age increases, the mortality rate also increases. When the age of the patient is more, the surgery and radiation is required at higher level to reduce the risk factor of the survival rate of the patient. Lung cancer is divided into two stages, namely limited stage and the extended stage. A sponge like structure available in the chest area helps the breathing easier. The air enters through the mouth and the windpipe is called as alveoli. Lung cancer begins in the bronchial lining of the cells. Lung cancer is divided into three categories: localised, regional, and distant. The TNM stage is used to indicate the cancer stages such as tumour, nodule, and metastasis.

Table 1 Cancer stages and lymph node

S.no	Stages	Description
01	Stage I	The malignancy in the lungs has not migrated to the lymph nodes
02	Stage II	The malignancy has migrated to the lymph nodes adjacent
03	Stage III	The malignancy has migrated to the nearby lymph nodes denoted by III A,B
04	Stage IV	The fluid has spread around all parts of the organs which is called as advanced stage of lung cancer.

2 Risk Factors of Lung Cancer

when the patient is diagnosed with lung cancer the cancer cells are identified using the radiotherapy. The Lung cancer is classified into two types namely small cell and non-small cell adrenal lung cancer. The different symptoms of the lung cancer include fatigue, chest pain, loss of appetite, and shortness of breath. The CT scan was used to determine the disease's diagnosis. The scan helps in indicating the spread level of the tumour which helps the radiologist and medical analyst to examine the disease (Table 1).

3 Materials and Methods

3.1 Patient Population

The Seer data set has 18 cancer registries across United States. The data is extracted from the SEER*Stat version software. The SEER data set helps in analysing the TNM stages of the disease. The system helps in analysing the cases in the local, regional, and distant throughout the body. The different clinical characteristics under examination were the gender, age, year of diagnosis, surgery, and radiation therapy. A multivariate observation was done based on the different set of attributes. Cancer-specific survival was analysed for each sub-group using the Kaplan–Meier method.

4 Statistical Analysis

The SEER data contains different information on patient information such as primary tumour size, tumour morphology information, stage at diagnosis, and first course of treatment. The SEER registries keep track of cancer patients' survival rates. The analysis of the SEER data was done using the cohort selection. Cohort selection for our experiments is performed with SEER*Stat, a dedicated statistical software for

Fig. 1 Survey on the SEER
population

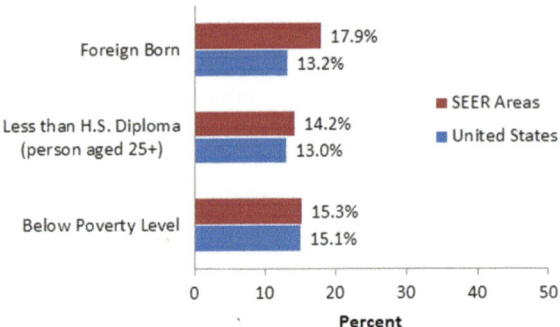

the analysis of the SEER data. SEER*Stat configuration can be stored into a session file and re-used by others for the cohort analysis data. The cohorts consist of 229,011 cases for lung cancer which is called as the SEER*Stat session files. The SEER data included the cases of the malignant tumours and the benign tumours. The American Survey is taken between the years 2016–2021.The analysis of the lung cancer data set was analysed based on the different set of cancer survival based on the each survival group (Fig. 1).

4.1 Tumour Analysis

The SEER data set consists of the lung cancer data which performs stage classification based on the size of the tumour. The nodules present in the contralateral lungs is called as M1a, and the tumours in the extra-thoracic organ are called as metastases (Fig. 2).

Fig. 2 Tumours in lung CT
image

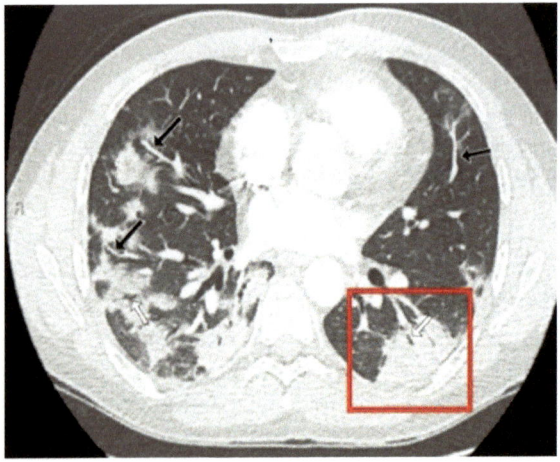

The SEER data indicates that the chemotherapy causes 10% increase in the survival rate. Two types of treatment are given namely chemotherapy and radiotherapy. These treatments reduce the risk of the disease. Surgery is the most effective treatment for patients with early deduction of the stages of lung cancer. Surgical removal of tumour is faster and more efficient method compared to the other type of treatments. Different kinds of survival analysis include clinical trials, cohort studies, and statistical analysis of the lung cancer patients.

5 Parametric Analysis

The survival analysis is used to predict the mortality rate of the patients. The different survival analysis metrics are parametric and non-parametric analysis. The Kaplan–Meier estimator is used to compute the mortality rate of the patients for a limited set of groups.

Table Data

Surveillance, epidemiology, and end results program: unique analyses and critical insights.

Analyses	Critical insights
1. Population-based cancer rates	1. Absolute risk of cancer occurrence
2. Rare cancer rates	2. Precise and comprehensive description
3. Cancer rates in minority groups	3. Healthy disparity assessments
4. Birth cohort effect	4. Risk factor exposure assessments
5. Calendar periodic effect	5. Benefits/harms of screening

6 Experimental Setup

6.1 Online Lung Cancer Outcome Calculator

An online tool is used to predict the lung cancer. The analysis of the tool was done using the five outcome variables to remove the redundant attributes. In order to compute the mortality rate, 13 variables are used.

The description of the variables is given as follows:

1. **Patient age**: The age of the patient is given as numeric value during the diagnosis of the lung cancer.
2. **Birth place**: The birth place of the patient is given as character value. There are totally 198 options available in the SEER database to select for the attribute.

3. **Cancer grade**: It is represented as attributes such as well-grown, poor, and well differentiable. It is the description of how the cancer cells grows from the initial level.
4. **Diagnostic confirmation**: The most commonly used attribute for the confirmation of the lung cancer is denoted by the laboratory test results such as positive histology and negative histology.
5. **Tumour extension**: The tumour spreads from the local region to the metastasis region is called as spread. There are 20 options available which is called as localised or the lymphatic region. The attribute name is represented as 'EOD extension'.
6. **Lymph node**: The most commonly used attribute for the lymph node is denoted by EOD lymph node. There are eight options available.
7. **Surgery Type**: It is the description of the technique used to remove the cancerous tissue from the lung. There are totally 25 options available in the SEER database to select for the attribute.
8. **No surgery attribute**: The reason should be represented as character type. The different options available are surgery performed (yes/no) and reasons.
9. **Surgery and radiation therapy**: It is denoted as the sequential procedure for the different operations such as surgery and radiation.
10. **Lymph node surgery**: It is the description of the surgical procedures used to remove the lymph nodes at the time of surgery and biopsy. There are totally eight options available in the SEER database to select for the attribute.
11. **Cancer stage**: The stage is denoted as the spread of cancer such as tumour, region analysis, and the detection of the disease spread.
12. **Malignant tumours**: It is denoted as the total number of tumours during the patient lifetime. It helps us in identifying the numeric, categorical tumours.
13. **Regional lymph nodes examination**: It is denoted as the total number of regional lymph nodes that were removed and examined by the pathologist. A total of the 63 attributes are available which provides a maximum accuracy of 91.4% (Fig. 3).

7 SEER Data Dictionary

The SEER data dictionary consists of different variables such as age, size, tumour size, T, N, M classifications. Age: The age is represented as a three-digit code for denoting the patient age in years.

Grade: The grades are represented in ranges ICD-O-2.The Grade 1 indicates the cell may look normal, and Grade 2 indicates abnormal growth of the cells. Tumour Size: The tumour size is measured in mm. The codes for representing the data are 991–995.

The SEER data dictionary uses the unlabelled data which uses data-driven process [1]. The most commonly used techniques are the clustering for lung cancer stage classification and prediction.

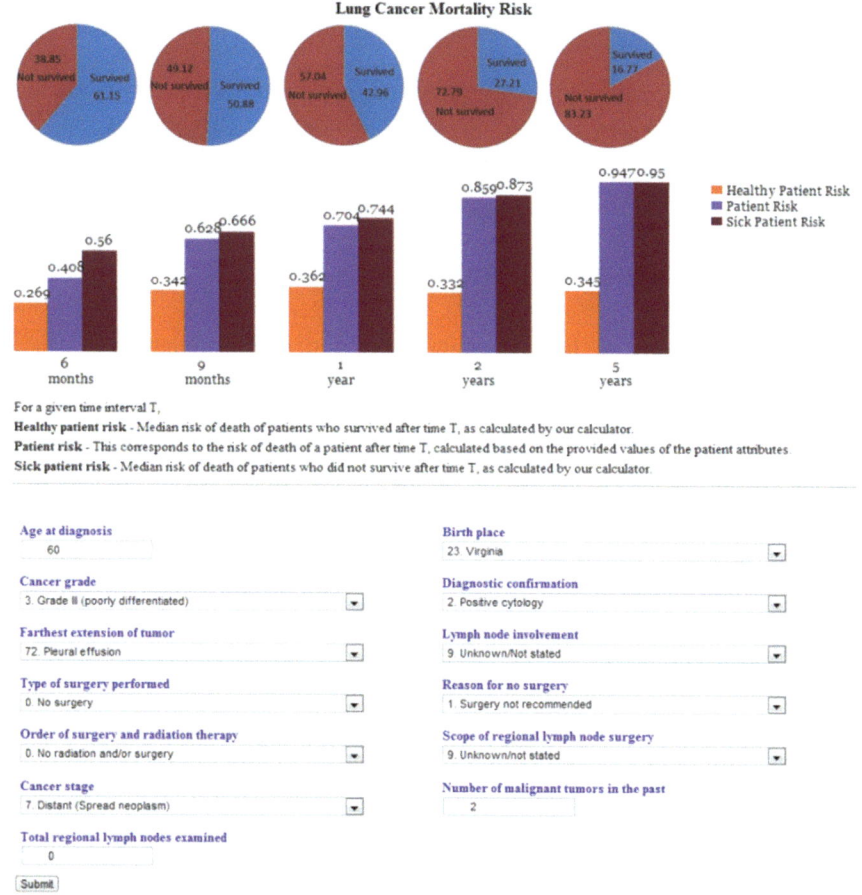

Fig. 3 Screenshot of the lung cancer outcome calculator

The ultimate goal of the SEER data model is the text classification which provides the best outcome for classification which operates on unlabelled data [2] for a given example.

The SEER data model refers the classification task having "true" or" false "value or "yes" or "no" [3].

The SEER data model is regarded as a class label for model prediction for lung cancer [4]. To predict the class for the given set of data points, it can be carried out in structures and unstructured data [5].

Another commonly used statistical model for lung cancer prediction and stage classification is the logistic regression (LR) [6].

7.1 Results

The SEER data helps in analysing the percentage of people affected with the lung cancer. It helps in calculating the mortality rate using the survival analysis. The statistical information can be computed from the cancer registry.

8 Conclusion and Future Work

The SEER population study has several limitations since it relies mainly on the cancer registry data. Our study of the SEER data also has several strengths. The patients are derived from population-based tumour registry, hence providing the accurate results. Because of the extensive data collected from the SEER program, we were able to analyse the cancer stage survival using the demographic information such as age, sex, and race. Lung cancer was one of the most spreading cancer type used for experimental analysis.

References

1. Magaji BA, Moy FM, Roslani AC, Law CW (20170 Survival rates and predictors of survival among colorectal cancer patients in a Malaysian tertiary hospital, pp 162–173
2. SEER* Stat Software. National Cancer Institute Surveillance, Epidemiology, and End Results Program (SEER). Available online: https://seer.cancer.gov/seerstat/. Accessed on 4 May 2018
3. Ung M, Rouquette I, Filleron T, Taillandy K, Brouchet L, Bennouna J et al (2016) Characteristics and clinical outcomes of sarcomatoid carcinoma of the lung. Clin Lung Cancer 234–245
4. Yendamuri S, Caty L, Pine M, Adem S, Bogner P, Miller A et al (2019) Outcomes of sarcomatoid carcinoma of the lung: a surveillance, epidemiology, and end results database analysis. Surgery 152(3):397–402
5. Imani F, Chen R, Tucker C, Yang H. Random forest mode ling for survival analysis of cancer recurrences, 9:183–192. Bentham publishers
6. Wongvibulsin S, Wu KC, Zeger SL (2020) Clinical risk prediction with random forests for survival, longitudinal, and multivariate (RF- SLAM) data analysis, 98:1–14
7. Pradeep KR, Naveen NC (2018) Lung cancer survivability prediction based on performance using classification techniques of support vector machines, C4.5 and Naive Bayes algorithms for healthcare analytics, 67:412–420
8. Nezhada MZ, Sadati N, Yanga K, Zhub D (2018) A deep active survival analysis approach for precision treatment recommendations: application of prostate cancer, 56:16–26
9. Fathima N, Liu L, Hong S, Ahmed H (2020) Prediction of breast cancer, comparative review of machine learning techniques, and their analysis, 8:173–174
10. Parikh RB, Manz C, Chivers, C, Regli SH (2019) Machine learning approaches to predict 6-month mortality among patients with cancer, 77, 1–7
11. Wongvibulsin S, Wu KC, Zeger SL (2020) Clinical risk prediction with random forests for survival, longitudinal, and multivariate (RF-SLAM) data analysis, 98:1–14
12. National Cancer Institute Surveillance, Epidemiology, and End Results Program (SEER) (2017) From electronic health-records, vol 56, Nov 2017, pp 37–56. Available online: https://seer.cancer.gov/. Accessed on 4 May 2018

13. Herbst RS, Morgensztern D, Boshoff C (2018) The biology and management of non-small cell lung cancer. Nature 553:446–454
14. Ettinger DS, Aisner DL, Wood DE et al (2018) NCCN guidelines insights: non-small cell lung cancer, version 5 2018. J Natl Compr Canc Netw 16:807–821
15. Roesel C, Terjung S, Weinreich G, Hager T, Chalvatzoulis E, Metzenmacher M et al (2016) Sarcomatoid carcinoma of the lung: a rare histological subtype of non-small cell lung cancer with a poor prognosis even at earlier tumour stages. Interact Cardiov Th 24(3):407–413

Data Analytics for Athlete Safety in Training

Chandra Prayaga, Lakshmi Prayaga, Aaron Wade, John Chamblee, and Kyle Rank

Abstract Data Analytics for Athlete Safety in Training (DFAST) is a system designed to improve performance and safety in athlete training. Wearable devices on the athlete send real-time data on movement, accelerations, rotations, heartrate, etc., to the system during workout. The system uses machine learning to analyze the data and send back real-time alerts to the athlete via the wearable device so that the athlete can correct posture and technique, thereby increasing safety.

Keywords Athlete training · Safety · Wearable devices · Machine learning

Physical fitness training is an important aspect of an athlete's overall performance. Athletic trainers (AT) engage with their clients and educate them on proper techniques to avoid or lessen the risk of injuries while training. The trainers also serve as medical responders to offer rehabilitation from injuries. The American Medical Association in fact recognizes athletic trainers as healthcare professionals who assist in the prevention, diagnosis, assessment, treatment, and rehabilitation of muscle and bone injuries and illnesses. Shanley et al. [10] describe AT as one who "is uniquely positioned to positively affect the overall health care of this population." Pike et al. [7] state that the presence of ATs is crucial in training athletes in secondary schools, given the levels of adolescents' participation in athletics and associated injuries. However, it is found that there is a major shortage of ATs, or they are just not being hired, which effects the quality of treatment and training received by the athletes.

C. Prayaga · A. Wade
Physics Department, University of West Florida, Pensacola, USA

L. Prayaga (✉)
Information Technology, University of West Florida, Pensacola, USA
e-mail: Lprayaga@uwf.edu

J. Chamblee
Cyber Security, University of West Florida, Pensacola, USA

K. Rank
Movement Sciences and Health, University of West Florida, Pensacola, USA

Post et al. [8] suggest that more than 53% of schools in CA did not employ ATs. Shanley et al. [10] report that athletic injuries among young adolescents account for almost 500,000 physician visits and over 50% of these injuries are preventable with proper athletic training. It is in this context that we present initial results of Data Analytics for Athlete Safety in Training (DFAST), a work in progress on a system which monitors athlete training to supplement or substitute ATs when they are not available and to provide feedback or training instructions.

1 Role of Wearables

Wearables are used by many to help track fitness progress with Q2 2021 showing a 34.4% increase in sales over the same quarter in 2020 [1]. In addition, ankle and waist wearables have been shown to be effective at monitoring the form while working out [2, 5, 6, 9]. Fuller et al. [3] showed that although wearables can be accurate in measuring steps and heart rate, the constant upgrading and redesigning to new models suggests the need for more current reviews and research.

The goal of DFAST is to use current technologies to provide a seamless interface for athletes to obtain feedback from their wearable devices on their training in real time. We use Internet of things (IoT) to establish the confluence between the wearable devices, sensors, and the cloud to collect data, analyze it, and provide meaningful feedback on their training and performance, so athletes can improve their performance and reduce the chances of injury. Hooren et al. [4] discuss the capabilities of existing current technologies, specifically wearables, in providing real-time feedback to assist in training athletes. Current wearables are equipped with sophisticated sensors that can be used to monitor several variables from the most basic vitals, such as heart rate, blood pressure, steps, speed, and quality of sleep, to more advanced data related to kinematics and mechanics of each task performed by an athlete. This kind of rich information obtained by the sensors impacts the quality of feedback that can be provided by the wearable device to the athlete in training. Hooren et al. [4] also suggest that it is this holistic feedback, which includes both physiological and psychological aspects such as motivation and personalization that will positively influence an athlete to improve performance and reduce injury.

In this paper, we describe our initial attempts to quantify data from specific workouts (Bicep Curls and Squats), so that an athlete may get real-time feedback from the system, if the performance deviates from a "baseline". The baseline itself will be derived, for each type of workout, by aggregating and processing data from several athletes, performing under the supervision of a coach or instructor. Presented in the rest of the paper is a description of the system architecture for DFAST, data collection, data cleaning, and initial experimental results for our attempts to quantify the quality of the workout.

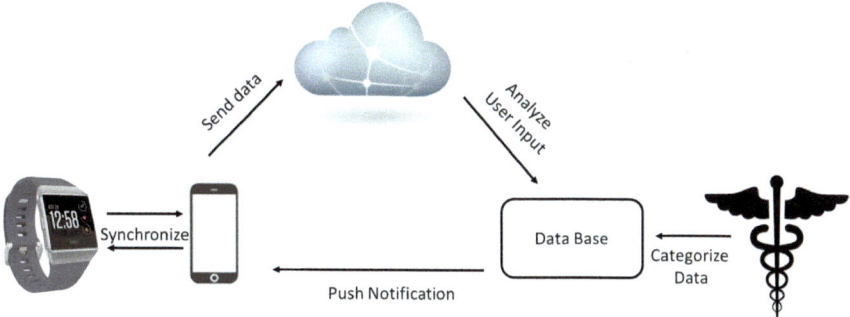

Fig. 1 Internet of things system architecture for DFAST

2 System Architecture

Figure 1 describes the architecture of DFAST. IoT is the underlying technology that connects several pieces of the application. Each individual athlete will use a mobile phone that is connected to a wearable device. The mobile phone + the wearable device, in this case a Fitbit, is used to allow the athlete to personalize his information and settings and collect data for individual activities. The data is transferred via IoT to the cloud for processing. Machine learning is used to check for the accuracy of each activity such as bicep curl, squats, etc., and provide real-time feedback to the athlete on the accuracy of the posture while training. The athlete can access this feedback on their wearable device and make appropriate changes per the feedback obtained improve performance and reduce the chances of injury.

3 Bicep Curls

Baseline exercise data was collected from six athletes and four non-athletes. Videos of each exercise were analyzed by a professional, and the data was categorized into correct or incorrect with subcategories for the type of error. Utilizing Fitbit watches, an application was designed to measure acceleration, rotation, orientation, and heart rate. With the measurements, the data is delivered from the watch to a centralized database stored in the cloud.

4 Feature Engineering and Data Cleaning

The athlete data is cast into a time series using a custom script that isolates each workout session. The new session is added to previous five. If there are no previous recorded sessions at the start of the first workout, blank regions are filled with zeroes.

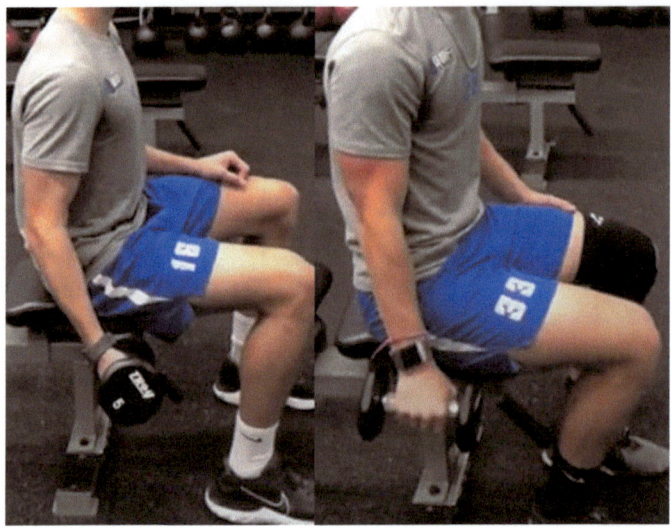

Fig. 2 Bicep curl at full extension. The form on the left is defined as good, and the one on the right as bad

Initially, the entire workout is randomly assigned the target label of good or bad, even if only segments of the workout are bad. One of the issues denoted in the bicep curls was the rotation of the wrist at full extension (Fig. 2). To save time, data points above horizontal were relabeled as good.

5 Methodology

The trained data consists mostly of good workouts with only a few bad workouts. A cosine similarity metric is used to determine how well a data point matches the good baseline workout data, and from this comparison, a performance rating is provided to the user. Records are matched with a label when the similarity score is greater than a threshold. The threshold ensures that irrelevant data does not influence the measurement. Matches are assigned a score based on the target label. If the record does not match any label, then it is assumed that the data point is bad and is assigned a value of zero. The scores are then averaged for the entire workout to get an overall rating for the workout session or averaged for each record to get a rating over time throughout the workout session.

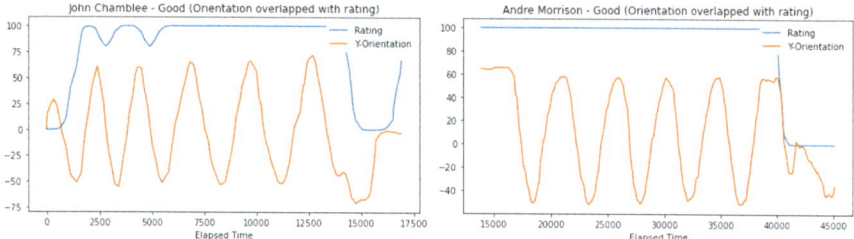

Fig. 3 Ratings for good bicep curls from the test set overlapped with the watch orientation

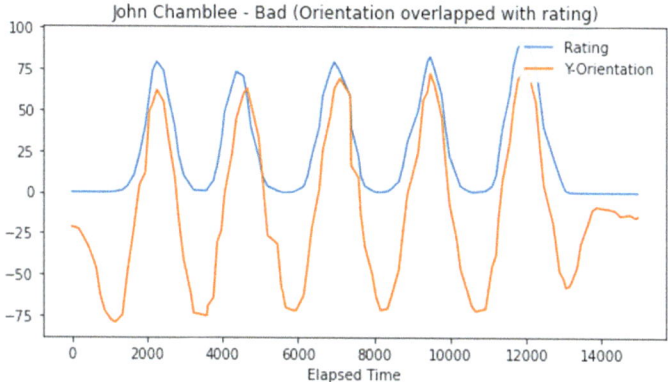

Fig. 4 Ratings for bad bicep curls from the test set overlapped with the watch orientation

6 Results

Figures 3 and 4 detail the rating over time of bicep curls with the wrist rotated at extension, as seen in Fig. 2. Because the wrist is not rotated at the peak of the bicep curls, it is rated higher at those points.

7 Squats

In the case of squats, data was collected for two separate workouts, one with "good form" and one in which the athlete was leaning too far forward. Fitbit data for both workouts show interesting features, which can be used to generate feedback to the athlete.

Figure 5 shows plots of acceleration (x, y, and z) readings for leaning forward too much (left) compared to keeping good form (right). We note that

1. Good form has majority of oscillations in z-direction. Learning forward shows oscillations in x-direction.

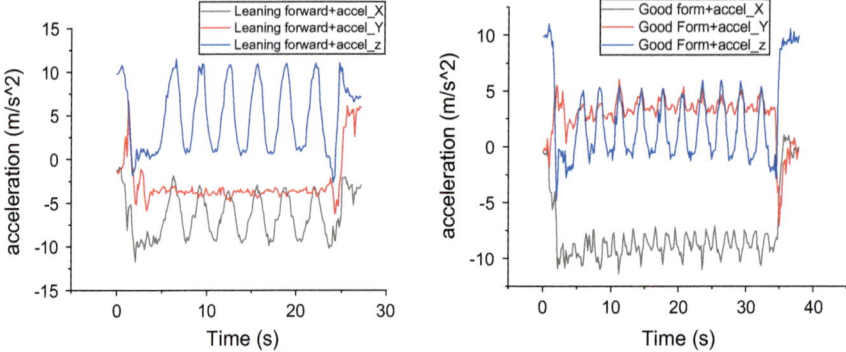

Fig. 5 *x*-, *y*-, and *z*- acceleration squat data (left) leaning forward, (right) good form

2. We can use the amplitude of the acceleration in the *x*-direction as an indicator that the person is leaning forward too much and provide feedback to adjust form.

Figure 6 shows a similar trend in the Euler orientations. The good form data has only small amplitude oscillations of the y-orientation compared to the leaning forward data.

As a test, we also performed FFT analysis of the rotation_y for both cases, for which the time series data is shown in Fig. 7.

FFT results of rotation about y-axis show that for the good form data, the y rotations show a narrow peak with small amplitude, but a much larger amplitude for the forward leaning case (Fig. 8).

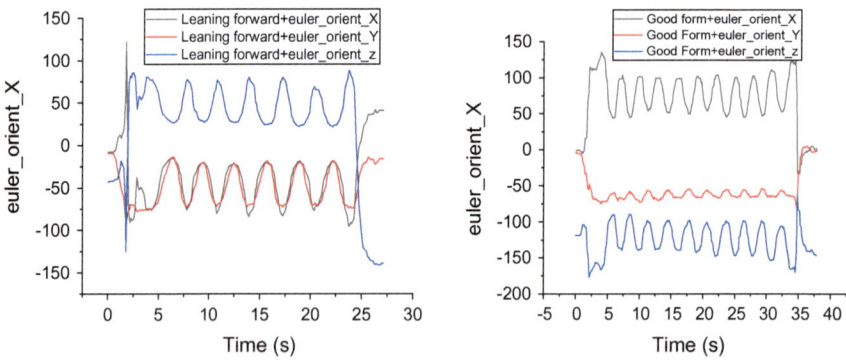

Fig. 6 Euler orientations (-*x*, -*y*, and -*z*) (left: leaning forward, right: good form)

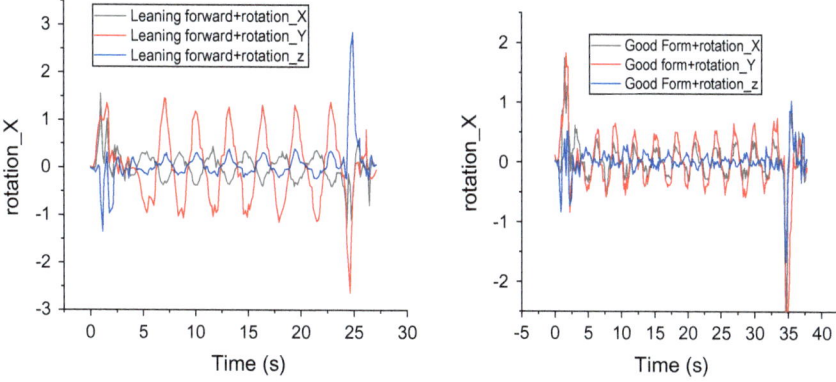

Fig. 7 Rotations (_x, _y, and _z) leaning forward (left) and good form (right)

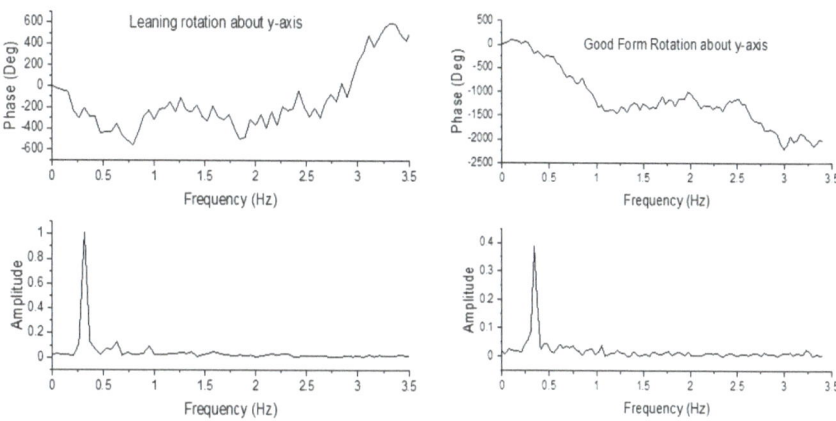

Fig. 8 FFT results of the *y*-axis data for squats: (left) leaning forward, and (right) good form

8 Conclusions and Further Development

Analytics of wearables data from workouts yields reliable indicators of "good form" workouts and deviations from the baseline. We continue to analyze all the data using cosine similarity, FFT, and other standard techniques of analyzing time series data, to produce quantitative, real-time feedback to athletes during workouts.

References

1. Axon S (2021) More people are buying wearables than ever before—and Apple is in the lead. Ars Technica. https://arstechnica.com/gadgets/2021/05/more-people-are-buying-wearables-than-ever-before-and-apple-is-in-the-lead/
2. Bauer CM, Rast FM, Ernst MJ, Kool J, Oetiker S, Rissanen SM, Suni JH, Kankaanpää M (2015) Concurrent validity and reliability of a novel wireless inertial measurement system to assess trunk movement. J Electromyogr Kinesiol 25(5):782–790. https://doi.org/10.1016/j.jelekin.2015.06.001
3. Fuller D, Colwell E, Low J, Orychock K, Tobin MA, Simango B, Buote R, Heerden DV, Luan H, Cullen K, Slade L, Taylor NGA (2020) Reliability and validity of commercially available wearable devices for measuring steps, energy expenditure, and heart rate: systematic review. JMIR Mhealth Uhealth 8(9):e18694. https://doi.org/10.2196/18694
4. Hooren BV, Goudsmit J, Restrepo J, Vos S (2020) Real-time feedback by wearables in running: current approaches, challenges and suggestions for improvements. J Sports Sci 38(2):214–230. https://doi.org/10.1080/02640414.2019.1690960
5. Kuenze C, Pfeiffer K, Pfeiffer M, Driban JB, Pietrosimone B (2021) Feasibility of a wearable-based physical activity goal-setting intervention among individuals with anterior cruciate ligament reconstruction. J Athl Train 56(6):555–564. https://doi.org/10.4085/1062-6050-203-20
6. O'Donovan KJ, Kamnik R, O'Keeffe DT, Lyons GM (2007) An inertial and magnetic sensor based technique for joint angle measurement. J Biomech 40(12):2604–2611. https://doi.org/10.1016/j.jbiomech.2006.12.010
7. Pike AM, Pryor RR, Vandermark LW, Mazerolle SM, Casa DJ (2017) Athletic trainer services in public and private secondary schools. J Athl Train 52(1):5–11. https://doi.org/10.4085/1062-6050-51.11.15
8. Post EG, Roos KG, Rivas S, Kasamatsu TM, Bennett J (2019) Access to athletic trainer services in California secondary schools. J Athl Train 54(12):1229–1236. https://doi.org/10.4085/1062-6050-268-19
9. Rantalainen T, Pirkola H, Karavirta L, Rantanen T, Linnamo V (2019) Reliability and concurrent validity of spatiotemporal stride characteristics measured with an ankle-worn sensor among older individuals. Gait Posture 74:33–39. https://doi.org/10.1016/j.gaitpost.2019.08.006
10. Shanley E, Thigpen CA, Chapman CG, Thorpe J, Gilliland RG, Sease WF (2019) Athletic trainers' effect on population health: improving access to and quality of care. J Athl Train 54(2):124–132. https://doi.org/10.4085/1062-6050-219-17

Analysis of Various Techniques of Fetal Growth Detection

G. Mohana Priya and P. Mohamed Fathimal

Abstract Internal organs and soft tissues can be observed using imaging, which is one of the most vital medical instruments used for diagnosis. Ultrasound screening is one of these tools, and it's widely used in the field of gynecology. Ultrasound scanning has many benefits, including the fact that it is non-invasive, radiation-free, cost-effective, and real time. The methods for detecting fetal growth are discussed in this article. However, new methods for identifying the fetus object are developed from time to time. This overview paper depicts, compares, and organizes numerous strategies of fetal development recognizable proof. Analyzing the fetal growth and features for any abnormalities using machine learning (ML) techniques helps in deciding the growth of the fetal period during the gestational period.

Keywords Fetal growth · Ultrasound · Doppler ultrasound · Cardiotocography · Fetal weight · Gynecology

1 Introduction

The period between conception and birth in which a baby grows and develops within the mother's womb is regarded as gestation. Since determining when a fetus is viable is difficult, gestational age is determined by counting from the first day of the mother's last menstrual period to the present day. Diagnosis during pregnancy: During the pregnancy, a number of tests are performed. To gain access to the developing fetus, we must use techniques that will enable us to do so. Ultrasound, Cardiotocography, and Doppler checks are various techniques (Fig. 1), which can be conducted securely at any period before childbirth and blood chemistry assessments utilizing maternal blood monitoring during the first few months of pregnancy.

G. Mohana Priya (✉) · P. Mohamed Fathimal
Department of Computer Science and Engineering, SRM Institute of Science and Technology, Vadapalani Campus, Chennai 600026, India
e-mail: mg8504@srmist.edu.in

© The Author(s), under exclusive license to Springer Nature Singapore Pte Ltd. 2023
A. Kumar et al. (eds.), *Proceedings of the 4th International Conference on Data Science, Machine Learning and Applications*, Lecture Notes in Electrical Engineering 1038,
https://doi.org/10.1007/978-981-99-2058-7_5

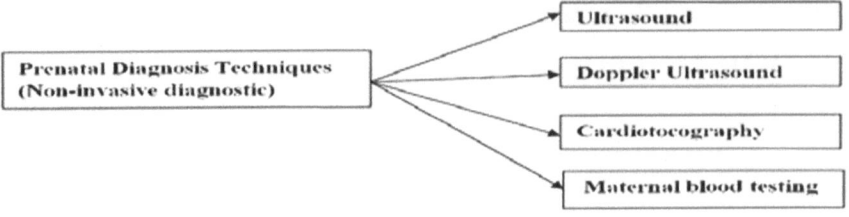

Fig. 1 Prenatal diagnosis techniques

1.1 Ultrasound

An ultrasonic scan is a medical procedure that captures to obtain real-time images of the body using high-frequency vibrations. Sonography is another name for it. Different tissue densities can be used to diagnose internal and exterior abnormalities. It works on the same principle as radars used during air traffic control and maritime navigation. Ultrasound, unlike other imaging procedures, does not utilize radiation.

1.2 Doppler Ultrasound

Doppler ultrasonography is a type of imaging technique that observes the change in waves of a sound objects to visualize blood flow through blood arteries. A conventional ultrasound can also make images of structures inside the body using sound waves, but it cannot reveal flow of blood.

1.3 Cardiotocography

The term Cardiotocograph (CTG) is a medical procedure used to monitor and assess the fetal heart rate (FHR) and uterine contractions during pregnancy. It is a common method employed to evaluate the well-being and health of the fetus during labor and delivery. The CTG is also known as electronic fetal monitoring (EFM) and plays a crucial role in obstetrics.

1.4 Maternal Blood Testing

The pregnant women can get a blood test called Maternal Serum Screening (MSS). It will assist them in determining if their unborn child has a neural tube defect and down syndrome.

2 Related Literature

Advancements in the early detection of structural abnormalities. Edwards et al. [1] address ultrasound's technological aspects, early anatomical survey, detection rates, and limitations. Increased nuchal translucency, skin edema, micrognathia, and omphalocele are all first trimester defects. Their paper discussed soft markers, identification rates in the mid-trimester, structural malformation detection, and the role of 3D ultrasound in the second trimester ultrasound. Echogenic lung lesion, micrognathia, and micrognathia are all found on second trimester scans. The anatomy of second trimester neurosonography differs from that of the first trimester. Beyond the routine morphology scan, second trimester neurosonography addresses the functions of magnetic resonance imaging (MRI) fetal brain assessment. The MRI diagnosis for prenatal abnormalities and the technical aspects and signs of fetal MRI as well as a comparison of MRI and ultrasound and the function of ultrasound during non-invasive prenatal testing are discussed. Models in Caradeux et al. [2] showed a significant improvement in sensitivity using meta-regression analysis for evaluating ultrasound done later in pregnancy. The third trimester ultrasound examination for early growth restriction focuses on abdominal circumference and is performed around 37 weeks of pregnancy.

The models proposed in Anggraini et al. [3] reduced the error than the current ultrasonic and clinical models, according to the findings. This inquire has brought about the creation of models to foresee approximate fetal weight during various stages of pregnancy in areas where ultrasound amenities are not available. Lu et al. [4] results show that their ensemble methods reduce the estimation error in fetus's weight by predicting the growth curve of an embryo based on changes in relevant parameters of pregnant women. Tao et al. [5] proposed a method which will help with birth weight estimation and the creation of clinical delivery care guidelines. It can also be used to construct a decision-making framework based on a temporal machine learning prediction model. In [6], eight out of ten suspected SGAs had a birth weight in the 10th 66th percent for gestational age, with the 5th centile category having a better estimate of real birth weight. Hiwale et al. [7] established models which have errors in the Indian population more than in their native populations. This highlights the limitations of the Indian population. As a result of this report, clinicians are advised to use caution when interpreting fetal weight estimates. Furthermore, this research emphasizes the need for models centered on Indian populations. Crockart et al. [8] proposed a predictive model whose performance inspires researchers to look at further birth abnormalities, paving the groundwork for more complex models and less studied topics. The knowledge available for this model was critical to its success. In this Akhtar et al. [9], support vector machine generated the highest precision 85% for prediction. The primary goal of this inquire is to assist physicians in the initial stages of the LGA prognosis to reduce the disease's negative consequences using a restricted range of biochemical markers. The inclusion of biological markers and defective placentation just slightly increases the prescient yield of maternal variables and prenatal biometrics between 35 and 36 weeks pregnancy for childbirth of small gestational age newborns [10].

Moreira et al. [11] compared various machine learning techniques on a database of expectant mothers who developed hypertension during pregnancy. The findings show that ensemble approaches are capable of accurately prophecy the fetus's predicted birth weight. Kalafat et al. [12] proposed the forecasting model which identifies six main risk factors linked to the likelihood of delivery for fetus decision in term small for gestational age (SGA) fetal. The model has good bias and fit, and it could be used to make decisions and advise women about their individual prenatal risk. Lappen et al. [13] demonstrated the effect of the systematic assumption inaccuracy of fetal weight on the diagnostic of fetal growth restriction (FGR). Rueda et al. [14] used five approaches to the fetal head sub-challenge focusing on either the appearance of the image or the details of the edges using graph-based techniques. The results showed that it is possible to obtain a high level of output that is comparable to manual delineation. One of the most important variables in determining a fetus's growth and health is the fetal head circumference (HC) (Table 1).

3   Conclusions

This paper presented various techniques of fetal growth detection and different methods pursuing various classifications. The analyses of the fetus characteristics such as fetus weight and health status are most important for the well-being of the fetus. This paper's major goal is to identify fetal growth using a variety of machine learning techniques: Hybrid-LSTM, ANN, CNN, RF, LR, Support Vector Machines, and Back Propagation Neural Network, Naive Bayes, SVM, Stochastic Gradient Descent and k-Nearest Neighbors, LightGBM, Random Forest, and XGBoost algorithms, as well as the Genetic algorithm. The future work shall include research using suitable deep learning techniques to achieve the desired outcome of the well-being of fetal and to improve the accuracy of fetal growth estimation.

Table 1 Comparison study

S.No	Author	Category	Approach	Analysis
1	Edwards and Hui	Fetal weight	At the 11–13 and additional 6 week ultrasound scan, anatomical assessment were performed. MRI is used to assess fetal structural anomalies, and chromosome testing is done after a structural abnormality ultrasound is performed at the time of delivery Non-Invasive Prenatal Testing (NIPT)	Ultrasound-based infant structural anomaly screening is an important aspect of standard hospital care since it enables for further testing such as prognostic knowledge, genetic testing, specialized imaging, and consideration of care alternatives before the baby is born
2	Caradeux et al.	Fetal weight	A meta-analysis using stratified summary receiver operating characteristic curves, quantitative data synthesis using random-effects models, and a meta-regression examination found a substantial improvement in sensitivity when ultrasound examination was conducted afterward in pregnancy	Finally, by limiting analyses to study on normal populations conducted within the following decade, researchers were able to quantify the efficacy of the present routine screening
3	Anggraini et al.	Fetal weight	Multivariate linear regressions were used	The retrospective longitudinal study's goal was to collect guideline data from designated primary hospitals. Women are likely to have used health care facilities other than those included in this assessment. While this may effect in information under rate, it is unlikely to have an influence on the authenticity of the analysis

(continued)

Table 1 (continued)

S.No	Author	Category	Approach	Analysis
4	Lu et al.	Fetal weight	The techniques include the LightGBM, Random Forest, and XGBoost algorithms, as well as the genetic algorithm. The genetic algorithm is used to refine a variety of features	The experimental findings show image processing between the estimated fetal weight limit at each Preterm delivery predicted by the ensemble model and that of ultrasonography. The machine learning technique utilized in this work was capable to estimate fetal weight with good accurateness at various embryonic ages in the nonappearance of an ultrasound exam
5	Tao et al.	Fetal weight	Hybrid-LSTM, ANN (artificial neural network), convolutional neural network (CNN), random forests (RF), LR, SVR, and back propagation neural network	A lengthy short-term memory creates a continuous model of factors relevant to pregnant mothers and prenatal medical exams
6	Mlynarczy et al.	Fetal weight	Hadlock's regression equation and statistical analyses were done	The fifth percentile of sonographic estimated unborn weight is related with a greater degree of cumulative newborn morbidity. It gives therapeutically important information for mothers in community hospitals with a sonographic estimated fetal weight (SEFW) of 5th vs. 5–9th percent in favor of GA, as well as the possibility to construct an interventional study toward reducing the morbidity associated by questionable SGA

(continued)

Table 1 (continued)

S.No	Author	Category	Approach	Analysis
7	Hiwale et al.	Fetal weight	Statistical analysis methodology are (a) The average of variation between estimated fetal weights (EFW) and actual birth weights (ABW), (b) standard deviation discrepancies, and (c) mean of percentage errors (MPE) were used to equate calculated fetal weights (EFW) from various models with real birth weights (ABW) (MPE)	MPE in models based only on AC or AC-BPD pairings statistically significantly lower than in estimates based on certain combos (p 0.05)
8	Crockart et al.	Fetal weight	LR, RF, stochastic gradient descent, and k-nearest neighbors	The final model outperformed all other models in all evaluation metrics, especially the stochastic gradient descent system
9	Akhtar et al.	Fetal weight	Naive Bayes, SVM, logistic regression(LR), random forest for gestational age	This paper presents three experiments that used 10-fold cross-validation to find graded risky variables associated with large for gestational age (LGA) baby prognostication. Including preference of characteristics in the development of a good classification technique capable of predicting an illness in real time, accurate. Computationally efficient manner is frequently important
10	Ciobanu et al.	Fetal weight	Multivariable logistic regression is used and Statistical analysis is done	A scan positive rate of 66% due to parental causes, 32% due to parental factors and successful fetus weight, and 30% due to the addition of markers resulted in a prediction model of 90% of SGA newborns delivering at any point after evaluation
11	Moreira et al.	Fetal weight	Various ML (Decision Tree, SVM, KNN, Boosted Tree, Bagged Trees, Subspace KNN). The confusion matrix was used to execute procedures under the ROC	Used an actual database of expectant mothers who had a hypertension condition throughout their gestation and prophesy the estimated weight of the fetus at delivery

References

1. Edwards L, Hui L (2018) First and second trimester screening for fetal structural anomalies. In: Seminars in fetal and neonatal medicine, vol 23, no. 2. WB Saunders, pp 102–111
2. Caradeux J, Martinez-Portilla RJ, Peguero A, Sotiriadis A, Figueras F (2019) Diagnostic performance of third-trimester ultrasound for the prediction of late-onset fetal growth restriction: a systematic review and meta-analysis. Am J Obstet Gynecol 220(5):449–459
3. Anggraini D, Abdollahian M, Marion K (2018) Foetal weight prediction models at a given gestational age in the absence of ultrasound facilities: application in Indonesia. BMC Pregnancy Childbirth 18(1):1–12
4. Lu Y, Fu X, Chen F, Wong KK (2020) Prediction of fetal weight at varying gestational age in the absence of ultrasound examination using ensemble learning. Artif Intell Med 102:101748
5. Tao J, Yuan Z, Sun L, Yu K, Zhang Z (2021) Fetal birthweight prediction with measured data by a temporal machine learning method. BMC Med Inform Decis Mak 21(1):1–10
6. Mlynarczyk M, Chauhan SP, Baydoun HA, Wilkes CM, Earhart KR, Zhao Y, Abuhamad AZ (2017) The clinical significance of an estimated fetal weight below the 10th percentile: a comparison of outcomes of< 5th vs 5th–9th percentile. Am J Obstet Gynecol 217(2):198-e1
7. Hiwale SS, Misra H, Ulman S (2017) Ultrasonography-based fetal weight estimation: finding an appropriate model for an Indian population. J Med Ultrasound 25(1):24–32
8. Crockart IC, Brink LT, du Plessis C, Odendaal HJ (2021) Classification of intrauterine growth restriction at 34–38 weeks gestation with machine learning models. Inf Med Unlocked 23:100533
9. Akhtar F, Li J, Azeem M, Chen S, Pan H, Wang Q, Yang JJ (2019) Effective large for gestational age prediction using machine learning techniques with monitoring biochemical indicators. J Supercomput 1–19
10. Ciobanu A, Rouvali A, Syngelaki A, Akolekar R, Nicolaides KH (2019) Prediction of small for gestational age neonates: screening by maternal factors, fetal biometry, and biomarkers at 35–37 weeks' gestation. Am J Obstet Gynecol 220(5):486-e1
11. Moreira MW, Rodrigues JJ, Furtado V, Mavromoustakis CX, Kumar N, Woungang I (2019) Fetal birth weight estimation in high-risk pregnancies through machine learning techniques. In: ICC 2019–2019 IEEE international conference on communications (ICC), pp 1–6. IEEE
12. Kalafat E, Morales-Rosello J, Thilaganathan B, Tahera F, Khalil A (2018) Risk of operative delivery for intrapartum fetal compromise in small-for-gestational-age fetuses at term: an internally validated prediction model. Am J Obstet Gynecol 218(1):134-e1
13. Lappen JR, Myers SA (2017) The systematic error in the estimation of fetal weight and the underestimation of fetal growth restriction. Am J Obstet Gynecol 216(5):477–483
14. Rueda S, Fathima S, Knight CL, Yaqub M, Papageorghiou AT, Rahmatullah B, Noble JA (2013) Evaluation and comparison of current fetal ultrasound image segmentation methods for biometric measurements: a grand challenge. IEEE Trans Med Imaging 33(4):797–813

Mr. Bot—A Survey on Arduino-Based Autonomous Robotic Vehicle

J. Karthiyayini, Chayanika Biswas, C. C. V. N. Ashish, N. B. Hrishikesh, and A. Ayesha Siddiqua

Abstract Robots have a rising contribution in real-world settings, such as schools, homes, hospitals, laboratories and workplaces. With the fast-moving world we like everything automated which is increasing the development of autonomous systems to perform both minor and major tasks which would save our time and effort. Mr. Bot acts as a helping hand and tries to reduce the human effort and is the first step in the involvement of robots in daily use. Mr. Bot is a robotic vehicle that has the capability to reach a particular destination to deliver small items and/or convey messages. Mr. Bot has sufficient intelligence (Joshi G, Kolhe P Intelligence spy robot with wireless night vision camera using wi-fi) to follow the shortest path in the provided space and detect the obstacles in both day and night visions (Manasa P, Harsha KS, Deepak DM, Karthik R, Nichal NO Night vision patrolling robot). Mr. Bot acts as helper, it can be summoned to any required place, and it can be sent to any destination to deliver small objects (such as cables, papers and books) and conveys messages via LCD.

Keywords Robot · Autonomous system · Intelligence · LCD

J. Karthiyayini · C. Biswas · C. C. V. N. Ashish · N. B. Hrishikesh · A. Ayesha Siddiqua (✉)
Department of Information Science & Engineering, New Horizon College of Engineering, Karnataka Bengaluru, India
e-mail: ayeshasiddiquaaman@gmail.com

J. Karthiyayini
e-mail: jkarthi1952@gmail.com

C. Biswas
e-mail: chayanika982001@gmail.com

C. C. V. N. Ashish
e-mail: cvnnashish@gmail.com

N. B. Hrishikesh
e-mail: hrishikeshnb007@gmail.com

© The Author(s), under exclusive license to Springer Nature Singapore Pte Ltd. 2023   47
A. Kumar et al. (eds.), *Proceedings of the 4th International Conference on Data Science, Machine Learning and Applications*, Lecture Notes in Electrical Engineering 1038,
https://doi.org/10.1007/978-981-99-2058-7_6

1 Introduction

The origin of robot's marks to the ancient world. The concept of robots came into existence as early as 3000 BC, and they were certain mechanical devices which would carry out particular functions involving physical tasks as per the instructions. The early built-in mechanical devices were Egyptian water clocks to hit the hour bells, wooden pigeon (400 BC) developed by Archytus of Taremtum could fly, hydraulically operated statues (second century BC) built in Hellenic Egypt could speak and show gestures, Petronius Arbiter made a doll (first century AD) that could move, and Giovanni Torriani developed a wooden robot (1557) that could fetch bread from the store, talking doll (nineteenth century) by Edison and steam-powered robot (nineteenth century) by Canadians. These inventions and discoveries were the seeds of inspiration in the field of robotics, then in the twentieth century the field of robotics took a huge leap and surpassed all the previous inventions. The recent inventions of robots include AMECA which has the capability of face and multiple voice recognitions, ARMAR-6 has the capability of moving objects and handing them to the desired person, DIGIT has fully functional limbs, JIAJIA can express certain emotions such as laugh and cry, SOPHIA can process visual, emotional and conversational data for better interaction with human beings, and many more to go.

The term robot has several definitions but all of them come down to the same concept of "a reprogrammable, multifunctional, manipulator device" that is designed intelligently to perform certain physical tasks such as moving materials, tools and delivering objects through various programmed motions for completion of the tasks. The word "Robotics" was penned by Russian born American Science fiction writer Isaac Asimov in his short story "Runabout" in the year 1942.

Asimov proposed three laws of robotics which are being followed till date, and those are:

1. A robot must not injure living beings via any form.
2. A robot must obey its master's commands except where it would disagree with the first law.
3. A robot must defend its own existence as long as it would not disagree with the first and second law.

The world of robotics has taken a huge leap from just being certain mechanical devices to the development of humanoid robots. Robots in today's world are very well developed and assist human beings in their day-to-day life in various fields such as hospitals, education and health care.

Our concept of Mr. Bot has a better improvised version from the early-century robots but less features compared to our current-day humanoid robot it tries to inculcate a lot of features such as detecting the obstacles, live streaming its path and can also capture and store them, read characters, it can be controlled by voice and respond to it, and it can be summoned via mobile.

2 Literature Survey

Arduino Based Voice Controlled Vehicle [1] by M Saravanan, Anandhu Jayan, B Selvababu, Aswin Raj and Angith Anand proposes the idea of controlling the robot via voice commands using mobile applications (mobile app). The robot has higher accuracy in voice recognition and is highly sensitive to surrounding noise, but the voice commands must be provided via the Android app.

Review on Optical Character Recognition [2] by Muna Ahmed Awel and Ali Imam Abidi proposes the idea of a robot which uses different approaches of character recognition system accuracy to understand the alphabets with higher accuracy, but it can only detect English, Arabic and Devanagiri characters.

Robot Voice—A Voice Controlled Robot Using Arduino [3] by Vineeth Teeda, K Sujatha and Rakesh Mutukuru proposes the idea of controlling the movement of robot using voice commands, and these voice signals are captured using inbuilt microphone. The robot takes voice commands, executes them and gives acknowledgment through speech output, but the impact of the distance between the mouth and microphone on the robot affects the performance of the robot, and it also impacts on the speech to text conversion.

Moving Obstacle Avoidance of a Mobile Robot Using a Single Camera [4] by Jeongdae Kim and Yongtae Do proposes the idea of robot which detects the object using single camera and tries to find a proper path, but it fails when the distance between objects is more, if the object's color is similar to the surroundings and too much reflection of light.

Optical Character Recognition based Auto Navigation of Robot by Reading Signboard [5] by Prof. Suneel K Nagavi, Mahesh S Gothe and Prof. Praveen S Totiger proposes the idea of a robot that reads characters and symbol and carries out navigation process using those characters and symbol. It permits a robot to find path consequently by distinguishing and reading textual information in signs located (sign board) by utilizing OCR, but it is standardized by using black color with character size from 34–48 and written in Ariel style.

An Abstraction Layer Exploiting Voice Assistant Technologies for Effective Human–Robot Interaction [6] by Ruben Alonso, Diego Reforgiato Recupero and Emanuele Concas proposes the idea of robot that uses voice assistant to convey messages directly to the user. It provides an effective communication between the robot and the user, but the language and pronunciation of certain words might create confusions and lead to different search results.

Night Vision Patrolling Robot [7] by Poojari Manasa, Deepak D M, K Sri Harsha, Karthik R and Naveen Nichal O proposes the idea of robot that uses the night vision camera to work in both day and night light and find a suitable path, but it can only be controlled by sound sensor and not by manually or by Wi-Fi.

Development of an Arduino—Based Obstacle Avoidance Robotic System for an Unmanned Vehicle [8] by Kolapo Sulaimon Alli, Moses Olluwafemi Onibonoje, Akinola S Oluwole, Michael Adegoke Ogunlade, Anthony C Mmonyi, Oladimeji Ayamolowo and Samuel Olushola Dada proposes the idea of a robot that detects the objects using IR sensors, and it can also be controlled utilizing an IR sensor and a remote controlled device, but the robot cannot detect the long distance objects as it is using IR sensors.

3 Objectives

- To develop a robot which has both manual control and automatic control to deliver small objects and/or convey messages from one place to another. The robot is always stationed at specific point and can be summoned in front of any room in its path and can be sent to any other room to deliver objects or conveys messages; after the work has been completed the robot returns back to its station.
- In order to do this task successfully the robot has to perform various operations such as obstacle detection [8], path detection, night vision [9] and character reading [2].
- Mr. Bot detects the path using path and obstacle detection mechanism by following the specified path marked on the floor by connecting all the rooms; when an obstacle is encountered it tries to avoid the obstacle if possible, else takes a different path, if a different path is not available it alerts with a buzzing sound, and it also reads the unique labels marked in front of each room using character [10] reading to reach its destination.
- Night vision [7] enables the robot to perform obstacle detection and path detection in absence of light.

4 Existing System

In the existing project, we will find robots that implement only a specific feature instead of implementing as a whole with multiple features. Our project, Mr. Bot, tries to sum up the existing project, add few additional features and create a single multifunctional bot.

5 Proposed System

Mr. Bot is a mini robot that is developed to perform minor tasks. Mr. Bot can assist human beings in their day-to-day life in various fields such as hospitals, schools and colleges. Mr. Bot has both manual control and automatic control to deliver small objects or convey messages by performing various operations inculcating a lot of features such as detecting the obstacles [8], live streaming its path and can also capture and store them and read characters [5, 10], and it can be controlled by voice and respond accordingly and can be summoned via mobile.

6 Proposed Methodology

Our suggested robot would follow simple commands provided via the mobile app [1] or voice commands, and with better future enhancements it can be used to perform tedious tasks.

The Major features are listed below:

Obstacle Detection [4]**:** It is achieved using IR [8] sensors and ultrasonic sensors [11] to detect the object size and decide whether it can pass over it or take a diversion and find a new path, and if it is not able to find a new path then it will stop and alert with a buzzing sound.

Character Recognition [5, 10]**:** The robot uses camera [5, 9] to read the characters and/or numbers to reach its destination.

Night Vision Camera [7]**:** It is designed using night vision camera which uses infrared light to detect path and obstacles in the night vision [7] where there is no sufficient lighting condition.

Voice Control [1, 3]**:** It is achieved using microphone to interact with the robot, where we can directly give commands to the bot instead of using the mobile app [1]. The bot not only accepts the command, but also responds to it with certain short messages.

LCD Screen: It displays the message that has to be conveyed to the desired person.

Mobile App to Monitor [2]**:** An app is developed to give commands to the robot. It is basically used to summon it, give instructions to it, type a message that needs to be conveyed and provide a path to it. The app also helps us to control the movement of the robot.

The diagram displays the basic working of the bot. Mr. Bot is initially in its idle state, stationed at a particular place; when it is summoned it reaches the desired place and alerts. The user can add any message to be conveyed or place any object that needs to be delivered; these instructions can be provided via voice command or app, and then the bot will be sent to the destination; after completion of the task, the bot goes back to its station.

7 Applications

- **Schools, Colleges and Universities:** The bot can be used to collect the attendance sheet from every classroom, call the teachers for the meetings, etc.
- **Hospitals:** To deliver the medicines to the required patient and guide any person to reach a particular destination such as pharmacy, emergency ward and laboratories.
- **Shopping mall:** Helps the customer to find any required object and can be used like a cart to carry small objects.

8 Expected Result

- A fully functional robot which can be used to deliver small objects and convey messages using a LCD screen, from the summoned point to the required destination.
- Detects and tries to avoid obstacles if possible, else alarms everyone about the obstacles.
- Performs character reading to identify room numbers which are marked on its path.
- The robot is enabled with night vision [12] to perform the tasks smoothly in absence of light.
- Mr. Bot can be summoned in front of any room in its path from its original station and can be sent to any other room its path, after which it will return to its original station.
- Mr. Bot can be controlled manually or can be automated, and certain voice commands are understood by the robot and responds accordingly.

9 Conclusion

Mr. Bot is a robotic vehicle that acts as a helper and can be used in schools, colleges and hospitals. Mr. Bot can be used to deliver small objects and display message via its LCD screen. It is always stationed at a particular place and can be summoned via mobile; it can be sent to any other place on its path to complete our desired task. With better future enhancements it can replace peons in any institutions and not only be limited to one floor.

10 Future Enhancements

- The robot can move freely without a specific path on the ground.
- AI technology for face and object recognition to identify the required person and the objects that are being delivered.
- Security surveillance feature [13] can be added for night safety.
- Complete speech recognition and voice control for performing all tasks.
- Movement between multiple floors by using lift or stairs (Figs. 1 and 2).

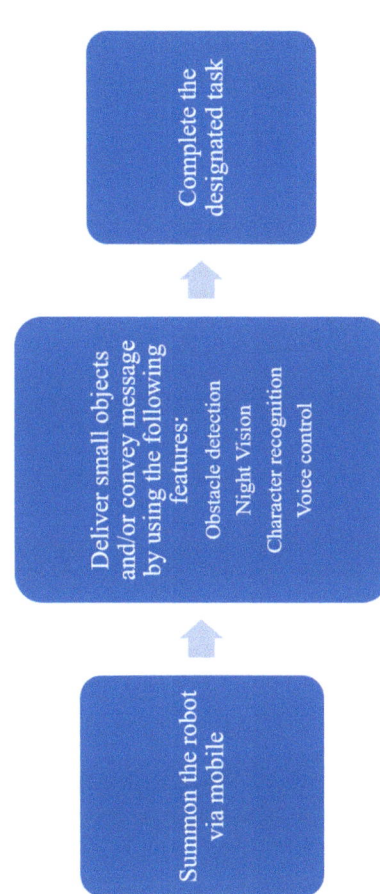

Fig. 1 System architecture

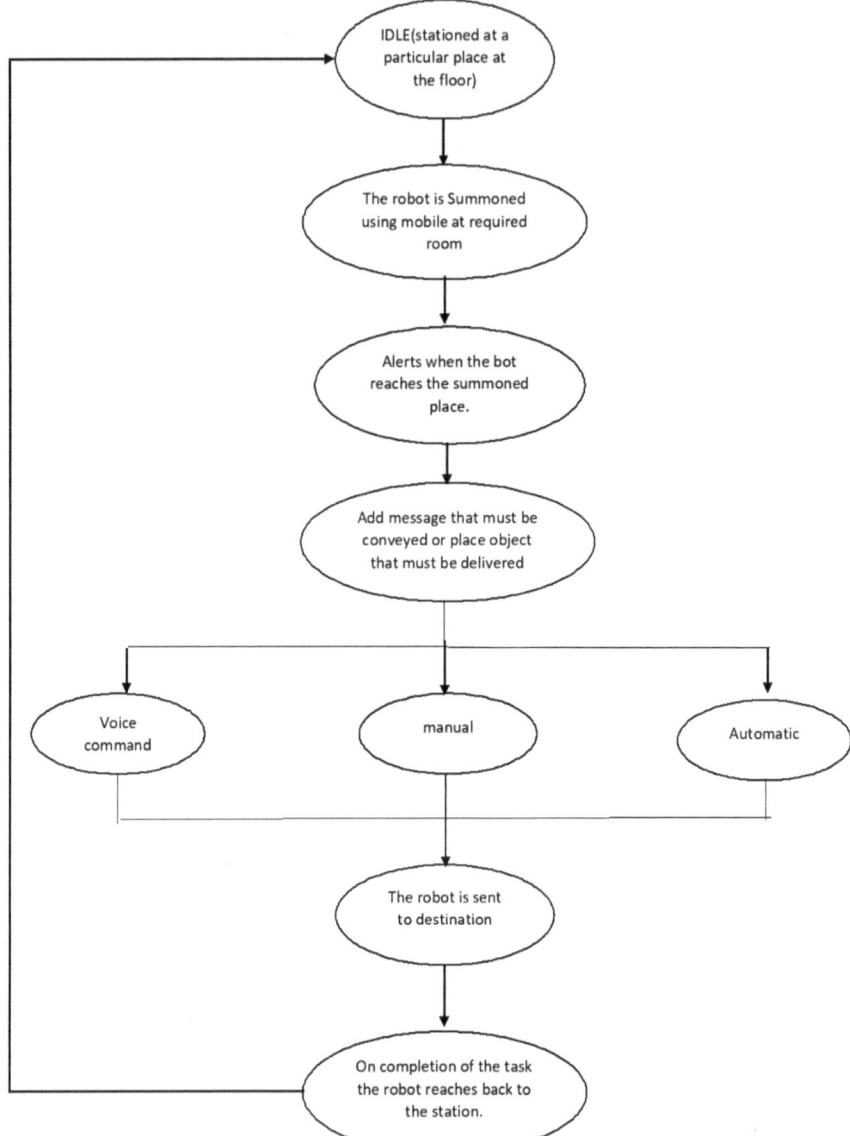

Fig. 6.2 Flow diagram

References

1. Saravanan M, Selvababu B, Jayan A, Anand A, Raj A (2020) Arduino based voice controlled robot vehicle
2. Muna Ahmed Awel, Ali imam Abidi (2019) AI Review on optical character recognition
3. Vineeth Teeda, K Sujatha, Rakesh Mutukuru (2016) Robot voice a voice controlled robot using Arduino
4. Jeongdae Kim, Yongtae Do (2012) Moving obstacle avoidance of a mobile robot using a single camera
5. Suneel K Nagavi, Mahesh S Gothe, Praveen S Totiger (2015) Optical character recognition based auto-navigation of robot by reading signboard
6. Ruben Alonso, Emanuele Concas, Diego Reforgiato Recupero (2021) -voice assistant technologies for effective human robot interaction
7. Poojari Manasa, K Sri Harsha, Deepak DM, Karthik R, Naveen Nichal O (2020) Night vision patrolling robot
8. Kolapo Sulaimon Alli, Moses Oluwafemi Onibonoje, Akinola S. Oluwole, Michael Adegoke Ogunlade, Anthony C. Mmonyi, Oladimeji Ayamolowo and Samuel Olushola Dada (2018) Development of an arduino—Based obstacle avoidance robotic system for an unmanned vehicle
9. Gayatri Joshi, Prashant Kolhe (2021) Intelligence spy robot with wireless night vision camera using wi-fi
10. Karthiyayini J, Pramod M, Vamsipriya A, Sheriff A, Anand G (2020) Assisting visually impaired for shopping using OCR(Optical character recognition)
11. Karthiyayini J (2020) Robot assisted emergency and rescue system with wireless sensors
12. N Hemavathy, Arun K, Karthick R, Srikanth AP, Venkatesh S (2020) Night vision patrolling robot with sound sensor using computer vision technology
13. Srinivasan L, Nalini(2019) Abadent object detection & IOT based multi-sensor smart robot for surveillance security system

Water Quality Monitoring and Controlling Systems for Aquaculture

Rajeshwarrao Arabelli and T. Bernatin

Abstract Due to large increases in demand for fish and seafood around the world, aquaculture has been a rapidly expanding sector. Shrimp, fish, and other aquatic crops are grown in aquaculture, where the water's qualities determine the crop. Dissolved oxygen, salinity, temperature, alkalinity, turbidity, hardness, pH, ammonia, water level, and nutrient levels are a few water quality factors that should be kept at ideal levels to enhance yield. Depending on the circumstances, these characteristics might change drastically throughout the day. In order to fully exploit their potential, these factors must be regulated at high frequencies. This paper affords an in-depth assessment of the diverse strategies utilized in aquaculture to Detect and manage water quality. This survey highlights the research gap on this area and presents higher scope for superior work. Keywords: Aquaculture, Water quality index, Water quality monitoring, Wireless Sensor Net-works.

Keywords AI (Artificial Intelligence) · IoT (Internet of Things) · Sensors · MCU · Wireless sensor networks

1 Introduction

Water is one of the major sources for human beings for drinking and other maintenance, agriculture, fisheries, industrial applications etc. In India and other countries in the world, the water resources like lakes, rivers, and ocean are polluted because of the increase of population, and industrialization, cor- responding wastages are

R. Arabelli · T. Bernatin (✉)
Department of ECE, Sathyabama Institute of Science and Technology, Chennai, Tamilnadu, India
e-mail: bernatin12@gmail.com

R. Arabelli
e-mail: rajeshwarrao432@gmail.com

R. Arabelli
Center for Embedded Systems and IoT, Department of ECE, SR University, Warangal, Telangana, India

© The Author(s), under exclusive license to Springer Nature Singapore Pte Ltd. 2023
A. Kumar et al. (eds.), *Proceedings of the 4th International Conference on Data Science, Machine Learning and Applications*, Lecture Notes in Electrical Engineering 1038,
https://doi.org/10.1007/978-981-99-2058-7_7

dissolved in the soil and water resources. They can pollute the ground and surface water.

In India, fisheries and aquaculture play a significant role in food produc- tion, ensuring the food basket's nutritional security, boosting agricultural exports, and employing over 14 million people in various occupations. The water of a high standard has to be kept in aquatic ponds to enhance production. The quality of water refers to its condition in respect to the requirements of one or more biotic species or any agricultural necessity or purpose. Water's quality is determined by its physical, chemical, and biological characteristics. Manual and automated devices are now used to measure the water quality for aquaculture. Farmers that use the manual approach must take water samples from the ponds and evaluate the water's properties in a lab. This strategy is tiresome. Aqua farms are monitored by wireless sensor networks in automated systems employing the unreliable GSM and Zigbee protocols, which are only effective for small distances and pond.

2 Literature Review

Different kinds of water quality monitoring techniques have been researched in the section of the literature review. The following is a discussion of some of those that are being used in aquaculture farms.

2.1 Water Quality Monitoring Using Wireless Sensor Networks

Wireless sensor networks are used to continuously monitor water quality parameters using GSM / Zigbee protocols, which are not reliable and limited to short distances and ponds [1], developed an inexpensive, open-source hardware system for tracking and recording water-quality indicators for aquaculture production. It has the ability to take records and transmit data using the ZigBee wireless protocol to a graphical user interface, which can show data graphically and save the data in a database [2], have tested a prototype smart system for monitoring the pH, DO, and temperature in an Eel fish aquarium. The system is designed by a single Raspberry-Pi3 system. They anticipated that the Raspberry Pi 3 system will use integrated Wi-Fi to connect data to the internet network. Additionally, cell phones may be used to remotely regulate an automated system that maintains water quality at the appropriate sensor value [3], developed an Arduino-based low-cost water quality monitoring system, in order to effectively monitor water quality parameters in distribution networks using the ZigBee protocol. According to [4], the focus is on using Wireless Sensor Networks (WSN) for real-time water quality monitoring in order to avoid the delay caused by using GPRS to handle the situation. In order to regulate the changes in water parameters like temperature, conductivity, turbidity, and the presence of an oil layer over the water as well as fish behavior during the feeding process using the fish swimming depth sensor and velocity sensor, optimized a low-cost WSN for aquaculture monitoring [5].

[6], established a fully autonomous system for water quality monitoring employing an improved Kalman filtering algorithm and route tracking algorithm to monitor the water quality parameters in a broad region of cultivation areas. The Android mobile app obtains the water quality characteristics of any place using GPS for decision making [7], designed online PH and DO monitoring system in Shrimp Aquaculture. The main aim of the system is to reduce the energy consumption and optimal usage of water. The PH and DO sensors continuously measure and send the master station and corresponding necessary action will be taken by the corresponding system. This system has a smaller number of sensors. Therefore, the complexity of the system is less, simple in operation and maintenance. [8], illustrated a wireless sensor network based on virtual instruments for aquaculture monitoring and management as shown in Figure 1. They also detailed the physical design of smart nodes that enable real-time alterations in water properties like as pH, humidity, and temperature.

[9], designed to monitor the various parameters of the water in large-scale fish farms like the potential of hydrogen, dissolved Oxygen, water level, temperature, dissolved ammonia and dissolved carbon dioxide. Because the fish development not only depends on food feeding but also depending on the water quality. In the proposed system the water sample is collected at the sampling and sensing chamber. The required parameters can be sensed by the sensors and they are displayed on the central display board. After each sample collection, the chamber is cleaned with fresh water. This cleaning is also monitored by the program-based central system. If the measured parameters are above or below the specified limits and the operator not taking any necessary action, then the alarm circuit will be activated and it will give the alarm. This system is more suitable for the large-scale fish forming unit. But this system installation cost is more. [10], developed an intelligent, networked aquaculture environment monitoring system with low cost, low power consumption, and high-reliability characteristics of the wireless sensor network based on the Zigbee

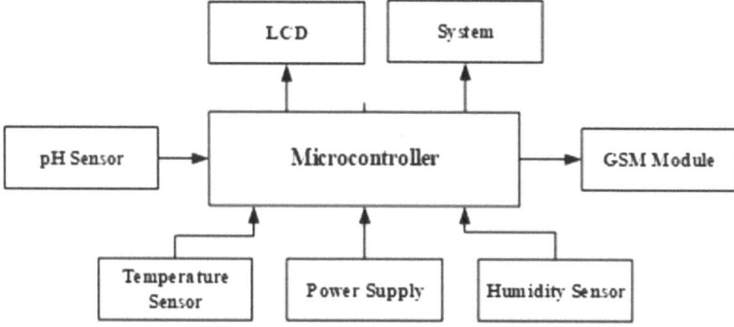

Fig.1 Block diagram of transmitter

protocol. [11], Present water recycling and reducing water waste with the assistance of a vertical aquatic system for energy management with electronic control. They also described the self-cleaning of organic waste in aquaculture. This design can easy to regulate water flow, and reduce labor costs. [12], reviewed various water quality parameters that are required to measure the water quality index during the summer and winter seasons of lake water and also offered strategies to enhance the lake water quality, such as a total ban on pollution-causing activities. [13], have designed a time-to- time fish feeding system to reduce labor cost and food wastage. This system can be designed by a microcontroller, DC motor, food storage system, and control circuit. The feeding time and food container opening time are adjusted as per the requirement. The program is stored in the microcontroller. Based in the program, the DC motor is operated through the PWM (Pulse Width Modulation) technique, and the food is fed to the fish through a food spreader. This device reduces the operation cost and as per the schedule, the food is spread in the fish container. But the productive efficiency is less be- cause some of the fishes will take more food than other fishes. In this method, food wastage is also more. [14], Since independence, the country has shown constant and sustained increases in fish output due to its diversified aquatic resources. In addition to producing about 6.3 percent of the world's fish, the sector contributes 1.1 percent to GDP and 5.15 percent to agricultural GDP. Currently, the inland sector contributes around 65 percent of the 10.07 mil- lion metric tonnes of overall fish production, with cultural fisheries making up roughly the same. However, disease recurrence has become a significant limitation to sustainable aquaculture output and product trade, hurting fishers' socioeconomic condition. Infection by opportunistic infections can be caused by a variety of stress conditions, including insufficient physicochemical and microbiological quality of culture water, low nutritional status, and excessive stocking density. [15], provide a survey on Wireless Sensor Network (WSN) applications in water quality monitoring. They also compare various sensor node topologies in terms of microcontroller units, wireless communication protocols used, data security implementation, and power supply topologies. The survey focuses on monitoring water quality metrics such as pH, electrical conductivity, oxidation-reduction potential (ORP), and turbidity, as well as the numerous obstacles connected with water quality assessment using WSN. They noted how the benefits of WSN stem from its low cost and capacity to execute measurements virtually and in real-time, as proposed by previous writers while taking into consideration their coverage, energy, and security is- sues. These networks, however, have resource constraints in terms of memory, processing power, energy/electricity, and communication bandwidth. [16], examined several approaches for analyzing lake water quality, such as water quality index, Hyperion, and hazard quotient. Analyzed the existence of contaminants in the lake that damage the aquatic ecosystem and proposed that pollution prevention and water re-use be implemented with nutrient re- cycling in regulated urban agri-culture. [17], reviewed comprehensively and discussed the use of new techniques in the detection of critical water quality parameters, such as pH, effective chlorination, dissolved oxygen, turbidity, fluoride, and biochemical oxygen demand (B.O.D.) and summarized the benefits and limitations of optical sensors, MEMS and Biosensors for indication of different water parameters. [18], describes the Guidelines for

Regulating Coastal Aquaculture for improving the production efficiency in various aqua- culture forms. [19], concentrated on identifying the most significant biological species necessary for certain processes, such as nitrification or fertilization of a specific algal species for nitrogen removal or oxygen replenishment for sustainable fish production. An autonomous system was developed to monitor water quality based on aquaculture size and temperature [20, 21].

2.2   Water Quality Monitoring Using IoT

[22], Presents an intelligent web-based control system to improve aquaculture. They designed an aquaponic cycle for recycling water between aquariums and plants. This cycle will convert fish wastage into fertilizers for plants to grow and maintain the water clean. [23], built an NB-IoT-based monitoring system to keep an eye on aquaculture ponds' water quality. It enables distributed monitoring and centrally managed management of aquaculture environment water quality parameters. It involves collecting data, distant transmission, data recovery, remote access, and advanced control functions. [24], have de- signed and implemented an online water monitoring system in a botanical garden for the measurement of pH, EC, ORP, and water Temperature using four sensors. The system consists of a microcontroller and all the sensors connected to the controller. The electrical power supply is given to the controller through the solar PV panel with a battery. The measuring data was continuously measured by the sensors and it sends to the data monitoring system through LoRa transmission. This system construction cost is low, real-time data will be collected and simple in data visualization. [25], has developed a smart IoT-based mini aquarium monitoring system. This system was designed by various sensors like Dissolved Oxygen (DO), temperature sensor, water level sensor, PH sensor, Electric Conductivity (EC) sensor, and total dissolved solids (TDS) sensors. This system can be used to closely monitor the fish behavior and sufficient food feeding of the system. This system's accuracy is more, but the installation cost is high. [26], developed a water level and quality monitoring system to monitor and report the parameters to mobile applications using MQTT protocol. A method suggested for real-time monitoring of water quality indicators uses wireless sensor networks, which monitor several water quality parameters remotely through distributed IoT [27–29].

[30], Present the findings of various European initiatives that use IoT to monitor water quality in aquaculture. The usage of PROTEUS novel sensors, which are based on cutting-edge carbon nanotube technology, was also discussed. to install high-performance, low-cost sensors to decrease maintenance costs. [31], showed the viability of a multi-parameter water quality monitoring system for floating harbors by gathering real-time high-frequency water quality data and displaying it online using wireless sensor networks and IoT. [32], have designed an IoT-based water level, salinity and pH in the water measurement device using the beat sensor. This system's power consumption is less and the device is suitable for agriculture and aquaculture. [33], have designed air and water quality monitoring system with the help of sensor

and IoT-based technology. It is portable and easy to install as per the requirements. It is used to measure temperature, humidity, volatile organic compounds in the air and temperature and pH level in the water. However, in this system power consumption is more and ethernet cable cables are required. [34], created an event-based Internet robotic system that was used to organize aquaculture chores. The robotic system is a dependable instrument for performing feeding and water quality measurement chores in an experimental shade house to aid aquaculture research and increase intense cultivation efficiency. It offers a closed loop event-referenced control system that may be expanded to a bilateral teleoperation system to deliver sensory input and improve user remote perception over the ponds. [35], have proposed a seawater quality monitoring system in the country of Fiji. This system is used to measure or sense the various parameters in the water and that data will send to the monitoring system using IoT and remote sensing technology. In the research work, 4 samples were tested under various conditions. The results were match the expected results. The system operation is depending on the GSM and cloud technology.

2.3 Water Quality Monitoring Using Artificial Intelligence

[36], employing wireless technologies, created a prototype water quality monitoring system The prototype system may be separated into two sections for implementation. The sensors in the hardware implementation may be used to test water quality in terms of PH, Oxidization-Reduction Potential (ORP), Dissolved Oxygen (DO), and Electrical Conductivity (EC). The self-healing method can be used to restore data if it is interrupted for an extended length of time. The data that is continually captured will be sent to the cloud. Continuous data monitoring is feasible in this system under all situations. In addition to a deep learning prediction model of water quality parameter content distribution based on multi-source feature fusion of spectral image and convolutional neural network, a technique for acquiring water quality parameters suitable for freshwater aquaculture was also created [37]. [38], have developed a drinking water quality measurement device using IoT, Machine Learning, and Cloud Computing. This device is helpful for the measurement of water quality in any area (like rural and urban areas). This measurement including required actions will reduce the number of people from various dis- eases and deaths. This system's operation and maintenance cost is less. [39], created an early warning system for recirculating aquaculture water quality monitoring. It assesses the interaction hazardous behavior of a combination of un-ionized ammonia, nitrite, zinc copper, and aluminum to Aliivibrio fischeri using linear independent action models and linear concentration addition models. [40], have proposed a fish behavior-based smart fish feeding system. In this system, the fish container is divided into two unequal parts. The larger part contained sand and plants. This chamber is comfortable for the fish to spend more time in. The smaller chamber is used for food feeding and when they want the food, then only they will enter the chamber. In this system, a webcam, interface circuit, and automatic dispenser with a stepper motor are used for food feeding. When food

is required, the fish entered the food feeding chamber. It is captured or recorded by the webcam. Based on the fish behavior, the stepper motor is operated by the interface circuit to open the nozzle of the automatic dispenser to release the food into the smaller chamber. This method reduces food wastage, decrease maintenance cost and increase productivity. But in this system, continuous monitoring is required because if there is any fault in any part of the smart system, detecting de- vices are not inserted. [41], created a prediction model for an online water quality monitoring system for intensive fish farming in China, which was integrated with a web server and mobile communications technologies. Based on past data saved on the server, it is intended to anticipate water quality using artificial neural networks (ANNs) and adjust water quality in real-time to minimize catastrophic losses. [42], implemented a wireless sensor network for collecting seawater temperatures with sensor nodes and uploading the data to a server platform using a LoRa and MQTT combination network, the results of which can be obtained instantly by logging in to the WEB. [43], introduced a new smart sensor system for water quality monitoring that employs spectroscopic techniques in conjunction with the measurement of physicochemical variables to estimate global pollution parameters in water samples, specifically the Chemical Oxygen Demand (COD). An artificial neural network technique is used to generate this estimation, which is based on a multisensor fusion approach. A machine learning model was developed to estimate the Water Quality Index Class (WQI) based on the parameters like temperature, dissolved oxygen, pH value etc [44–47]. [48], a machine learning method was developed to predict the Water Quality Class (WQC) based on characteristics such as temperature, dissolved oxygen, pH value, turbidity, and nitrates.

3 Discussion

The summary of various kinds of water quality monitoring techniques for aquaculture are shown in Table 1.

Which includes the various water quality parameter for aquatic crops growth and the communication of those parameters through wireless net- works, Internet of Things and artificial intelligence.

4 Conclusion

Aquaculture in India is a growing business with diverse aquatic resources and poten- tial, employing millions at the primary level and many further up the value chain. The crop (shrimp, fish, etc.) in aquaculture is determined by the parameters of the water. Water quality is a measure of the state of water in relation to the needs or purposes of one or more biotic organisms. To increase production, water's phys- ical, chemical, and biological characteristics should be kept at ideal levels. Climate

Table 1 Comparison of Water quality monitoring systems using WSN, IoT and ML

Item	Parameters	Control Board	Technology Used	Operating Frequency band (MHz)	Data rate (Kbps)	Range	Comparative Power consumption
[1]	Temp, pH, DO	Arduino MEGA2650	Wireless ZigBee Network	2400, 915, 868	250, 40, 20	100 m +	Low
[3]	Temp, pH, DO, EC, ERP	Arduino Uno, Raspberry Pi 3	ZigBee	2400, 915, 868	250, 40, 20	100 m +	Low
[4]	Turbidity, Temp	Raspberry Pi	GPRS	800, 900, 1800, 1900	56–114	Depends on Internet access	Low
[5]	Temp, conductivity, Turbidi-ty, fish presence	Arduino MEGA2650	Wi-Fi	2400	1000	46 m – 92 m	Low
[6]	Temp, pH, DO	STM32	GPRS, GPS	800, 900, 1800, 1900	56–114	Depends on Internet access	Low
[23]	Temp, pH, DO, Aerator	STM32	IoT	758–960	125–150	15 km	Low
[24]	Temp, pH, ORP, EC	Arduino Mega2560	Lora	865–867	0.3 - 50	5–15 km	Low
[25]	pH, EC, DO, TDS, Water level, Temp	ESP8266, IoT kit	IoT	758–960	125–150	15 km	Low
[31]	pH, DO, EC, turbidity, ORP	Microcontroller	Wi-Fi	2400	1000	46 m – 92 m	Low
[36]	pH, EC, Temp, DO, ORP	Microcontroller	GPRS, IoT, ML	800, 900, 1800, 1900	56–114	Depends on Internet access	Low

change consequences must be evaluated with strategies to reduce air pollution. Water purification procedures should exist in current systems from filtering operations that should be carried out be- fore introducing any foreign material into the water body. Existing systems simply monitor water parameters such as pH, dissolved oxygen, temperature, and turbidity and communicate data to remote locations via wireless sensor networks and the Internet of things. The majority of these systems do not monitor the chemical and biological properties of water, which impact yield output. These issues, however, may be overcome with considerable research, advancements, and the application of appropriate current approaches. The existing systems can be improved by taking into account a large set of physicochemical and bacteriological parameters of water, storing the database in servers, and developing risk assessment algorithms using internet of things and machine learning tools for integrating many sensors to evaluate the risk of water degradation and improving production. With more study, these approaches may be able to give a "One-Stop Solution" for monitoring and maintaining the entire water quality management system, allowing for a very bright and good future in the forthcoming science and technology period.

References

1. B´orquez L´opez RA, Martinez Cordova LR, Gil Nun˜ez JC, Gonzalez Galaviz JR, Ibarra Gamez JC,Casillas Hernandez R (2020) Im- plementation and evaluation of open-source hardware to monitor water quality in precision aquaculture. Sensors 20(21):6112
2. Salim TI, Haiyunnisa T, Alam HS (2016) Design and implementation of water quality monitoring for eel fish aquaculture. In: 2016 Interna- tional symposium on electronics and smart devices (ISESD), IEEE, pp 208–213
3. Khatri P, Gupta KK, Gupta RK (2019) Smart water quality moni- toring system for distribution networks. In: Proceedings of international conference on sustainable computing in science, technology and man- agement (SUSCOM), Amity University Rajasthan, Jaipur-India
4. Doshi S, Dube S (2019) Wireless sensor network to monitor river wa- ter impurity. In: International conference on computer networks and communication technologies, pp 809–817
5. Parra L, Sendra S, Garc´ıa L, Lloret J (2018) Design and deployment of low-cost sensors for monitoring the water quality and fish behavior in aquaculture tanks during the feeding process. Sensors 18(3):750
6. Zhu X, Liu H, Chen Tian X (2018) Automatic cruise system for water quality monitoring. Int J Agric Bi ol Eng 11(4):244–250
7. Wiranto G, Maulana YY, Hermida IDP, Syamsu I, Mah- mudin D (2015) Integrated online water quality monitoring. In: 2015 Interna- tional conference on smart sensors and application (ICSSA), IEEE, pp 111–115
8. Chandanapalli SB, Reddy ES, Lakshmi DR et al. (2014) Design and deployment of aqua monitoring system using wireless sensor networks and iar-kick. J Aquac Res Dev 5(7)
9. Kamisetti SNR, Shaligram AD, Sadistap S (2012) Smart electronic system for pond management in fresh water aquaculture. In: 2012 IEEE Symposium on industrial electronics and applications, IEEE, pp 173–175
10. Ding W, Ma Y (2011) The application of wireless sensor in aquaculture water quality monitoring. In: International conference on computer and computing technologies in agriculture, Springer, pp 502–507
11. Shin KJ, Angani AV, Akbar M (2017) Fully automatic fluid flow control system for smart vertical aquarium. In: 2017 International con- ference on applied system innovation (ICASI), IEEE, pp 424– 427
12. Gorde S, Jadhav M (2013) Assessment of water quality parameters: a review. J Eng Res Appl 3(6):2029–2035
13. Noor M, Hussian A, Saaid MF, Ali M, Zolkapli M (2012) The design and development of automatic fish feeder system using pic microcon- troller. In: 2012 IEEE control and system graduate research collo- quium, IEEE, pp 343–347
14. Mishra SS, Rakesh D, Dhiman M, Choudhary P, Debbarma J, Sa-hoo S, Mishra C (2017) Present status of fish disease management in fresh- water aquaculture in india: state-of-the-art-review. J Aquacul Fish 1(003):14
15. Pule M, Yahya A, Chuma J (2017) Wireless sensor networks: a survey on monitoring water quality. J Appl Res Technol 15(6):562–570

16. Bhateria R, Jain D (2016) Water quality assessment of lake water: a review. Sustain Water Res Manag 2(2):161–173
17. Bhardwaj J, Gupta KK, Gupta R (2015) A review of emerging trends on water quality measurement sensors. In: 2015 international confer- ence on technologies for sustainable development (ICTSD), IEEE, pp 1–6
18. Guidelines for regulating coastal aquaculture. caa.gov.in/uploaded/doc/Guidelines-Englishnew.pdf, accessed: 2022–08–24
19. Naughton S, Kavanagh S, Lynch M, Rowan NJ (2020) Synchronizing use of sophisticated wet-laboratory and in-field handheld technologies for real-time monitoring of key microalgae, bacteria and physicochemical parameters influencing efficacy of water quality in a freshwater aquaculture recirculation system: a case study from the republic of Ireland. Aquaculture 526:735377
20. Dolan A (2015) The effects of aquarium size and temperature on color vi- brancy size and physical activity in bettasplendens. Maryville College, Maryville, TN, USA, Tech. Rep 53309811
21. Chen J-H, Sung W-T, Lin G-Y (2015) Automated monitoring system for the fish farm aquaculture environment. In: 2015 IEEE international conference on systems, man, and cybernetics, IEEE, pp 1161– 1166
22. Elsokah MM, Sakah M (2019) Next generation of smart aquaponics with internet of things solutions. In: 2019 19th international confer- ence on sciences and techniques of automatic control and computer engineering (STA), IEEE, pp 106–111
23. Huan J, Li H, Wu F, Cao W (2020) Design of water quality monitoring system for aquaculture ponds based on nb-iot. Aquacul Eng 90:102088
24. Ngom B, Diallo M, Gueye B, Marilleau N (2019) Lora-based mea- surement station for water quality monitoring: case of botanical garden pool. In: 2019 IEEE sensors applications symposium (SAS), IEEE, pp 1–4
25. Lin Y-B, Tseng H-C (2019) Fishtalk: an iot-based mini aquarium sys- tem. IEEE Access 7:35457–35469
26. Sapkal R, Wattamwar P, Waghmode R, Tamboli U (2019) A review- water quality monitoring system
27. Encinas C, Ruiz E, Cortez J, Espinoza A (2017) Design and implemen- tation of a distributed iot system for the monitoring of water quality in aquaculture. In: 2017 wireless telecommunications symposium (WTS), IEEE, pp 1–7
28. Tseng S-P, Li Y-R, Wang M-C (2016) An application of internet of things on sustainable aquaculture system. In: 2016 International con- ference on orange technologies (ICOT), IEEE, pp 17–19
29. Raju KRSR, Varma GHK (2017) Knowledge based real time moni- toring system for aquaculture using Iot. In: 2017 IEEE 7th international advance computing conference (IACC), IEEE, pp 318–321
30. Dupont C, Cousin Y, Dupont S (2018) Iot for aquaculture 4.0 smart and easy-to-deploy real-time water monitoring with Iot. In: 2018 Global internet of things summit (GIoTS), IEEE, pp 1–5
31. Chen Y, Han D (2018) Water quality monitoring in smart city: A pilot project. Autom Constr 89:307–316
32. Manyvone D, Takitoge R, Ishibashi K (2018) Wireless and low-power water quality monitoring beat sensors for agri and acqua-culture Iot applications. In: 2018 15th international confer- ence on electrical en- gineering/electronics, computer, telecommunications and information technology (ECTI-CON), IEEE, pp 122–125
33. Simi´c M, Stojanovi´c GM, Manjakkal L, Zaraska K (2016) Multi-sensor system for remote environmental (air and water) quality monitoring. In: 2016 24th telecommunications forum (TELFOR), IEEE, pp 1–4
34. Luna FDVB, de la Rosa AE, Naranjo JS, Jagu¨ey JG (2016) Robotic system for automation of water quality monitoring and feeding in aquaculture shadehouse. IEEE Transa Syst Man Cybern Syst 47(7):1575–1589

35. Prasad A, Mamun KA, Islam F, Haqva H (2015) Smart water qual- ity monitoring system. In: 2015 2nd Asia-pacific world congress on computer science and engineering (APWC on CSE), IEEE, pp 1–6
36. Ariffin SH, Baharuddin MA, Fauzi MHM, Latiff NM, Syed- Yusof SK, Latiff NA (2017) Wireless water quality cloud monitoring system with self-healing algorithm. In: 2017 IEEE 13th Malaysia international conference on communications (MICC), IEEE, pp 218–223
37. Wang L, Yue X, Wang H, Ling K, Liu Y, Wang J, Hong J, Pen W, Song H (2020) Dynamic inversion of inland aquaculture water quality based on uavs-wsn spectral analysis. Remote Sensing 12(3):402
38. Koditala NK, Pandey PS (2018) Water quality monitoring system using iot and machine learning. In: 2018 International Conference on Research in Intelligent and Computing in Engineering (RICE), IEEE, pp 1–5
39. Da Silva LF, Yang Z, Pires NM, Dong T, Teien H-C, Store-bakken T, Salbu B (2018) Moni- toring aquaculture water quality: design of an early warning sensor with aliivibrio fischeri and predictive models. Sensors 18(9):2848
40. AlZubi HS, Al-Nuaimy W, Buckley J, Young I (2016) An intelligent behavior-based fish feeding system. In: 2016 13th international multi- conference on systems, signals & devices (SSD), IEEE, pp 22–29
41. Zhu X, Li D, He D, Wang J, Ma D, Li F (2010) A remote wireless system for water quality online monitoring in intensive fish culture. Comput Electron Agric 71:S3–S9
42. Huang A, Huang M, Shao Z, Zhang X, Wu D, Cao C (2019) A prac- tical marine wireless sensor network monitoring system based on lora and mqtt. In 2019 IEEE 2nd international conference on electronics technology (ICET), IEEE, pp 330–334
43. Charef A, Ghauch A, Baussand P, Martin-Bouyer M (2000) Water quality monitoring using a smart sensing system. Measurement 28(3):219–224
44. Veeramsetty V, Shadamaki N, Pinninti R, Guduri N, Ashish G (2022) Water quality index estimation using linear regression model. In: AIP conference proceedings, vol 2418, no 1. AIP Publishing LLC, p 040033
45. Rupal M, Tanushree B, Sukalyan C (2012) Quality characterization of groundwater using water quality index in Surat city, Gujarat. India. Int Res J Envir Sci 1(4):14–23
46. Ahmed U, Mumtaz R, Anwar H, Shah AA, Irfan R, Garc´ıa JN (2019) Efficient water quality prediction using supervised machine learn- ing. Water 11(11):2210
47. Smith DG (1990) A better water quality indexing system for rivers and streams. Water Res 24(10):1237–1244
48. Veeramsetty V, Shadamaki N, Pinninti R, Mohnot A, Ashish G (2022) Water quality classifi- cation using support vector machine. In: AIP Con- ference Proceedings, vol 2418, no 1, AIP Publishing LLC, p 040022

IoT-based Fire Analyzer and Fire Fighting System

Shaik Fayaz Begum, K. Yaswanthi, R. Yogitha, V. Sreenath Reddy, and S. Mohammad Maaz

Abstract Assuring minimum safety of the workers at the work places as fire accidents are occurring in factories and other workplaces which had been one of the major issue that the workers are facing in current days. In this paper, an Internet of things (IoT)-based fire detection system is intended to keep individuals from fire by giving an alarm message in the crisis, and analysis of a brilliant IoT framework will be done. Fire identifiers are used to perceive the fire or smoke, and automatic fire extinguisher can help in saving lives. At the present time, IoT-based caution has been arranged using temperature and smoke sensor. It wouldn't simply signal the closeness of fire in a particular explanation yet will in like manner send-related information to convenient through IoT. By using the fire sensor, smoke sensor and there is an easy to cutting edge convertor. This project can encourage various new experts to do research in the impending space of IoT.

Keywords Internet of things (IoT) · Arduino etc.

S. F. Begum (✉) · K. Yaswanthi · R. Yogitha · V. S. Reddy · S. M. Maaz
Department of ECE, AITS, Rajampet, Andhra Pradesh, India
e-mail: fayazbegums@gmail.com

K. Yaswanthi
e-mail: yaswanthireddy16@gmail.com

R. Yogitha
e-mail: yogitharagala@gmail.com

V. S. Reddy
e-mail: vennamsreenath@gmail.com

S. M. Maaz
e-mail: mohammadmaaz2125@gmail.com

1 Introduction

A fire is a state of consuming that conveys the bursts and warmth. The fire might perhaps make hurt its occupants and serious mischief to the property. Mechanical security overview magazine communicates that there are 25,000 individuals passed on due to fire setbacks in India in the hour of 2001–2014. The damage of designs and loss of human existence can be happened due to fire disasters in the endeavors. This current assessment tries for to find the staff qualities of business factors and work factors that incorporate which prompts fire accident in the business. Balance of fire accident and fire risk level control inconvenience is extended bit by bit. Extinguishing fires and noticing conditions are extraordinary today. They focus on work on the science and development in contradicting fire calamities. They are stressed over the utilization of new development, for instance, IoT and far off sensor orchestrate in extinguishing fires and noticing field. IoT is really proper for extinguishing fires with wide degree close by far off sensor network. A critical piece of fire protection in the business is to develop the prosperity structure by using caution sign to the relationship by strategies for IoT development to the enveloping domain in the business. The possible profound damages and troublesome costs on both condition and organization require extra improvement of prosperity procedure and choice of legitimate methods in dealing with risks in industry and quick action on standard working methodology in the event that there ought to be an event of appearance of dangers detected.

The fire disaster from Hong Kong shipyard is a critical episode which gave to search for respect for the impossible capacity of hazards like fire shoot in term of ruin of human existence, their prosperity and assets, and getting through impacts. The woods fire can be constrained by utilizing IoT-based alert system which is used in every one of the organizations to keep from the fire mishaps. A fire incident has wounds offers to the workers eliminated eliminate a sad setback's very own fulfillment to a basic degree. Second and seriously roasted regions routinely leave dreadful scars, and if these scars are instantly clear, then the settlement will generally be higher as a result of excited hurts. Consume wounds can leave a harmed person with endless distress or loss of conveyability as well, the two of which will require advancing activity-based recovery. It is gigantic to realize the issue state of fire risk in the business present to the earnestness of accident and effect the authority measures to control. The effects will be capable assuming fire will make change state to smother themselves due any human or mechanical disillusionment causes and clearly impact the human existence nearby and working condition. The place of this assessment about risk presents in condition which can be lead to fire setback in the affiliation. The fire accidents occurred in adventures should be recognized by using this way of thinking, and sensible exercises in expected to control the fire disaster in the ventures are ought to be made. The backwoods fire risk must be distinguished and give the answer for firemen utilizing the IoT innovation, the untamed life sanctuary and the creature lives in the timberland are to be saved. The greatest part of the significant that fire detection system is new widely used in various safety and security applications. The major amount of fire accidents occurs due to electric short circuits. It leads to damage the property. To avoid this damage, we are using this IoT-based fire detector and extinguisher system. This can be implemented in many places like colleges, apartments, companies, factories, etc. Normal traditional fire detection systems will only detect fire and blow alarm. In IoT-based, it sends message to the head of the property and its automatically fire extinguishers.

2 Literature Review

2.1 Automatic Fire Detection System Using Arduino

Automatic fire detection using Arduino's paper was published by R. Angiline, Asst Prof CSE Aditya' s, B.TECH Abhishek B.TECH in year 2009.

A. Outcomes
     Rings alarm if fire or smoke is detected.
B. Limitation.
     Rings alarm only, it will not send any message to head. There is no extinguisher system.

2.2 Fire Detector and Extinguisher System Using Arduino

Fire detector and extinguisher system using Arduino's paper was published by AV Duraivel, S. Naveen in year 2017.

A. Outcome
     It will detect fire and automatically it gets extinguisher using fan or sprinklers.
B. Limitations
     Using external interface like Wi-Fi module or GSM to send message to head.

3 Methodology

3.1 Existing System

- In existing system, we had used normal fire detection alarms which are there from the early 2000s.
- But there is a disadvantage with this detection alarm which is that it only rings the alarm, but no extinguishing measures have been taken.

There are many fire security systems which only make the alarm or buzzer sound and LED to glow when the fire is detected and smoked (Fig. 1).

Fig. 1 Fire detection alarm
circuit

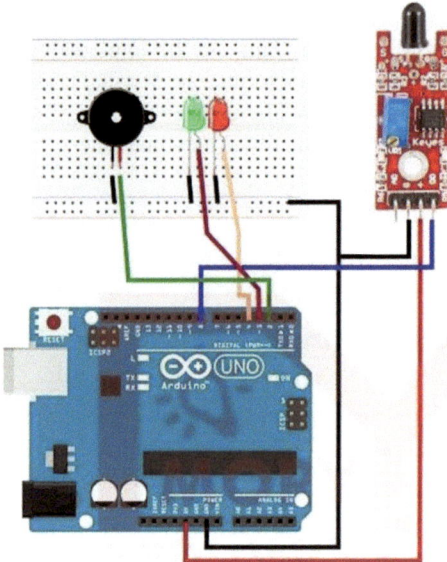

- Its waste of alarm when there is no one at the site when the fire is detected. The alarm will be beeping, but no one will be at the site.
- So, then, we updated to fire detection and automatic extinguishing system. In this, it will detect the time using sensors and sends the signal to microcontroller board, and it sends some signals to DC load fan to extinguish.
- But the problem is, there is no communication with head of property and who is staying in that property.
- So we have updated to send message and automatically time extinguish.

3.2 Proposed Methods

As seen in existing system, the projects and fire detectors only detect fire which is of no use when we are not in the spot or sight. In our project, we are going to detect fire and also extinguish it by using12V DC fan so that we can make fire get blowed off. And we are going to add a new feature where you get a notification to your mobile phone or any electronic gadget you are using by the use of the application called Blynk application. Blynk application is IoT-based application where you can monitor or get warnings, alerts based on your programming during app settings. Here, we are using NodeMCU instead of Arduino uno. Arduino uno is a microcontroller which does not have a Wi-Fi connectivity. As NodeMCU consists of in built Wi-Fi module where we can connect to the Internet and trigger its pins which they are connected to the sensors externally. We will be connecting flame sensor where we will be connected to the NodeMCU. Whenever the flame is detected, it triggers the DC fan which is

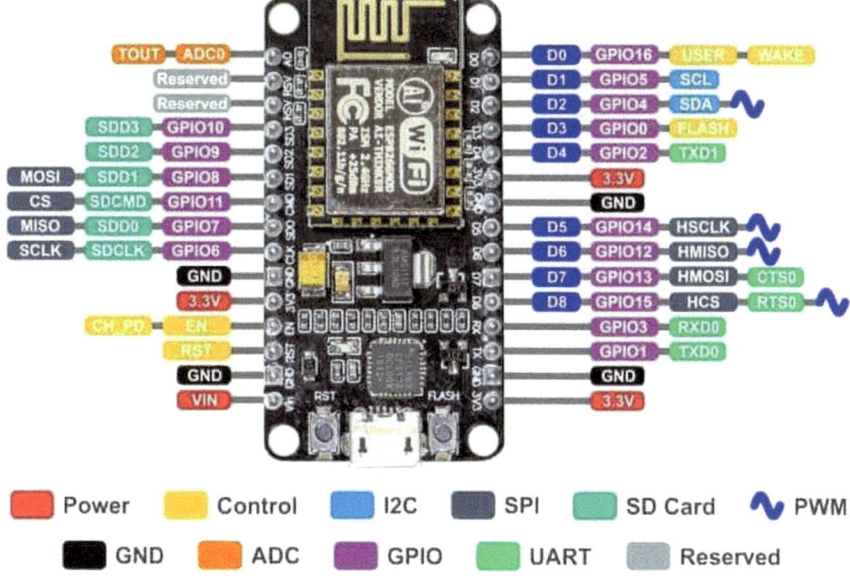

Fig. 2 NodeMCU feature

connected to the other pins and make relay module as open switch. So that DC fan gets started, and flame gets blowed off. And also it sends the notification to the Blynk application that all need to be done during programming. Because of this, we can alert the site or area which is going to be made bad because of this fire accidents (Fig. 2).

3.3 Block Diagram

We are using IR sensor or flame sensor for detecting the fire and ESP8266 (NodeMCU), it is a Wi-Fi module actually, and relay acts as switch and 12 V DC fan for extinguishing the fire (Fig. 3).

4 Results and Discussion

5 Result

Fire detection and controlling system using NodeMCU ESP8266 are made.

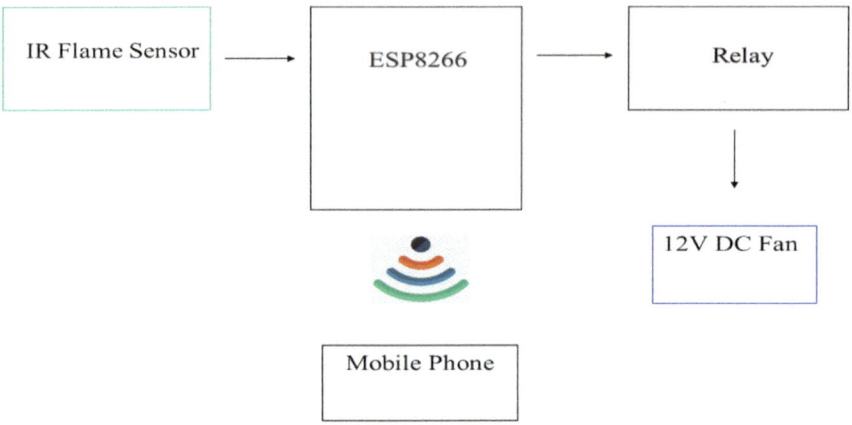

Fig. 3 Block diagram

This IoT-based project detects the nearby flame using an infrared flame sensor, and then, NodeMCU triggers the relay to extinguish the fire automatically.

It also informs the authority using IoT Blynk application (Fig. 4).

Fig. 4 Result

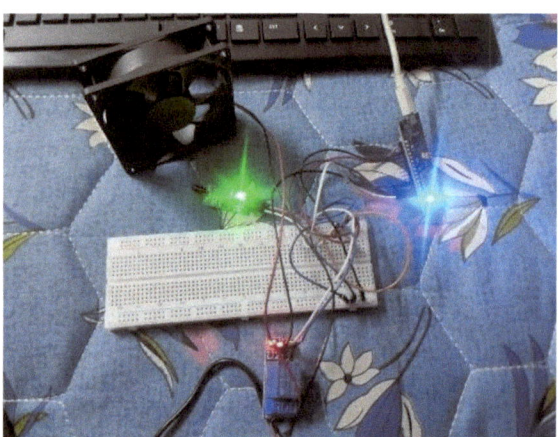

6 Future Scope and Conclusion

6.1 Future Scope

Mainly, the identifiers were for the recognition of intensity. As time and innovation progressed, identifiers additionally utilized for the fixed temperature, pace of rise, rate expectation, and direct. Till today, these finders are being used for various applications stay a suitable method for location, however, not with the end goal of life well-being. Using the thermistors and product/firmware of the finder and the framework, we really observe that the response time index (RTI) of an intensity locator will be decreased so the recognition of the warm occasion would be all the more immediately recognized. Alarm frameworks, not withstanding, were planned and introduced in the most of the uses for life well-being. A main finder is utilized for this application which is the smoke alarm. Smoke alarms and smoke cautions are the absolute strategy for the early recognition of the fire and would have saved incalculable lives. This gadgets anyway have a guideline issue that is a hotspot for undesirable cautions. Starting from the original of smoke alarms were delivered, there have been various headways to both diminishing the hour of discovery while simultaneously decline the actuation of the indicator when these results of ignition are absent. Smoke alarms and cautions are relocating from only the discovery of smoke, to mix finders and multiple indicators. In future, the finder would be even more a effective sensor, with the location something else for the results of ignition, for example, carbon monoxide (CO), carbon dioxide (CO_2), sulfur dioxide (SO_2), nitrogen oxides (NO) not withstanding intensity and particulate matter. Sensors will likewise can detect or follow when a room is involved or not and can be coordinated with inhabitant warning and departure. The improvement of further developed calculations and computerized reasoning, both inside the actual sensor and in the frontend control unit would diminish the time from the outset of a occasion to the warning of the occasion.

It is not unlikely that the recognition innovation will be actually want to identify a beginning stage of the fire instead of the blazing stage. This simultaneously could lessen the probability of an undesirable initiation from happening. Inside the following 10 years, video picture recognition (VID) will turn out to be more standard in which, through examination, the picture of one or the other smoke or fire will actually want to be segregated and distinguished from inside a room or space. The VID framework would likewise have the option to identify on the off chance that an individual is inside the space and through the mix with the notice machines, give a way of exit.

As per the notice of the inhabitants, inside the (US) United States, we are still fundamentally giving an alert all through the inhabitance and believing that the tenants will regard the admonition and go to the closest exit.

Consider this to be for future with notice inside a reason, in which the identification of fire framework will, through the sensors, know where the tenants are corresponding to where the caution is being produced from and have the option to direct them from the occasion to an exit. This might be through informing by means of warning machines yet could likewise be through the point of interaction of the discovery and notice framework to the shrewd gadget that the structure inhabitants would have on them.

Inhabitant area is likewise indispensable data for specialists on call. Right now, in the event that there is a functioning fire inside a construction, the principal obligation is to play out an essential hunt, and afterward an optional pursuit of the structure to verify that nobody is still inside.

7 Conclusion

In this project, we designed and implemented a fire detection and controlling system using NodeMCU ESP8266 which is very much necessary and helpful for fire detection and security purpose.

References

1. https://theiotprojects.com/iot-fire-detector-automatic-extinguisher-using-nodemcu/
2. https://www.researchgate.net/publication/353205967_IOT_Based_Fire_Detection_System
3. https://www.irjet.net/archives/V4/i1/IRJET-V4I1190.pdf
4. National Fire Protection Association (2001, February) Chapter 3 fundamental fire protection program and design elements. NFPA 805 Performance-based standard for fire protection for light water reactor electric generating plants. National Fire Protection Association. Standard: Gaseous Fire Suppression Systems 3.10.7
5. National Fire Protection Association (2011) Chapter 4 Annex A. NFPA 12 standard on carbon dioxide extinguishing systems. National Fire Protection Association

Multifocus Image Fusion Based on Convolutional Simultaneous Sparse Approximation

G. Obulesu, M. Sai Padma Thulasi, K. Kusuma, G. Bhavya, and N. Kesava Naidu

Abstract Convolutional simultaneous sparse approximation algorithms are based on the alternating direction approach of multipliers with various sparsity structures. Outdoor VL photos are improved using the NIR images. The NIR images utilised in this study are shown in grayscale and can be combined with the intensity parts of the VL photos, which are typically accessible in red, green, and blue (RGB) format. Using the sparse feature maps with the same supports, we try to approximate the input signals. Utilising two multimodal NIR-VL dictionaries that have already been learned and thus suggested CSSA approach for the problem concerned. An approach for fusing NIR and VL images is based on CSSA and CDL. On the basis of the SSA model, we address the CSSA problem with various sparsity structures as well as the convolutional feature learning problem in multimodal data/signals. We assess the suggested algorithms.

1 Introduction

Image processing is one of the technologies that has recently experienced significant development. Digital image processing, which is utilised in digital computers, comprises of the modifications that are included in it [1]. Though it primarily focuses

G. Obulesu (✉) · M. S. P. Thulasi · K. Kusuma · G. Bhavya · N. K. Naidu
Department of ECE, Annamacharya Institute of Technology and Sciences, Rajampet, India
e-mail: obulesuyadav66@gmail.com

M. S. P. Thulasi
e-mail: padmathulasisai@gmail.com

K. Kusuma
e-mail: kummithikusuma@gmail.com

G. Bhavya
e-mail: gunjibhavya@gmail.com

N. K. Naidu
e-mail: kesavanaidu463@gmail.com

on images, the subfield of image processing is known as signals and systems. A common procedure that includes both digital signal processing and image-specific techniques is called digital image processing [2]. The most typical illustration is Adobe Photoshop which is considered as Currently, the most popular tools for processing digital photos.

Digital image processing is the act of processing digital images using a computer. Digital image processing offers fundamental and sophisticated principles for an image processing tutorial [3]. Digital photographs are primarily processed in Adobe Photoshop.

The image fusion technique collects all the essential information from numerous photos and combines it into fewer, usually single, images [4]. This single image is more accurate and instructive than any other image from a single source since it contains all the necessary information. In addition to lowering the amount of data, image fusion strives to provide visuals that are more pertinent and understandable for both human and machine perception [5–7]. The process of combining pertinent information from two or more images into one image is known as multisensor image fusion in computer vision. The input photographs will all be less instructive than the finished image [8, 9]. The expanding array of space-based sensors accessible for uses in remote sensing. Picture fusion can be done in a variety of ways.

2   Literature Survey

2.1   High Dynamic Range Imaging

A remarkably easy way for considerably boosting the dynamic range of practically any imaging system was put out by the researchers [11, 12]. The fundamental idea is to sample the image irradiance's spatial and exposure dimensions simultaneously. Placing an ideal mask next to a traditional image detector array is one of several ways to accomplish this. Adjacent pixels on the detector receive various exposures to the scene due to the mask's pattern of spatially variable transmittance. An effective image reconstruction technique maps the collected image to a high dynamic range image 13. The ultimate result is an imaging system that can produce a significantly greater number of brightness levels and measure a very wide range of scene irradiances, with a minor drop in the number of brightness levels.

2.2   Multi-Exposure Image Fusion

A. Goshtasby developed a technique for combining multiple exposure photos of a still scene shot with a stationary camera to create an image with the most amount of information possible. The technique divides the image domain into uniform blocks and chooses the image inside each block that has the greatest information. Next, monotonically decreasing blending functions with a sum of 1 throughout the entire image domain and centres at the blocks are used to combine the selected images. A gradient-ascent approach is used to calculate the ideal block size and width for the blending functions in order to optimise the amount of information included in the combined image. Finding the image at a specific location that has the most information is the key issue at hand.

2.3 Exposure Fusion

A method for combining a bracketed exposure sequence into a high-quality image without requiring HDR was put out by T. Mertens, J. Kautz, and F. Van Reeth. The acquisition workflow is made easier by skipping the physically-based HDR assembly. This saves time on computation by avoiding camera response curve calibration. Additionally, it permits the sequence to include flash pictures. Utilising straightforward quality indicators like saturation and contrast, this approach combines many exposures. To account for the brightness change in the sequence, this is done in a multi-resolution manner. Using exposure fusion, the desired image sequence is computed. A scalar-valued weight map, which is composed of a collection of quality measurements, directs this procedure. It helps to visualise the input sequence as a collection of pictures.

2.4 Random Walks for Multi-Exposure Image Fusion

The multi-exposure picture fusion problem was approached from a new angle by Rui Shen, Irene Cheng, Jianbo Shi, and Anup Basu. It is thought to be best to strike a balance between the two quality metrics of local contrast and colour consistency using a probabilistic approach. Based on these two metrics, it is computed pixel-by-pixel probabilities that a pixel in the fused image originates from various input images, which are then utilised as fusion weights in the composition step. Local contrast is indicated by the pixel, which is how the fused image is displayed. The colour consistency measure mandates both uniformity with the surrounding natural landscape and consistency within a broad community. This measurement is based on the supposition that neighbouring pixels with comparable colours in the majority of the input images will suggest similar colours.

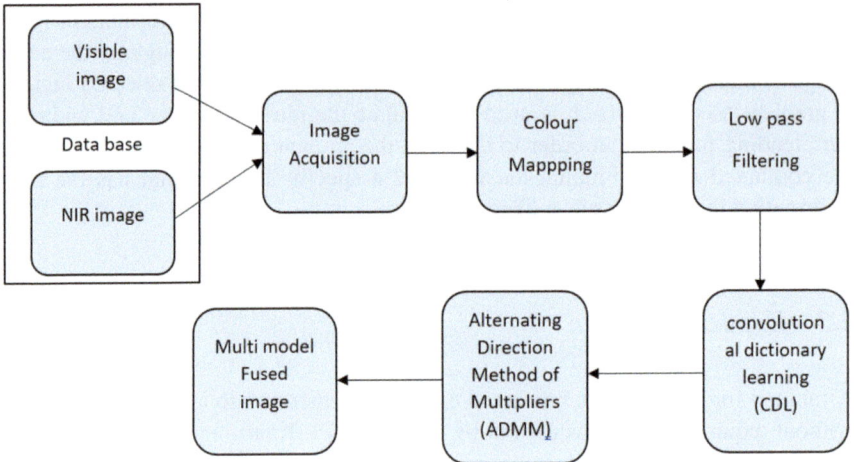

Fig. 1 Block diagram

3 Methods

3.1 Existing Method

The CDL problem can be expressed as follows: minimise D s.t. Dk, D, k = 1 K, where D. Given T sets of N dependent input signals and associated concurrent SRs (and). It is possible to solve problem utilising batch or online CDL methods because it is a common CDL problem. When using online CDL, training samples must be observed sequentially over time, whereas batch CDL requires all training data to be accessible at once. When there are more filters in the dictionary (here K) than there are training samples (here T N), online CDL is also more computationally efficient. The CDL issue can be extended to include learning multimodal convolutional dictionaries if the input signals are multimodal, and the order of modalities is fixed throughout all T sets of training samples. This can be written as n = 1 = 1 = 1 = 1 s.t. D, which has N distinct CDL issues that can be solved. Using the corresponding filters in the multimodal dictionaries, problem can be understood as learning correlated (coupled) characteristics in multimodal data (Fig. 1).

4 Proposed Method

The NIR images are distinguished by high contrast resolutions, which are useful for imaging in low-visibility atmospheric circumstances like fog or haze and for capturing scenes with vegetation. These qualities are utilised to enhance outdoor VL photos using NIR photographs. We provide an NIR-VL image fusion approach

Fig. 2 Visible image

based on CSSA and CDL in this section. Both '1' and '2,1' regularizations as well as multimodal dictionaries are used to perform the CSSA. The following list explains each stage of the suggested procedure for merging two identical-sized NIR and VL images (sn and sv, respectively). Due to the NIR images' greyscale presentation, they can be combined with the intensity components of VL images, which are often provided in using the proposed CSSA technique and two pre-learned multimodal NIR-VL dictionaries, the convolutional SRs Xn and Xv are generated for shn and shv,g, respectively (designated as Dn and Dv). The convolutional SRs are fused using the maxabsolute-value fusion criterion. Fnk and Fvk are fused convolutional SRs that contain only the most significant representation coefficients at each entry. Fvk if |Xvk(I, j)| |Xnk(I, j), otherwise Fvk if |Xnk I j) | > |Xvk I j), otherwise furthermore, the points I j) represent the locations of each pixel in the images shn and shv, whereas | | represents the absolute value of an integer and k = 1 in this example (number of filters in the dictionaries). Then, using the fused greyscale high-resolution component (Figs. 2, 3 and 4).

5 Results

First, a pair of NIR-VL pictures are sparsely approximated using the suggested CSSA algorithms with various sparsity structures. In order to do multifocus and multimodal image fusion tasks, we then apply the suggested methods. The 32 filters of size 8 × 8 in the convolutional dictionaries utilised in the experiments are learned using the online CDL approach. A multifocus picture dataset and an NIR-VL image dataset, each with 10 pairs of images, make up the training data. The RGB-NIR scene dataset and the Lytro dataset, respectively, are used to capture the NIR-VL and multifocus

Fig. 3 NIR image

Fig. 4 Fused image

photos. Both visually and through objective evaluation indicators, the fusion results are assessed. The average peak signal-to-noise ratio (PSNR), the structural similarity index (SSIM), average entropy (EN), and the average are the five metrics utilised for objective evaluations.

From the Figs. 5, 6, 7, there are two images: a visible image and an NIR image. The visible image is more amenable to human perception and contains a plethora of textural data. By acquiring considerable thermal radiation data, infrared pictures can be used to emphasise significant targets like cars, people, and other objects even in low light or other extremely hostile situations. An NIR light source is required to create an image since, like visible light, NIR is a reflected energy. During the day, the sun supplies plenty of IR light, but at night, an IR light source is needed to illuminate an area. A final image with more information is obtained when we merge the two input photos, a visible image and an NIR infrared image (Fig. 8, 9, and 10).

To create the fused convolutional SRs, we fuse the convolutional SRs (with identical supports) using the elementwise maximum absolute value rule. The other steps of the two algorithms are identical. The acquired fusion findings show that using CSSA results in significant gains in terms of greater contrast resolutions and better fusing of multifocus edges (boundaries where one side is in focus, and the other side

Fig. 5 Visible image

Fig. 6 NIR image

Fig. 7 Fused image

Fig. 8 Visual image

is out of focus). The figure depicts an example of fusion results obtained using the two procedures. The objective evaluation results in Table II also show that CSSA increases the overall performance of the CSA-based multifocus image fusion approach.

Fig. 9 NIR image

Fig. 10 Fused image

6　Conclusion

Based on the alternating direction approach of multipliers, algorithms for convolutional simultaneous sparse approximation with diverse sparsity structures have been presented. We tested the efficacy of the suggested approaches by applying them to two distinct types of picture fusion challenges and comparing the results to those of current image fusion methods. A novel near-infrared and visible light picture fusion approach based on convolutional simultaneous sparse approximation was suggested in particular.

References

1. Veshki FG, Vorobyov SA (2022) Convolutional simultaneous sparse approximation with applications to RGB-NIR image fusion, arXiv:2203.09913
2. Shaik F (2022) An enhanced image processing model for earlier detection and analysis of diabetic foot hyperthermia through cognitive approach. In: Gunjan VK, Zurada JM (eds) Modern approaches in machine learning & cognitive science: A walkthrough. Studies in Computational Intelligence, vol 1027. Springer, Cham. https://doi.org/10.1007/978-3-030-96634-8_48
3. Tropp JA, Gilbert AC, Strauss MJ (2006) Algorithms for simultaneous sparse approximation. part I: greedy pursuit. Signal Process 86(3):572–588
4. Gunjan VK, Prasad PS, Fahimuddin S, Bigul SD (2019) Experimental investigation to analyze cognitive impairment in diabetes mellitus. In: Kumar A, Mozar S (eds) ICCCE 2018. Lecture Notes in Electrical Engineering, vol 500. Springer, Singapore. https://doi.org/10.1007/978-981-13-0212-1_79
5. Tropp JA, Gilbert AC, Strauss MJ (2006) Algorithms for simultaneous sparse approximation. part II: convex relaxation. Signal Process 86(3):589–602
6. Jaya Krishna N, Shaik F, Harish Kumar GCV, Naveen Kumar Reddy D, Obulesu MB (2021) Retinal vessel tracking using Gaussian and Radon methods. In: Kumar A, Mozar S (eds) ICCCE 2020. Lecture Notes in Electrical Engineering, vol 698. Springer, Singapore. https://doi.org/10.1007/978-981-15-7961-5_37
7. Boßmann F, Krause-Solberg S, Maly J, Sissouno N (2022) Structural sparsity in multiple measurements. IEEE Trans Signal Process 70:280–291
8. Fahimuddin S, Lavanya D, Manasa T, Maruthi Praveen S, Raveendra Babu M (2023) Image dehazing using improved dark channel and Vanherk model. In: Kumar A, Senatore S, Gunjan VK (eds) ICDSMLA 2021. Lecture Notes in Electrical Engineering, vol 947. Springer, Singapore.https://doi.org/10.1007/978-981-19-5936-3_80
9. Zheng B, Zeng C, Li S, Liao MG (2022) The MMV tail null space property and DOA estimations by tail-'$_{2,1}$ minimization. Signal Process 194:108450
10. Yang B, Li S (2021) Pixel-level image fusion with simultaneous orthogonal matching pursuit. Inf Fusion 13(1):10–19
11. Veshki FG, Ouzir N, Vorobyov SA, Ollila E (2021) Coupled feature learning for multimodal medical image fusion. arXiv:2102.08641
12. Fahimuddin S, Subbarayudu T, Vinay Kumar Reddy M, Venkata Sudharshan G, Sudharshan Reddy G (2023) Retinal boundary segmentation in OCT images using active Contour model. In: Kumar A, Senatore S, Gunjan VK (eds) ICDSMLA 2021. Lecture Notes in Electrical Engineering, vol 947. Springer, Singapore. https://doi.org/10.1007/978-981-19-5936-3_82
13. Li J, Zhang H, Zhang L, Ma L (2015) Hyperspectral anomaly detection by the use of background joint sparse representation. IEEE J Sel Top Appl Earth Obs Remote Sens 8(6):2523–2533

Smart Reflecting Surface Approach for Reconfigurable Wireless Signal Transmission

K. Shankar, V. Yashwanth, A. Sumana, G. Sree Sandhya, and S. Vinay Kumar

Abstract In Wireless communication, when direct route is insufficient, wireless channels can be Reconfigure with an effective Intelligent Reflecting Surface (IRS) by utilizing the software control metasurfaces that serve as mirrors to reflect signals from source to target. In this analysis, compare IRS supported supported transmission with the repetition coded DF relaying in terms of channel gains and distance. When operating within rate constraint, IRS-supported transmission reduces transmitting power. Energy efficiency will rise as a result of the power reduction, and targets can be detected outside of the non-line-of-sight (NLos) using IRS-assisted radars. communication between a source with a single antenna and a destination with a single antenna. Capacity of wireless channels can be increased by involving an IRS surface configured to reflect signal towards target. Replacing the in place of relay with IRS minimizes total power minimization and improve energy efficiency.

Keywords IRS · DF relay · NLos target · Total power

1 Introduction

A reflect array is a layer that "reflects" an imposing plane wave in the type of a beam [1]. Real-time reconfigurable reflective surfaces, sometimes referred to as software-control metasurfaces and intelligent reflecting surfaces (IRS), have recently gained popularity in communications technology. By adjusting the propagation environment, the main objective is to make transmission possible from a source to a target. To achieve this, the IRS set up beamform its received signal in the direction of the destination. A relay actively evaluates the received signal prior retransmission an amplified signal, as opposed to an IRS, which passively reflects the signal no amplification but also with beamforming. [2] This particular use of IRS enhances the received signal energy at the remote users and broadens the BS coverage. An IRS

K. Shankar (✉) · V. Yashwanth · A. Sumana · G. S. Sandhya · S. V. Kumar
Bengaluru, India
e-mail: shan87.maddy@gmail.com

© The Author(s), under exclusive license to Springer Nature Singapore Pte Ltd. 2023   87
A. Kumar et al. (eds.), *Proceedings of the 4th International Conference on Data Science, Machine Learning and Applications*, Lecture Notes in Electrical Engineering 1038, https://doi.org/10.1007/978-981-99-2058-7_10

is made up of a sizable periodic array of subwavelength scattering meta-material components, which together create an electrically thin two-dimensional surface and reflect the incoming signal by adding a predetermined phase shift. The IRS can be used in integrate sensing and communication, 3D object detection, IRS-aided wireless systems. IRS -aided radars offer NLos target detection. The total transmits power minimization under rate constraints and effective energy efficiency. When the IRS are deployed power is minimized IRS beampattern gain with communications SNR constraints. The signal is transmitted from source of the single input and single output system and receives at the destination with the blocking objects the channel capacity is increased by involving the additional component which retransmits the signal directing towards the destination. The is done using the DF relaying and IRS. Using the IRS is efficient, it reduces the transmitting power and energy efficiency [3].

2 Literature Survey

Ozdogan G. Larsson Emil Bjornson Ozgecan: Comparison of repetition coded with IRS-supported transmission. The goal of DF relaying is to ascertain the size of surfaces required to defeat relaying. Adding the extra equipment in the communication may increase the capacity. The power consumptions will also be reduced with this technique.

Qingqing Wu, Xinrong Guan, Rui Zhang: The IRS technology for constructing cost-effective, energy- and spectrum-efficient wireless networks. Efficient designing of WIT as well as WET systems using IRS assistance systems with both. The answers to operating system-specific problems including channel estimation and placement, as well as IRS inactive reflection optimization.

A.M. Elbir, K Vijay Mishra, Symeon Chatzinotas and M. R. Bhavani Shankar: IRS modifies the behavior of wireless media, which has the potential to enhance the efficiency and dependability of wireless systems such radar remote sensing as well as communication. IRS deployment benefits, including as coverage expansion and interference suppression, were introduced in addition to ISAC scenario with common and specialized surfaces.

Qingqing Wu & Rui Zhang: The wave propagation for improved performance is done by IRS. A single IRS is implemented in the IRS-enhanced point-to-point multiple input single output (MISO) wireless system to aid in communication between many access points and a user using a single antenna. The signal sent direct from of the access point and the one reflected by the IRS are concurrently received by the user. By considering the global channel state information at IRS, the suggested methods are semidefinite relaxations. The proposed methods demonstrate significant performance improvements.

3 Methodlogy

Considering transmission from a source with one antenna to a destination with one antenna. Establishment of the deterministic flat-fading channels. By including the extra equipment in this communication, the channel capacity might be improved. We take into account two configurations: an IRS that is arranged to "reflect" signals in the direction of the destination and a relay that runs in DF mode.

3.1 Existing Method

A DF relay is far more sophisticated than an AF unit due to its full processing power, whereas an AF relay only amplifies and re - transmits that signal without decoding. The DF procedure additionally demands a complex media access control layer that the AF protocol does not require. A DF relay is almost as complicated as a base station overall [1] (Fig. 1).

We deploy a quarter relay in this configuration. This configuration is shown in figure, which uses the traditional repetition-coded DF relaying protocol with two equivalent phases for transmission. The source transmits during the first phase, and the destination receives the signal. Deterministic fading channels channel refers to the channel that has been constructed in between source and its destination. The extra equipment is used so that the capacity may be increased when this channel is weak. This DF relay is furnished. When a signal is sent from a source towards a destination, the DF relay receives it and retransmits it [2].

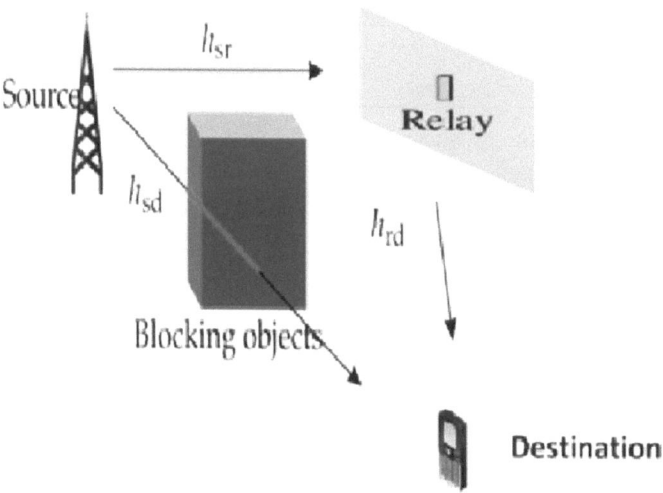

Fig. 1 Transmission supported by a relay

h_{sr}—The source to DF Relays deterministic channel is denoted by the symbol.

h_{rd}—The channel between DF relay and destination is represented.

h_{sd}—Representing deterministic flat-fading channel.

3.2 Proposed Method

Each reflecting elements on the intaelligent surface is connected to a tunable chip that allows it to change its load impedance, such as a PIN or varactor diode. The intelligent reflecting surface is made up of numerous subwavelength reflecting elements (small antennas like microstrip patches). In the setup, signal is transmitted from the source antenna to the destination. The same signal is received at the IRS with N elements, the software controlled metasurfaces which reflects the which is received from the source to the destination with help of passive reflecting elements, whereas the DF relays decodes the information and re modulates and re transmitted but IRS [4] elements just reflects the signal from source to destination (Fig. 2).

h_{sr}—Deterministic channel from source to IRS is represented.

$h_{rd,k}$—Channel between the IRS and destination represented.

$h_{sd,\ k}$—Representing deterministic channel with blocking objects to destination.

The IRS requires large number of small reconfigurable elements to overcome the DF relaying, the DF relaying consists of 2 phases where in the first phase the signal transmits from source to destination and in the second phase the same signal is given to DF relay and in this method also similar some portion of the signal is received to the k elements present in the IRS [5, 6] which passively just reflects the signal to the targets.

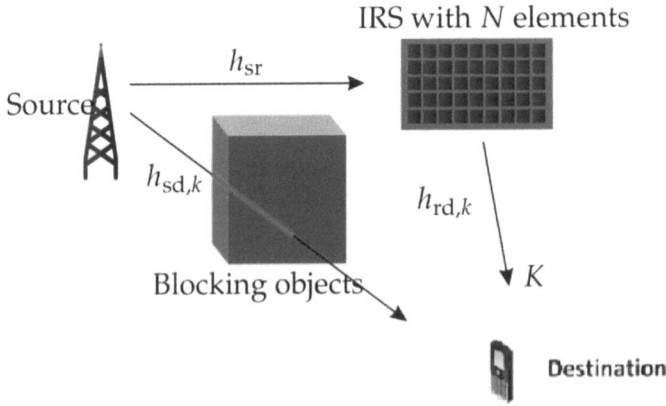

Fig. 2 IRS supported transmission

4 Result

Signal that travels at least 10 m from source to destination. the line of sight and non-line-of-sight channel gain comparisons. The channel gain is entirely dependent on the carrier frequency; as that of the carrier frequency changes, the channel gain shifts downward because the destination is close to the source or IRS.

The carrier frequency is of 3GHZs the channel gain of the targets presents on the line-of-sight detection increase as the distance varies, the channel gain varies from the −50db to the −70db and for the non-line-of-sight target channel gain varies using the −61db to the −108db. The IRS uses to detect NLos of targets similarly, when the carrier frequency is changed to the 5GHZs the channel gain is shifted of nearly -8dbs here also the channel gain is large when the destination is near to the source or IRS. As the distance varies from the 10 m to 100 m the channel gain varies from −55db to −75db for line-of-sight targets and for Non-Line-of -Sight channel gain varies from the −68db to −118db nearly.

In calculating the channel gain the gains of the transmitting antenna and receiving antenna and the channel gain calculating of at a certain distance represented with d[m]. Gains of the antennas are used at the source and the destination are of equal gains of −5db (Fig. 3).

Transmitting power is represented in dB. The single input single output system uses the high transmitting power to minimize the transmitting power the DF relaying transmitted power is less for the low rates, hear the transmitting power is of -3db, achieving the transmitting power the number of elements must be > 164. As the distance increase and the higher rates, the IRS over comes the DF relaying at the distance of 80 m, noise is 10 and is tis is changed to the high to 12 or 14 the transmitting power is increased to 14db to 5db and the noise lowered less than 10db the transmitted power is reduced.

The noise figure is changed the transmitted power in dB is shifted to above and below nearly of 5db,so the use of IRS supported system give better results in minimization of transmitted power, when compared to the SISO,DF relaying (Fig. 4).

F = frequency, N = Noise figure

The energy efficiency of the single input single out system is same as the IRS system and the efficiency of the DF relaying is high for the small rates. As the rates increases the DF relaying energy efficiency is high up to the rates of 8 as the rate is higher than the 8.48 the DF relaying efficiency is low. But the IRS system is highly efficient for the high data rates. As the rates are greater than the 8.48 the IRS is more efficient than the two techniques DF relaying and SISO (Fig. 5).

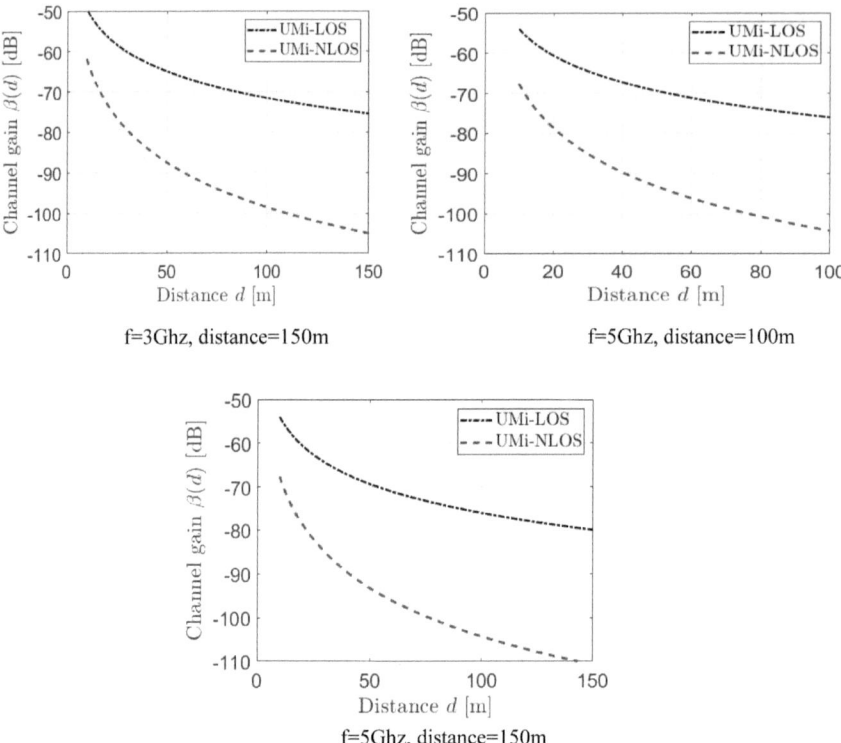

Fig. 3 For antenna gains Gt = Gr = 5db Channel gains for Los, NLos **a** f = 3Ghz, distance = 150 m, **b** f = 5Ghz, distance = 100 m, **c** f = 5Ghz, distance = 150 m

5 Conclusion

Comparison has been done with DF relaying with new technology IRS, as number of elements is used to be competitive with the DF relaying. The IRS needs hundreds if elements when compared to DF relaying nearly 164 elements. The main advantage of the IRS is no need of any power amplifiers and any decoders as it passively reflects the signal coming from the source to destination. Whereas the DF relay demodulates and decode and amplifies the signal and retransmits to destination as it becomes more power consumptions. The main key is the IRS achieves energy efficiency for higher rates also IRS needs large number of surfaces to achieve low rates.

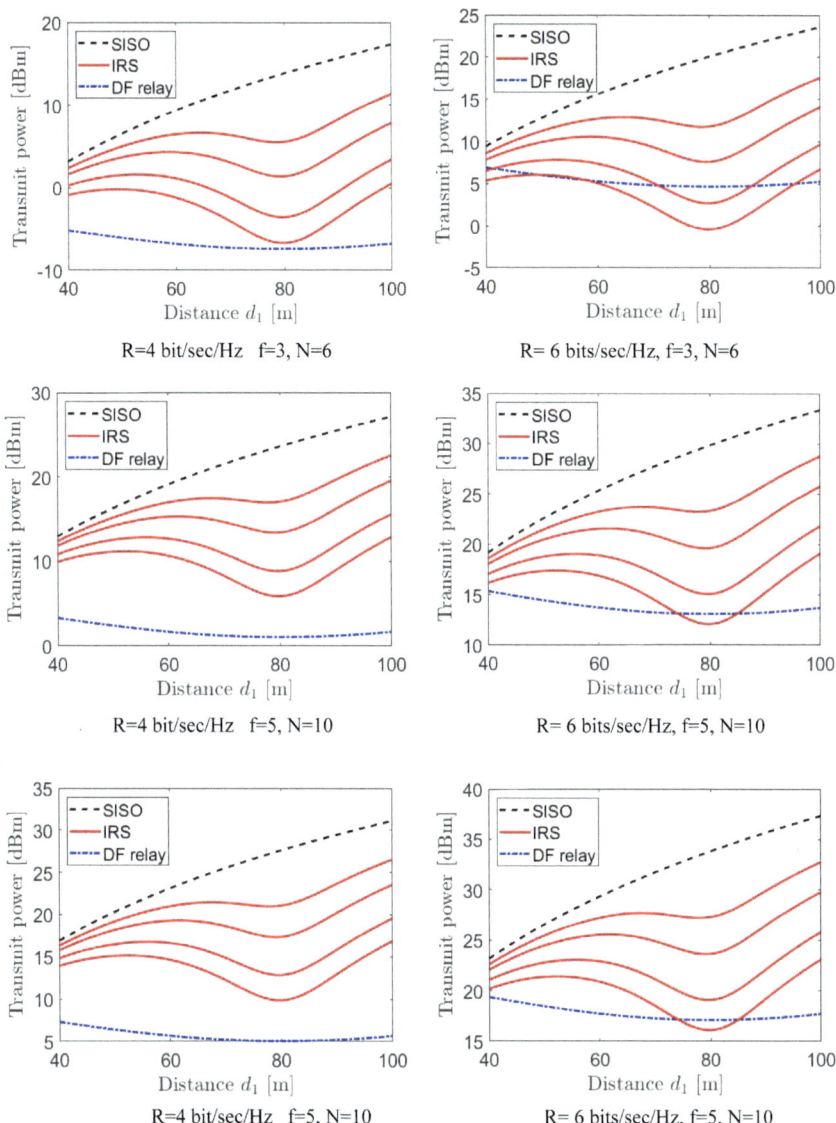

Fig. 4 Simulation set up distance as variable. **a** R = 4 bit/sec/Hz f = 3, N = 6 **b** R = 6 bits/sec/Hz, f = 3, N = 6. **c** R = 4 bit/sec/Hz f = 5, N = 10. **d** R = 6 bits/sec/Hz, f = 5, N = 10. **e** R = 4 bit/sec/Hz f = 5, N = 10. **f** R = 6 bits/sec/Hz, f = 5, N = 10

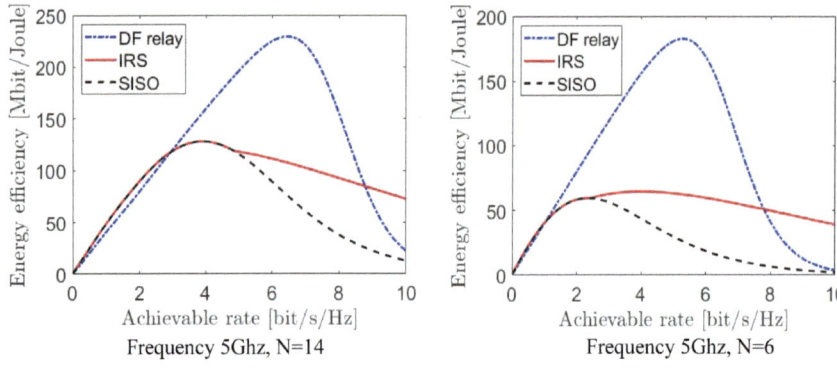

Fig. 5 Energy efficiency of DF relay, SISO, IRS. **a** Frequency 5Ghz, N = 14 **b** Frequency 5Ghz, N = 6

References

1. Arun Prakash S, Sumithra MG, Shankar K, Grover A, Singh M, Malhotra J (2021) Performance investigation of spectral-efficient high-speed inter-satellite optical wireless communication link incorporating polarization division multiplexing
2. Latif G, Saravanakumar N, Alghazo J, Bhuvaneswari P, Shankar K, Butt MO (2020) Scheduling and resources allocation in network traffic using multiobjective, multiuser joint traffic engineering
3. Reddy YP, Vaishnavi A, Devi MS, Prasad MS, Reddy BS (2023) Multimodal medical image fusion approach using PCNN model and shearlet transforms via Max Flat FIR Filter. In: Kumar A, Senatore S, Gunjan VK (eds) ICDSMLA 2021. Lecture Notes in Electrical Engineering, vol 947. Springer, Singapore. https://doi.org/10.1007/978-981-19-5936-3_73
4. Reddy YP, Hemadri G, Nithya UJ, Basha DH, Krishna YH (2023) Reader for Blind Using the Raspberry Pi. In: Kumar A, Senatore S, Gunjan VK (eds) ICDSMLA 2021. Lecture Notes in Electrical Engineering, vol 947. Springer, Singapore. https://doi.org/10.1007/978-981-19-5936-3_69
5. Du L, Zhang W, Ma J, Tang Y (2021) Reconfigurable intelligent surfaces for energy efficiency in multicast transmissions. IEEE Trans Veh Technol 70(6):6266–6271. https://doi.org/10.1109/TVT.2021.3080302
6. Abdelhady AM, Salem AKS, Amin O, Shihada B, Alouini M-S (2021) Visible light communications via intelligent reflecting surfaces: metasurfaces vs mirror arrays. IEEE Open J Commun Soc 2:1–20. https://doi.org/10.1109/OJCOMS.2020.3041930

ML Based Intelligent Transport System Through GPS and Traffic Analysis

K. Shankar, P. Guru Hema Sree, D. Harsha Vardhini, C. Abhinavi, S. K. Abdul Kalam, and P. Gowtham

Abstract Machine learning and feature extractions are playing a vital role in internet and health domain. The aim of this paper is to develop a system for forecasting precise and timely data on traffic flow. Everything that could affect how much traffic is moving along the road, such as traffic lights, accidents, protests, and even road repairs that could cause a delay, is referred to as the "traffic environment." If a motorist or rider has prior awareness of all of the above as well as the many other real-world situations that can impact traffic, they can make a better decision. In recent decades, traffic data has increased significantly, and big data concepts for transportation are becoming more prevalent. The present traffic flow prediction methods utilize a few prediction models. They still fall short when handling applications that are used in the real world, though. Since there is an excessive amount of data available for the transportation system, it is difficult to anticipate the traffic flow accurately. The aim of this paper is to significantly reduce the complexity of the big data analysis for the transportation system using machine learning, genetic, soft computing, and deep learning methods. Additionally, image processing algorithms are used to recognize traffic signs, which ultimately aid in the proper training of autonomous vehicles. Finally, the goal of this project is to develop an appropriate machine learning tool which can predict the traffic for intelligent transportation system using GPS, speed, direction, and start–end junction features' data. The algorithms used are KNN and Random Forest.

Keywords Machine learning · Traffic prediction · Genetic algorithm · Big data · Soft computing

K. Shankar (✉) · P. Guru Hema Sree · D. Harsha Vardhini · C. Abhinavi · S. K. Abdul Kalam · P. Gowtham
Department of ECE, Annamacharya Institute of Technology and Sciences, Rajampet, India
e-mail: shan87.maddy@gmail.com

© The Author(s), under exclusive license to Springer Nature Singapore Pte Ltd. 2023
A. Kumar et al. (eds.), *Proceedings of the 4th International Conference on Data Science, Machine Learning and Applications*, Lecture Notes in Electrical Engineering 1038,
https://doi.org/10.1007/978-981-99-2058-7_11

1   Introduction

These days, the globe is incredibly busy, which causes a lot of problems in society. Traffic congestion is one of the key issues due to population growth. Communication technologies for road transportation are managed by intelligent transportation systems to improve efficiency and safety. The intelligent transportation system's many applications are utilized to gather information, lessen environmental effect, improve traffic management, and boost the benefits of transportation. Through the use of the most recent traffic management systems, several issues, such as traffic congestion and low safety, can be resolved. ITS is preferred for a variety of applications and the transportation industry due to their numerous advantages. Additionally, because cellphones have a variety of sensors, they can be used to track and identify the pace and density of traffic due to the reduced connection of traffic flow. Nowadays, drivers utilize smartphones, which are tracked to determine the condition and speed of the road. Traffic flow information is gathered both in the past and in the present via inductive loops, radars, cameras, mobile Global Positioning Systems, crowdsourcing, social media, and other kinds of sensors. We have reached the era of big data transportation thanks to the widely used classic traffic sensors and new coming traffic sensor technologies. Data is being used in transportation management and control more and more. Although there are numerous methods and models for predicting traffic flow, the majority of them still fall short and use superficial traffic models. This encourages us to reconsider the challenge of traffic flow prediction with deep learning models and the vast amount of traffic data. Recently, both academic and commercial interests have been focused on deep learning, a type of machine learning technique. Application areas include categorization tasks, motion modeling, dimensionality reduction, natural language processing, and others.

2   Literature Review

This study intends to build a system employing the deep neural network method, a subtype of artificial intelligence, to add intelligence to the existing traffic control system in place at a four-way intersection. The main objective of this technology is to substitute artificial intelligence for the traffic control system's timing [1].

One of the key concerns with an intelligent transportation system is how to estimate short-term traffic. In order to effectively manage traffic, commuters need to be able to choose the best transport modes, routes, and departure times. It makes sense to enhance traffic data analysis in order to boost forecast accuracy. We are motivated to use deep learning techniques to increase the accuracy of short-term traffic forecasts due to the recent emergence of plentiful traffic data and computing power. Long short-term memory (LSTM) network-based innovative traffic forecasting methodology is proposed. The suggested LSTM network, which differs from traditional forecast models in that it uses a two-dimensional network made up of numerous memory units, takes into account temporal-spatial correlation in the traffic system. The suggested LSTM network's superior performance is confirmed by comparison with other comparable forecast models. [2].

Geometric mobile data can be handled by deep learning effectively, but this presents a challenge for other ML algorithms. Multivariate data that is represented by coordinates, topology, metrics, and order is referred to as geometric data. Point clouds and graphs, which have significant geometric qualities, can naturally depict mobile data, such as mobile user position and network connectivity. Devoted deep learning architectures, like Point Net++ and Graph CNN, can effectively model this data. The geometric mobile data analysis could be transformed by using these structures [3].

3 Methodology

3.1 Existing Method

The code was created using the logistic regression algorithm. This algorithm is one of the most regulated and under strict control machine learning algorithms. This algorithm makes predictions about the actual valued output based on inputs, such as location and days. These inputs form the basis of traffic forecast. The most significant benefits of traffic management are the decreases in time consumption and pollution [1]. The device can be useful in an emergency by showing the driver another route (Fig. 1).

Logistic regression is one of the machine learning algorithms that is most frequently employed in the supervised learning category. It is used to forecast the categorical dependent variable using a specified set of independent variables. Logistic

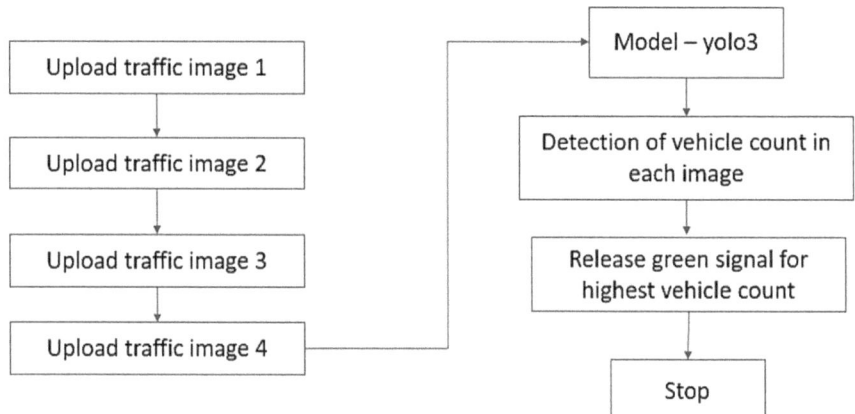

Fig. 1 Flowchart

Fig. 2 Logistics regression accuracy

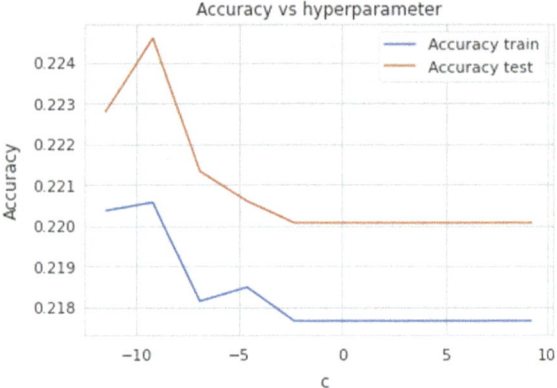

Fig. 3 Confusion matrix on logistic regression

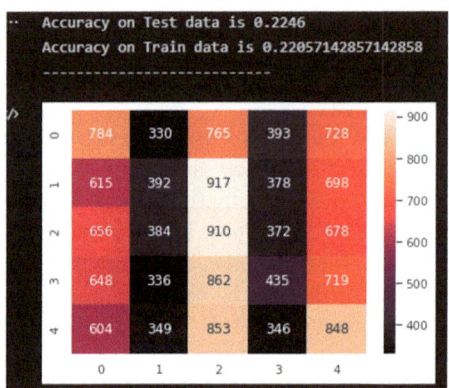

regression is used to predict the output for a dependent variable that is categorical. The outcome must therefore have a discrete or categorical value [4]. It gives probabilistic numbers between 0 and 1, rather than the precise values of 0 and 1. It can be either true or false, 0 or 1, or yes or no (Figs. 2 and 3).

3.2 Proposed Method

Although various traffic prediction models are used in the current methods for estimating traffic flow, they are still insufficient to deal with real-world scenarios. This reality drove us to focus on the problem of traffic flow forecasting utilizing traffic data and models [1]. Since there is an incredible amount of data available for the transportation system, it is difficult to anticipate the traffic flow accurately. We use ML models such as RF and KNN to anticipate the occurrence of traffic. Performance for each model is compared in order to demonstrate greater accuracy (Fig. 4).

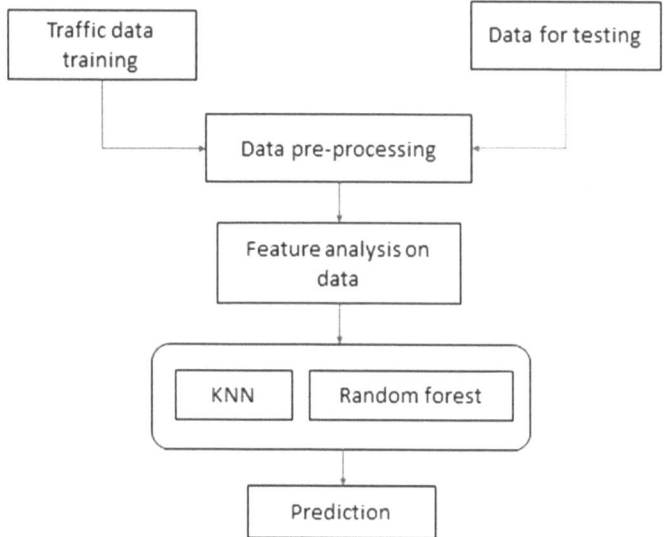

Fig. 4 Flow diagram

Popular machine learning algorithms that are a part of the supervised learning technique include Random Forest and K-Nearest Neighbor. Based on the supposition that the new data and the previous cases are comparable, KNN assigns the new case to the category that matches the existing cases the closest. In ML, RF can be applied to Classification and Regression issues [5]. The RF classifier aggregates the outcomes from numerous decision trees applied to distinct subsets of the input dataset in order to improve the predicted accuracy of the input dataset. Instead of relying exclusively on one decision tree, the RF uses predictions from each tree and predicts the result based on the votes of the majority of projections. The greater number of trees in the forest prevents higher accuracy and overfitting. The KNN algorithm stores all of the available data and classifies new data points based on similarity. This implies that new data can be instantly and precisely categorized using the KNN approach as it is generated.

4 Results and Analysis

See Figs. 5, 6, 7, and 8.

Fig. 5 Feature importance

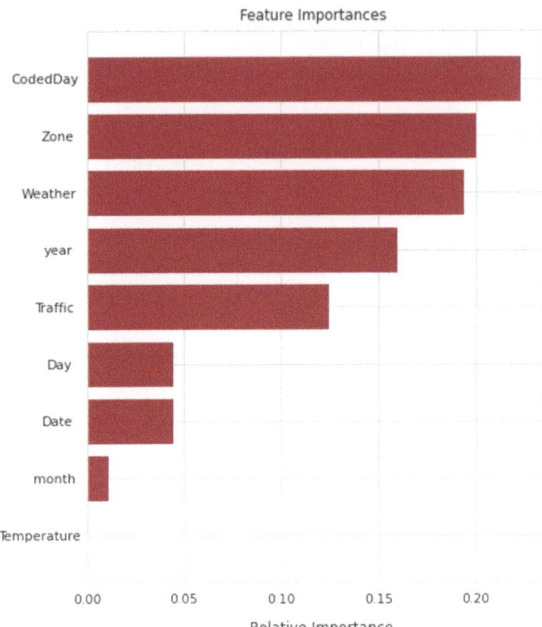

Fig. 6 KNN accuracy

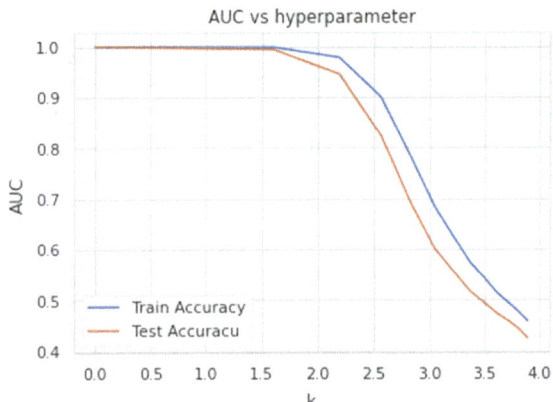

Fig. 7 KNN and Random Forest confusion matrix

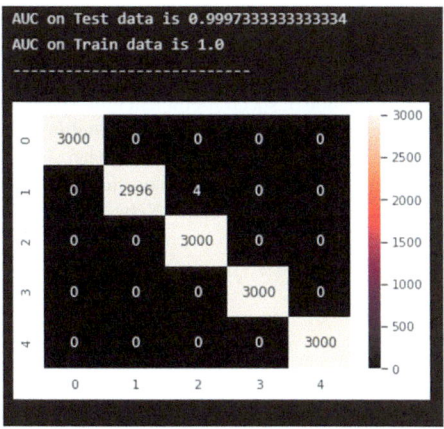

Fig. 8 Total performance table

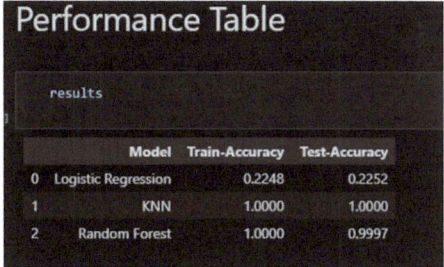

5 Conclusion

This paper suggests a system for analyzing and predicting traffic in real time. For prediction, this model employs the Random Forest algorithm and the KNN algorithm. Compared to the current system, the proposed method offers a significantly improved efficiency. It modifies the difficult issues from the dataset. By contrasting them with two sets of data, this forecast can support in evaluating the traffic flow. People or users may find it easier to judge the flow of traffic with the help of every prediction, and they may even choose their route using a navigation. Both algorithms have been considerably enhanced. Python programming is used to carry out the proposed work.

References

1. Geetha V, Gomathy CK, Harshitha T, Vijay Nagendra Varma P (2021) A traffic prediction for intelligent transportation system using machine learning
2. Mahrishi M, Morwal S (2020) Index point detection and semantic indexing of videos-a comparative review. Advances in intelligent systems and computing

3. Zhao Z, Chen W, Wu X, Chen PCY, Liu J (2017) LSTM network: a deep learning approach for short-term traffic forecast. IET Intelligent Transport Systems
4. Arun Prakash S, Sumithra MG, Shankar K, Grover A, Singh M, Malhotra J (2021) Performance investigation of spectral-efficient high-speed inter-satellite optical wireless communication link incorporating polarization division multiplexing. Opt Quant Electron 53:270
5. Latif G, Saravanakumar N, Alghazo J, Bhuvaneswari P, Shankar K, Butt MO (2020) Scheduling and resources allocation in network traffic using multiobjective, multiuser joint traffic engineering. Wireless Netw 26:5951–5963

Monitoring the Farming Conditions Using IoT

K. Riyazuddin, R. Anitha, A. Devika Chowdary, P. Durga Prasad,
and C. Girish Varma

Abstract The Internet of Things will transform every sector and everyone's lives by giving everything a sentient existence. It is a group of various devices that function as a self-configuring network. The use of Internet of Things in smart farming is revolutionising conventional agriculture by enhancing its productivity, expanding its accessibility to farmers, and lowering crop loss. The objective is to develop a system that can communicate with farmers in a number of ways. With real-time information from the fields, the tool will support farmers in their efforts to engage in smart farming (temperature, humidity, soil moisture, UV index, and IR).

Keywords ESP32s · DHT11 temperature and humidity sensor · Internet of Things (IOT) · Smart farming

1 Introduction

One of the key industries in India is agriculture. The support of human life is dependent on agriculture. Agriculture production rises in lockstep with population growth. Agriculture output essentially depends on the seasonal conditions if there are not enough water sources. IoT-based smart agriculture systems are used to improve agricultural outcomes and solve difficulties.

The term "Internet of Things" refers to a collection of hardware, software, and networking-enabled devices that allow for data sharing and communication between items. Farmers are gaining a plethora of advantages by implementing the IOT programme. Farmers have benefited from cost savings and higher agricultural yields. The irrigation system's primary goal is to create and maintain the right temperature and soil moisture conditions for the best possible development of crops [1].

In order to provide current information about food production, global and regional agricultural monitoring systems are being developed. An Internet of Things-based

K. Riyazuddin (✉) · R. Anitha · A. Devika Chowdary · P. Durga Prasad · C. Girish Varma
Annamacharya Institute of Technology and Sciences, Rajampet, Andhra Pradesh, India
e-mail: Shaik.riyazuddin7@gmail.com

© The Author(s), under exclusive license to Springer Nature Singapore Pte Ltd. 2023
A. Kumar et al. (eds.), *Proceedings of the 4th International Conference on Data Science, Machine Learning and Applications*, Lecture Notes in Electrical Engineering 1038, https://doi.org/10.1007/978-981-99-2058-7_12

smart farming system is developed to monitor the agricultural field using sensors for light, humidity, temperature, soil moisture, etc. The state of the fields is always accessible to the farmers. Internet of Things-based smart farming is far more efficient than conventional farming methods. A DHT11 Sensor and an ESP32S Node MCU Module are used in the proposed Internet of Things-based irrigation system.

2 Literature Review

A technology known as the "Internet of Things" enables connections and communication between things. This helps to improve the procedures and methods utilised in both business and agriculture. In order to enhance planting production, a system that explains smart farming is offered. A control system and a sensor system are the two main components of smart farming. Two Arduino boards are used to set up the sensor and control systems. The system's controls are programmed in Python. An LCD display and a serial monitor, respectively, show the values from the numerous sensors that were seen [2].

It is impossible for our world to exist without agriculture. It satisfies every requirement that a person has to live in this world. Automation is replacing outdated ways as technology advances with the rise of the Internet of Things, leading to substantial gains in a number of industries. In a wide range of sectors, including smart homes, waste management, automobiles, industries, farming, health, grids, and other areas, smarter technologies are constantly being enhanced at this time. Through the use of automation, the Internet of Things development has supported advancements in farming [3].

By making everything smart and intelligent, the Internet of Things (IoT) has revolutionised every industry and raised the standard of living for the average individual. An autonomous network of devices is referred to as "IoT." IoT-based technology is enabling the development of "intelligent smart farming," which improves agriculture production while lowering waste and increasing cost-effectiveness. This project aims to develop an innovative, clever IoT-based agriculture stick that will let farmers get real-time data (temperature, soil moisture) for effective environment monitoring. They will be able to practise smart farming as a result, increasing their total productivity and product quality [4].

This paper's major objective is to present cutting-edge innovation and talk about how it can advance agriculture. In the last century, agriculture has made some essential advancements, similar to how machinery has. Despite state-of-the-art innovation, growers and collectors perform better than their predecessors or have experienced only modest alterations. Agribusiness must use innovation effectively to increase productivity and employee employability. This exam's major objective is to decide how agricultural technology should be used. There are numerous methods to leverage innovations to boost productivity [5].

Physical things that are part of the Internet of Things can speak to one another when they are online. The agricultural industry, which by 2050 will be able to feed

9.6 billion people worldwide, depends on IoT agriculture techniques. In this work, a system is developed to monitor crop fields and manage irrigation using sensors (soil moisture, temperature, humidity, and light). Notifications are periodically sent to farmers' mobile devices. Farmers may monitor the condition of their fields from any location. This strategy will work better in areas with limited water supplies. The effectiveness of this strategy is 92% higher than the conventional approach [6].

3 Methodology

3.1 Existing Method

In the current system, agricultural land is controlled by a few sensors and micro-controllers. Below is a list of various sensors, including a soil moisture sensor and a UV sensor. In the current arrangement, the water motor will switch on based on the measurement of the moisture content of the soil by the soil moisture sensor. There is no automatic control on the water motor, though. The current system does not employ PIR sensors for either motion or animal detection. There is no such thing as cloud-based, motorised, remote pesticide spraying [7].

3.1.1 Limitations of Existing Method

This system is not secure. Motion detection is not present to protect agricultural crops. No automatic system exists.

3.2 Proposed Method

The soil moisture sensor measures the amount of soil moisture. Data from various sensors is transmitted and received by the node's MCU, which provides automatic controllability. We can continuously monitor a crop's growth using an ultrasonic sensor. The former can use this equipment to monitor and control the environmental conditions in their industry. Farmers need not physically visit their fields because they can remotely monitor and manage them using the cloud. The three techniques we employ to notify the farmers are the Blynk smartphone app, which also tracks live feeds, the various alert sounds produced by a little buzzer, and the LED visual alert. When using this product, farmers are urged to move swiftly. Nevertheless, there is still room for development, and the next task can be given more priority.

Block Diagram of Smart Farming:

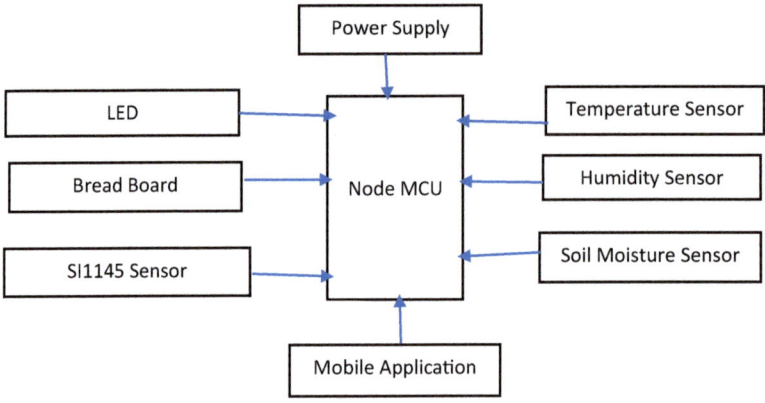

A. ESP32s Node MCU

The Node MCU ESP32s is one of the development boards created by Node MCU to test the ESP-WROOM-32 module. The ESP32 microcontroller, a single chip that includes Wi-Fi, Bluetooth, Ethernet, and low-power support, serves as its central component. An Internet of Things (IoT) platform called Node MCU uses the Lua programming language (Fig. 1).

B. Bread board

A breadboard is essentially a board for designing or building circuits. Without soldering, circuits can be built by placing components and connections on the board. The holes in the breadboard handle your connections by firmly holding on to the wires or components where you insert them and electrically connecting them inside the board [8] (Fig. 2).

Fig. 1 ESP32s Node MCU

Fig. 2 Breadboard

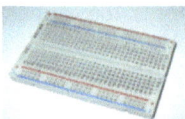

C. DHT11 Temperature and Humidity Sensor

A calibrated digital output is produced by the DHT11 temperature and humidity sensor. A cheap temperature and humidity sensor with great long-term stability and dependability is the DHT11 (Fig. 3).

D. Soil Moisture Sensor

It establishes the soil's moisture content. Using the open-circuit theory, the sensor produces both analogue and digital data. To indicate whether the output is high or low, this system uses an LED (Fig. 4).

E. SI1145 sensor for UV/IR and visible light index

The SI1145, a brand-new sensor from SiLabs, calculates the UV index using a calibrated light-detecting algorithm. It simulates UV sensing by using the sun's visible and IR rays in place of a real UV sensor component (Fig. 5).

F. LEDs

The light-emitting diode is a typical kind of standard light source in electrical equipment (LED). It can be applied in a wide range of contexts, including mobile phones and enormous billboards for advertising. They are used most frequently in devices that show different types of data and the current time.

G. KY-006 passive buzzer

Based on the input signal's frequency, the KY-006 Passive Piezoelectric Buzzer Module can emit a variety of tones. The KY-012 Active Buzzer can be used to produce single-tone noises [9] (Fig. 6).

Fig. 3 Temperature and humidity sensor

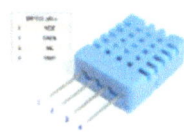

Fig. 4 Soil moisture sensor

Fig. 5 SI1145 sensor

Fig. 6 Passive buzzer

H. Power Supply–Power Bank

The capacity of a power bank is measured in milliampere hours, or mAh. The more mAh a power bank has, the more charge it can store for you to draw from when you use it to charge your devices. Once the power bank's storage space has been depleted, it must be replenished.

Implementation:

Our aim was to develop a prototype model that is simple to use and can be quickly installed in the field because farmers may lack technical expertise. The system is automated thanks to the Internet of Things (Figs. 7 and 8).

(1) The wireless and Wi-Fi-enabled ESP32s node MCU was used.
(2) We connected the ESP, the DHT11 temperature and humidity sensor, the soil moisture sensor, the buzzer, the LEDs, and the SI1145 digital UV index, IR, and visible light sensor to the breadboard using jumper wires.
(3) Every 18 min, the ESP32 goes to sleep, wakes up, takes a reading, uploads it to the cloud of the Blynk app to offer real-time data, and then goes back to sleep.
(4) The farmer can check the LEDs and take the necessary action even if he did not hear the alert or get a phone notification because the LEDs save the status, where multiple signs are provided by changing red, blue, or violet. If one buzzer indicates something, two imply something else.
(5) The soil moisture sensor is located at the bottom of the lid, together with temperature and humidity sensors, a digital UV index sensor, and a buzzer.
(6) We provide electricity using a 6000 mAh battery bank, so the system starts up automatically after the code has been uploaded.

Fig. 7 Circuit of the prototype

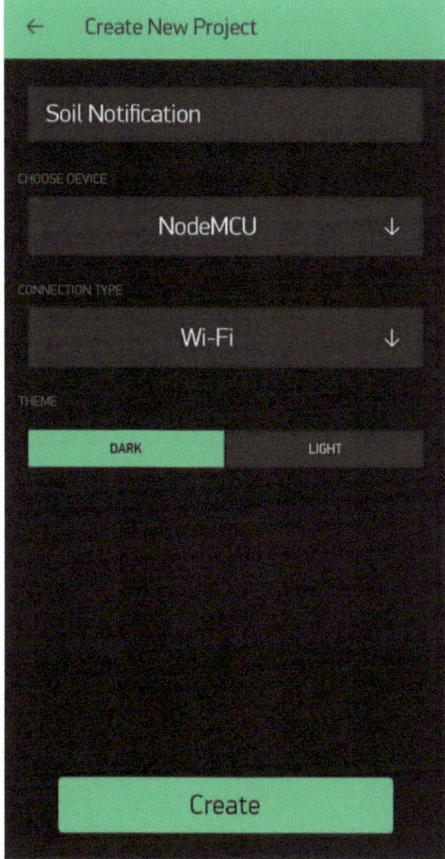

Fig. 8 Notification of Blynk mobile app

4 Result and Analysis

By giving farmers' real-time information on temperature, humidity, soil moisture, UV index, and infrared radiation using the Blynk mobile app, this article will help farmers. With the equipment and materials we used to build our prototype, we were able to provide farmers with a solution that was accurate, effective, and cost-effective. For farmers, this was also inexpensive and easy to implement. Thus, we can infer that this prototype will surely help farmers with little available land adequately to monitor their crops using the user-friendly software (Figs. 9, 10, and 11).

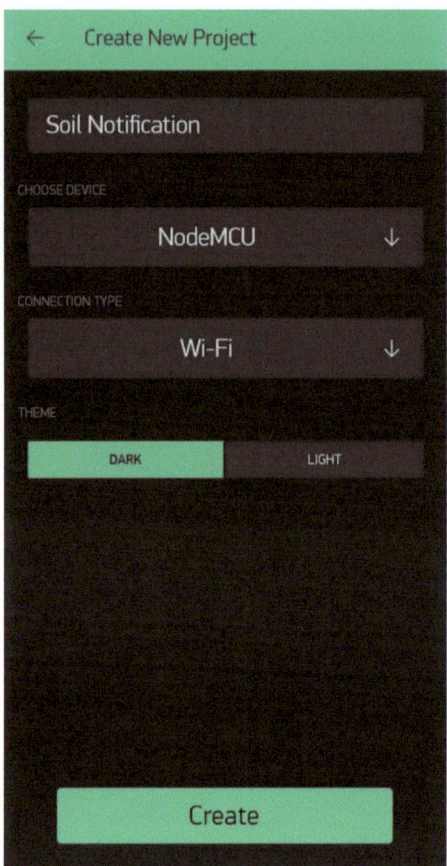

Fig. 9 Connecting to blink mobile app

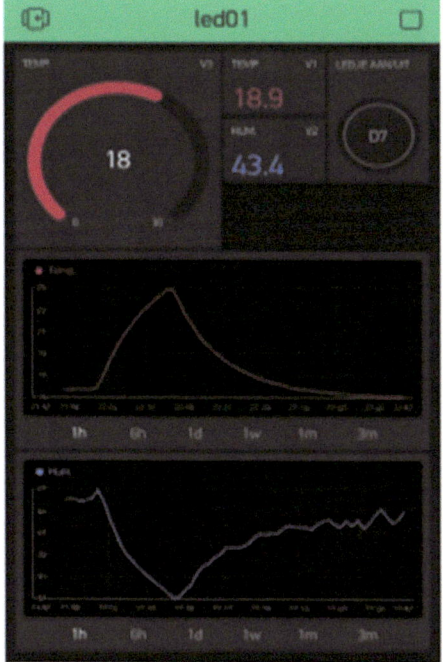

Fig. 10 Output of farm field

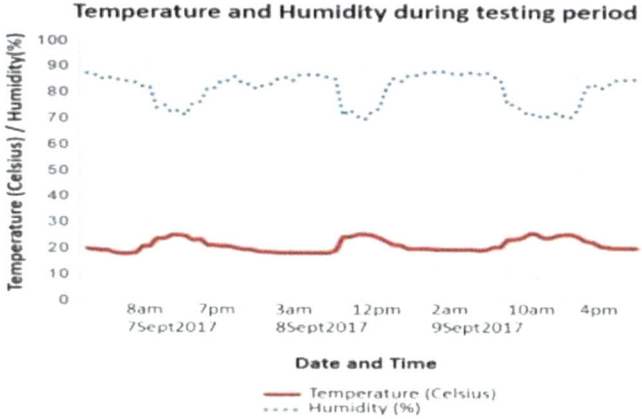

Fig. 11 Waveform of temperature and humidity

5 Conclusion

In IOT-based smart farming, a system is developed for remotely monitoring the agricultural field using sensors (light, humidity, temperature, soil moisture, and so on) and automating the irrigation system. Farmers can monitor the condition of their fields remotely. Smart farming powered by IoT is substantially more efficient than traditional farming.

References

1. de Wilde S (2016) The future of technology in agriculture, STT 81
2. Anushree MK, Krishna R (2018) A smart farming using Arduino based technology. Int J Adv Res Ideas Innov Technol 4(4):850–856
3. Gorli R, Yamini G (2017) Future of smart farming with internet of things. J Inf Technol Appl 2(1):27–38
4. Nagaraju CH, Sharma AK, Subramanyam MV (2018) Reduction of PAPR in MIMO-OFDM using adaptive SLM and PTS technique. International Journal of Pure and Applied Mathematics, Special issue 118(17):355–373. ISSN: 1311-8080 (printed version); ISSN: 1314-3395
5. Nayyar A, Puri V (2016) Smart farming: IoT based smart sensors agriculture stick for live temperature and moisture monitoring using Arduino, cloud computing & solar technology. The international conference on communication and computing (ICCCS-2016)
6. Shaik F, Sharma AK, Ahmed SM (2016) Hybrid model for analysis of abnormalities in diabetic cardiomyopathy and diabetic retinopathy related images. Springerplus 5:507. https://doi.org/10.1186/s40064-016-2152-2
7. Karimullah S, Vishnuvardhan D (2020) Iterative analysis of optimization algotithms for placement and routing in Asic design. In: Kumar A, Paprzycki M, Gunjan V (eds) ICDSMLA 2019. Lecture notes in electrical engineering, vol 601.Singapore, Springer. https://doi.org/10.1007/978-981-15-1420-3_199
8. Rajalakshmi P, Devi Mahalakshmi S (2016) IOT based crop field monitoring and irrigation automation. In: 10th international conference on intelligent systems and control (ISC0)
9. Kais M, Eddy B, Chaxel F, Fernand M (2019) A comparative study of LPWAN technologies for large-scale IoT deployment. ICT Express 5:1–7

Design of OTFS Modulation by Superimposed Pilot-Based Channel Estimation and Embedded Pilot-Aided Estimation

B. Rakesh Babu, Y. Gowthami, K. Anusha, C. Hari Rama Subba Reddy, and M. Guna Sekhar Reddy

Abstract By utilizing orthogonal time–frequency space modulation, we offer the first-ever analysis of cell-free massive MIMO performance. We investigate the trade-off between performance and overhead using embedded pilot-aided and overlay pilot-based channel estimation techniques. The factors taken into account while estimating the delay-Doppler-domain channels are the quantity of APs, users, and uplink and downlink spectral efficiencies for each user. We also propose new scaling rules that APs and users transmit power which must follow to maintain the desired level of service quality in light of these analytical results. As the number of APs, Ma, increases without bound, it has been discovered by reducing the broadcast power of each user and APs during the uplink and downlink phases to 1/Ma and 1/M2a, respectively. In situations with high mobility, we compare the effectiveness of OTFS with orthogonal frequency division multiplexing (OFDM). According to our research, OFDM is outperformed by OTFS modulation with integrated pilot-based channel estimate by a factor of 30 in terms of the 95%-likelihood per-user downlink rate. Not to note, at the median speeds over the associated shadowing channels, the increase in per-user throughput with overlay pilot-based channel estimate is quite noticeable.

Keywords Spectral efficiency · Cell-free massive multiple-input multiple-output · And orthogonal time–frequency space modulation

1 Introduction

Beyond 5G wireless communication networks must be capable of supporting both the constant increase in the number of wirelessly linked devices and the rising demand for a variety of network services. To satisfy these requirements [1]. This system

B. Rakesh Babu (✉) · Y. Gowthami · K. Anusha · C. Hari Rama Subba Reddy · M. Guna Sekhar Reddy
Department of ECE, Annamacharya Institute of Technology and Sciences, Rajampet, India
e-mail: rakesh777babu@gmail.com

A. Kumar et al. (eds.), *Proceedings of the 4th International Conference on Data Science, Machine Learning and Applications*, Lecture Notes in Electrical Engineering 1038, https://doi.org/10.1007/978-981-99-2058-7_13

consists of a large number of inexpensive, low-power access point antennas that are dispersed and linked to a network controller. Compared to the number of users, there are a lot more antennas. Increased data capacity—through spatial and multiplexing techniques, MIMO can increase data carrying and capacity without needing more bandwidth. Diversity—it is feasible to encrypt the signal because MIMO provides the capacity to discriminate transmission over several pathways. A 2D modulation method called OTFS transforms data that is transmitted using the delay-Doppler coordinate system [2]. Due to its durability in high-speed vehicular settings, it was initially employed for fixed wireless and is now a strong contender for 6G technology. In OTFS, multiple closely spaced narrowband sub-channel frequencies are used to transmit a single information stream rather than a single wideband channel frequency. OFDM is a widely used technology in satellite radio (DVB-T/H), digital audio transmission, and digital radio Mondale [3]. The channel bandwidth, covered area, and the number of base station sites are multiplied by the total number of system users to determine the maximum throughput, also known as goodput. UNITS—bits/hertz/cell (the number of bits sent per cell per sec per HZ) (the number of bits transmitted per cell per sec per HZ). The modulation effectiveness of a transmission method that uses a bandwidth of one kilohertz to send 1000 bits per second is one bit per Hz. The reflectors that make up the domain mirror the geometry of the wireless channel, which fluctuates much more slowly in the DELAY-DOPPLER domain than it does in the rapidly varying-time–frequency domain [4]. It is also referred to as a "Mobile haul network," which is a phrase for the connectivity of cloud radio access networks using fiber (C-RAN). By reducing complexity, the fronthaul network aims to reduce data rate requirements between radio equipment and radio equipment control [5]. Uplink signal aggregation and downlink symbol pre-coding are accomplished by, the APs can implement basic beamforming techniques. Additionally, channel hardening enables sufficiently excellent performance while decoding signals using channel statistics. We take into account a massive MIMO system without cells that consist of Ku users and Ma APs. Each APs and user has a single antenna, which is dispersed at random throughout a vast area. A fronthaul network connects the APs to the CPU. Assuming that users will move quickly, double-selective fading will occur on the channels between the APs and users [6].

Need and Significance:

In OTFS, modulation techniques have enabled the connections of billions of people to the internet so that they can reap the benefits of today's economy. Maintenance and installation costs have reduced compared to other forms of networks. In these Access points, major components act as transmitters and receivers like radio, two-way data transfers, vehicle-to-vehicle communications.

2   Literature Review

Improved conversion of OTFS signals from the time domain to the delay-Doppler domain [7] is achievable in circumstances with exceptionally high mobility. Initially proposed for high-mobility wireless applications, the orthogonal time–frequency space technique, which modulates data in the delay-Doppler domain as opposed to the conventional time–frequency domain, is now widely acknowledged as a game-changing technology and enables future wireless communications. Estimating the downlink channel for high mobility across a huge MIMO system with uplink assistance [8]. Although MIMO over orthogonal time–frequency space modulation is frequently utilized in OFDM systems, there may be a significant training overhead in situations with high mobility. Pilot-aided channel estimates for OTFS using delay-Doppler channels [9]. It has been shown that orthogonal time–frequency space modulation gives a significant improvement in error performance over orthogonal frequency division multiplexing in delay-Doppler channels. To recognize orthogonal time–frequency space-modulated data, the receiver needs to be aware of the channel impulse response. A threshold approach is employed at the receiver to estimate the channel, and the estimated channel information is used by a message-passing algorithm to detect data.

3   Methodology

3.1   Existing Method

By using numerous sub-carriers within a single channel, orthogonal frequency division multiplexing, a digital multi-carrier modulation technology, expands the concept of single sub-carrier modulation. OFDM uses a greater number of closely spaced, orthogonal subcarriers that are delivered in tandem rather than simply one subcarrier to provide a high-speed data stream [10]. A traditional digital modulation approach is used to modulate each subcarrier at a low symbol rate (e.g., QPSK, 16QAM, etc.). However, the use of a numerous sub-carriers in conjunction allows for data rates which are comparable to those of traditional single-carrier modulation methods within comparable bandwidths [11]. Frequency division multiplexing, a widely used method, is the foundation of OFDM (FDM). In FDM, several information streams are converted into numerous parallel frequency channels. In order to minimize interference between them, a frequency guard band is used to separate each FDM channel from the other channels.

The following linked differences between the OFDM system and conventional FDM:

1. The information stream is carried by a number of carriers (referred to as subcarriers).

2. The subcarriers are opposite to each other.
3. To lessen inter-symbol interference and channel delay spread, a guard interval is introduced to each symbol.

The fundamental ideas of an OFDM signal and the connection between the frequency and time domains are depicted in the diagram below. Separate modulation of many successive tones or subcarriers using complex frequency domain [12]. The frequency-domain sub-carriers are transformed into the OFDM symbol in the time domain using an inverse FFT transform. Then, to prevent inter-symbol interference at the receiver caused by multi-path delay spread in the radio channel, guard intervals are inserted between each symbol in the time domain [13]. To produce the final OFDM burst signal, several symbols can be concatenated. The original data bits are recovered from the OFDM signals at the receiver using an FFT.

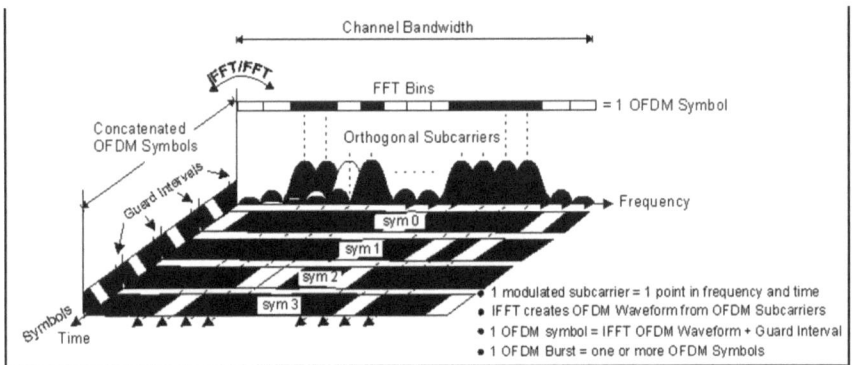

Frequency-Time Representative of an OFDM signal

3.2 Proposed Method

OTFS utilizes a set of basis functions that are orthogonal to both time and frequency changes to operate in the delay-Doppler coordinate system. In this coordinate system, both data and reference signals or pilots are carried. In the delay-Doppler domain, where changes happen much more gradually than in the rapidly changing time–frequency domain, the wireless channel's architecture is mirrored. Strong Doppler conditions cause OTFS symbols to experience the full variety of the channel across time and frequency, trading delay for performance. Figure 1 shows the modulation and demodulation processes as well as the modulation effects of the channel. Using the two-dimensional Simple Fourier Transform, the transmit information symbols (QAM symbols) are translated from the delay-Doppler to the time–frequency domain. The transmit information symbol is laid out on a grid or lattice in this domain (shown top left in green). In case you forgot, OFDM QAM symbols are located in

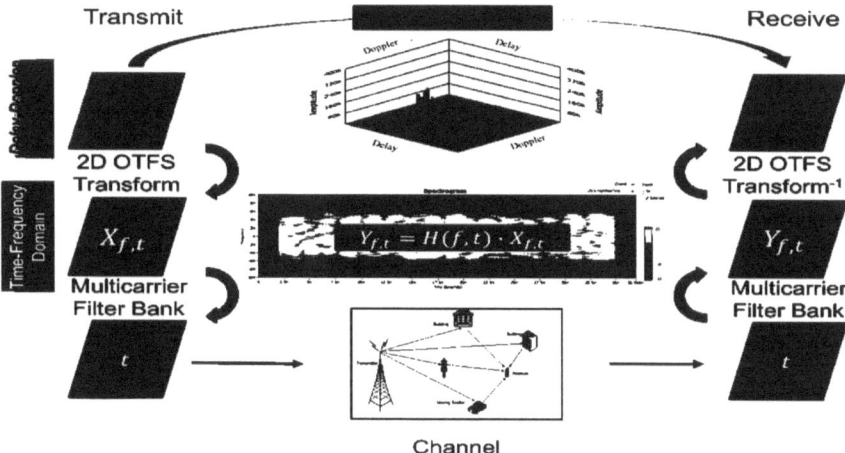

Fig. 1 OTFS processing

the well-known time–frequency domain. As opposed to this, each QAM symbol in OTFS is scattered over the time–frequency plane using a distinct basis function (i.e., across the chosen signal bandwidth and symbol time). The same channel and SNR are consequently experienced by all symbols of the same power. The implication is that when a suitable frequency and time observation window is utilized, frequency- or time-selective fading of QAM signals does not exist. After then, the signal is sent using a multicarrier filter bank, which provides the same advantages for filter shaping as other forms of filtered OFDM. On the receiving end, inverse processing is performed.

Reference signals (RSs) are frequently used in the time–frequency domain, typically in coarse (regular or irregular) grid, as in LTE, to help the receiver estimate the channel. Using various (preferably orthogonal) signature sequences, many antenna ports are multiplexed on the same coarse grid. To test the channel, the OTFS sends pilot or reference signals as impulses in the delay-Doppler domain. To enable the channel's maximum delay and Doppler scatters, each pilot has a specified area surrounding it.

4   Result and Analysis

We use the assumption that the OTFS system has a sub-carrier spacing of 15 kHz and an operating carrier frequency of fc = 4 GHz. The maximum movement speed of 300 km in the scenario results in a maximum Doppler index of 9, at most. We take into account the I extended vehicular A, with Lpq = 9 and max = 2.5 s, and the II extended vehicular B (EVB), with Lpq = 6 and max = 10 s, of the 3GPP vehicle models. In a square of size DXD km2 with its edges wrapped around to prevent boundary effects, we consider the distribution of Ku users and Ma APs to be random and predictable.

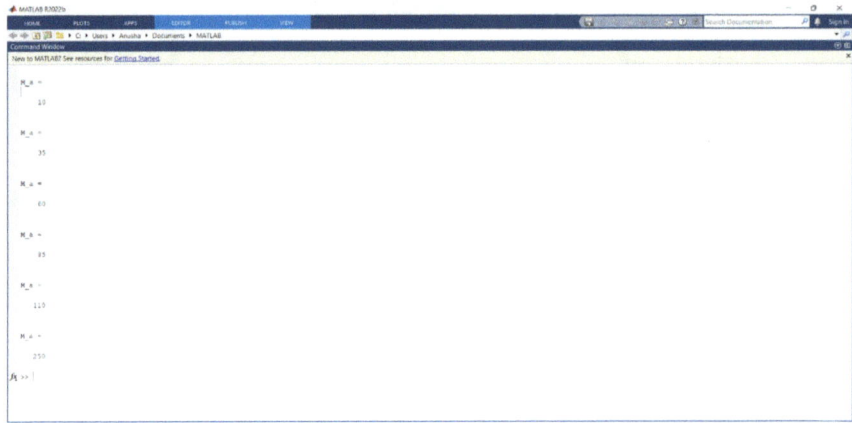

5 Conclusion

Cell-free massive MIMO has been introduced as a unique design to answer the growing demands for high SE, coverage probability, green output, and spread uniformly throughput for all network users. In beyond 5G and wireless communication systems, dense APs can be deployed throughout a substantial network region to cooperate with a few UEs using CF-massive MIMO. Giving a general overview of the design, application situations, and potentials was the aim of this study, as implementation difficulties of these cutting-edge concepts. Its motivation came from the numerous benefits of CF-massive MIMO. The CF-massive MIMO system paradigm, which includes UL/DL pilot-aided channel estimate, UL/DL training, and channel hardening, is thoroughly researched.

References

1. Mohammed SK (2021) Time-domain to delay-Doppler domain conversion of OTFS signals in very high mobility scenarios. IEEE Trans Veh Technol 70(6):6178–6183
2. Karimullah S, Vishnuvardhan D, Arif M, Gunjan VK, Shaik F, Siddiquee KN (2022) An improved harmony search approach for block placement for VLSI design automation. Wireless Communications and Mobile Computing, vol 2022, Article ID 3016709, 10 pages. https://doi.org/10.1155/2022/3016709

3. Albu A, Precup R-E, Teban T-A (2019) Results and challenges of artificial neural networks used for decision-making and control in medical applications. Facta Univ Ser Mech Eng 17(3):285–308

4. Karimullah S, Vishnuvardhan D (2022) Pin density technique for congestion estimation and reduction of optimized design during placement and routing. Applied Nanoscience

5. Shaik F, Sharma AK, Ahmed SM (2016) Hybrid model for analysis of abnormalities in diabetic cardiomyopathy and diabetic retinopathy related images. Springerplus 5:507. https://doi.org/10.1186/s40064-016-2152-2

6. Karimullah S, Basha SJ, Guruvyshnavi P, Sathish Kumar Reddy K, Navyatha B (2020) A genetic algorithm with fixed open approach for placements and routings. ICCCE, Springer, pp 599–610

7. Nagaraju CH, Sharma AK, Subramanyam MV (2018) Reduction of PAPR In MIMO-OFDM using adaptive SLM And PTS technique. International Journal of Pure and Applied Mathematics, Special issue, 118(17):355–373. ISSN: 1311-8080 (printed version); ISSN: 1314-3395

8. Karimullah S, Vishnuvardhan D (2020) Experimental analysis of optimization techniques for placement and routing in Asic design. ICDSMLA 2019, Lecture notes in electrical engineering 601, Springer Nature Singapore Pte Ltd.

9. Bai M, Urtasun R (2017) Deep watershed transform for instance segmentation. In: Proceedings of the IEEE conference on computer vision and pattern recognition, pp 5221–5229

10. Karimullah S, Vishnuvardhan D, Basha SJ (2020) Floorplanning for placement of Modulesin VLSI physical design using harmony search technique. ICDSMLA 2019, Lecture notes in electrical engineering 601, Springer Nature Singapore Pte Ltd.

11. Karimullah S, Vishnuvardhan D (2020) Iterative analysis of optimization algorithms for placement and routing in Asic design. ICDSMLA 2019, Lecture notes in electrical engineering 601, Springer Nature Singapore Pte Ltd.

12. Liu Y, Zhang S, Gao F, Ma J, Wang X (2020) Uplink-aided high mobility downlink channel estimation over massive MIMO-OTFS system. IEEE J Sel Areas Commun 38(9):1994–2009

13. Raviteja P, Phan KT, Hong Y (2019) Embedded pilot-aided channel estimation for OTFS in delay-Doppler channels. IEEE Trans Veh Technol 68(5):4906–4917

Morphology and EMD-Based Patch-Wise Image Fusion

Y. Pavan Kumar Reddy, Y. Sunanda, S. Charitha Adena, Sharmila Eruru, Naveen Chennamsetty, and Naveen Reddy Pothireddy

Abstract The process of integrating various source photos into a single image that is more informative than all the source images is known as image fusion. It is an efficient way of retrieving the information from the multiple sources into single image. The main purpose of image fusion is to not only decrease amount of data but also construct images that are more appropriate and comprehensible for human and machine perceptions. This paper describes morphology and empirical mode decomposition (EMD)-based image fusion strategy. The goal of this technique is to minimise the spatial distortions caused by noisy attributes of pixel-wise maps and to construct fusion images of high quality. Initially, we design a multi-channel, bidimensional empirical mode decomposition (EMD) algorithm that divides the image data into IMFs of different scales and a residue utilising morphological dilation as well as erosion filters. While retaining the decomposing quality, it further increases the computing efficiency of EMD. Additionally, we create a patch-based fusion method that merges the IMFs as well as the residue with intersecting partitions to reduce noisy attributes.

Keywords Empirical mode decomposition · Image fusion · Morphological filters · Patch-based fusion · Intrinsic mode functions

Y. Pavan Kumar Reddy · Y. Sunanda (✉) · S. Charitha Adena · S. Eruru · N. Chennamsetty · N. R. Pothireddy
Department of Electronics and Communication Engineering, Annamacharya Institute of Technology and Sciences, Rajampet, India
e-mail: sunanda.bujji@gmail.com

© The Author(s), under exclusive license to Springer Nature Singapore Pte Ltd. 2023   121
A. Kumar et al. (eds.), *Proceedings of the 4th International Conference on Data Science, Machine Learning and Applications*, Lecture Notes in Electrical Engineering 1038,
https://doi.org/10.1007/978-981-99-2058-7_14

1 Introduction

A wide range of image acquisition sensors are available as a result of technological advancements. The information collected with a single image acquisition sensor is insufficient, irrespective of the fact that each sensor offers characteristics that cannot be replaced in its optimum operating environment and range. Image fusion is a technique for creating a composite image from several different source images. It is an effective method of combining significant data from various sources into a single image [1]. Image fusion objects to generate images that remain more relevant and understandable for both human and machine perceptions, as well as lower the amount of data necessary to hold numerous images. Image fusion is widely cast off in computer apparition, medical imaging, remote sensing, military, etc. Various benefits of image fusion include image sharpening, feature extraction, replacement of defective data.

Different fusion techniques like (Intensity Hue Saturation), high-pass filtering, pyramid techniques, wavelet transform, discrete cosine transform are developed so far. Huang proposed a classical EMD algorithm that can decompose the one-dimensional (1D) time series signal into different IMFs and residue using iterative sifting process. It is further developed for 2D images known as bidirectional empirical mode decomposition (EMD) which is applied to image fusion by many researchers to extract the feature and to overcome the distortion introduced by pre-defined functions based on transformation techniques like wavelet transform and Fourier transform [2]. Still, the known EMD methods have some drawbacks. They are highly time consuming with the increase in image size which in turn reduces computation efficiency. While merging IMFs, the spatial distortions are caused by noisy attributes of pixel-wise maps. So, these fusion methods will produce inappropriate outcomes.

To enhance the efficiency of EMD-centred fusion technique, this research paper proposes a morphology and EMD-based fusion technique. Initially, to enhance computation efficiency, we design a multi-channel bidirectional EMD algorithm using morphological dilation and erosion filters which can moulder the source images obsessed by IMFs of various scales in addition to a residue. Further, to reduce spatial distortion, we design a patch fusion technique with overlapping partitions, where maximum selection rule based on energy levels is developed to merge IMFs and residue, and the final output image is obtained by accumulating all the IMFs and also the residue collected.

The assistances in this paper include developing a morphological filter-based empirical decomposition algorithm for multi-channel images and patch-based fusion technique to fuse IMFs and the residue which can further minimise the decomposition time and maximise computation efficiency.

2 Literature Survey

Huang proposed a classical 1D EMD algorithm [3] for processing non-stationary and nonlinear 1D time signals. Through an iterative sifting process, any complex set of data is divided into a pre-defined and indeed small number of intrinsic mode functions and residue. It is confined to only 1D signals. Later, Nunes and their fellows developed it for 2D images and proposed image [4] analysis by bidirectional empirical mode decomposition which lacks stability. In order to improve the dissolution by averaging algorithms of all noise-added images, Wang [5] created a BEEMD approach, that also consists of highly costly time expenses. By using automatic vehicle-selected selective restoration and improved fast empirical mode decomposition, Trusiak's [6] advanced computation of optical fringe patterns was proposed. While using segmentation method instead of order statistics filters, this will increase the efficiency of the algorithm of envelops, but it is only applicable to single-channel images.

A wide range of fusion techniques have been developed in spatial, frequency transform, deep learning, and neural network-based domains. The decomposition algorithms are mainly suited for transform-domain fusion techniques. The input source images' transformed coefficients, which were gathered using transform-domain techniques, are integrated, and restoration step with a conforming inverse transform results in the creation of the fused image [7, 8]. Certainly, in these techniques, the choice of the transform domain is crucial. The Laplacian pyramid, empirical mode decomposition (EMD), multi-scale geometric analysis, wavelet transform, fast Fourier transform, and other transformations have all been used till now to conduct image fusion. Unlike traditional transform techniques that rely on pre-explained basis functions. EMD is completely flexible, and data dependent. Qin [9] created the decomposition. The extreme selection criteria dependent on two saliency parameters and a pixel-based algorithm were used to combine the residual and all IMFs, which may have caused some distortions. The multivariate 1D EMD is cast off to dissect source images in order to equalise the quantity and properties of the decompositions of various source images [10]. A variance-based weighted averaging method can then be used to aggregate each component pixel by pixel. In order to gain the multi-scale breakdown, Xia [11] used the MBEMD based on surface projection, which may progress the fusion excellence of the multivariate one-dimensional empirical mode decomposition-based fusion technique. To handle Zhu's [12] ground-breaking fusion technique, sparse representation (SR) and bivariate bidimensional empirical mode decomposition (B-BEMD) are used, by properly combining the common and novel characteristics of two patterns of pictures. In order to successfully keep the fine qualities of the source pictures, the high-frequency components are combined using the "max-absolute" method as the activity level measurement. Then, in order to emphasise the common features and reserve the innovation features, the common and innovative features among low-frequency components are extracted by the deftly devised SR-based approach and fused, respectively, by the appropriate fusion rules. The fused picture is then rebuilt using the inverse B-BEMD procedure. Sufyan [13] proposed a new MMAI fusion method constructed on structure extraction and contrast which eliminates distortions from source images and then fuses the images based on local contrast and salient structure [14]. This paper presents a morphological filter-based empirical mode decomposition algorithm for multi-channel images and then the extracted IMFs and residue are fused with the help of patch-based fusion technique, where a maximum selection rule based on energy levels is employed [15]. This method increases the computation efficiency and minimises the decomposition time [16].

3 Methodology

3.1 Existing Method

IHS (Intensity Hue Saturation) Transform

The three characteristics of a colour—intensity, hue, and saturation—provide a regulated visual representation of an image. The IHS transform method is the most traditional picture fusion technique. Because the IHS space carries the majority of the spectral information, hue and saturation need to be carefully regulated [17].

High-pass filtering (HPF)

High-pass filtering is used to create high-resolution multispectral photographs. The high-frequency data from the high-resolution panchromatic image and the low-resolution multispectral image are combined to create the final image [18]. Either a high-pass filter is employed to filter the high-resolution panchromatic image or the original HRPI is used and the LRPI is removed from it.

Wavelet Transform

The wavelet transform is an alternative to the rapid Fourier transforms. The Fourier transform only provides the proper resolution in the frequency domain, but this method supplies it in both the time domain and the frequency domain [19]. In contrast to the wavelet transform, which scales and shifts versions of the mother wavelet or function, the Fourier transform separates the signal into sine waves of various frequencies.

Discrete Cosine Transform

It has become important for the MPEG, JVT, and other compressed picture formats. The spatial domain image is transformed into frequency-domain image using the discrete cosine transform [20]. Low frequency, medium frequency, and high frequency are three categories used to divide the images. The DC value reflects average illumination, whereas the AC values are the high-frequency coefficients. The RGB picture is divided into 8 × 8 pixel blocks for segmentation. The picture is then turned into a greyscale image after being separated into groups based on the red, green, and blue matrices [21].

3.2   Proposed Method

The bidirectional EMD method, which is based on morphological dilation and erosion filters, first divides the contribution source images hooked on numerous IMFs and also a residue. Second, it uses an overlapping patch-based fusion technique to fuse the residue and IMFs separately. A maximum selection method based on energy levels is constructed about the fusion of the IMFs, along with two separate rules which are built for the fusing of the residue based on the key information collected by IMFs from the input images. The intrinsic mode functions and also fused residue are ultimately used to reconstruct the required fused image.

The block diagram below depicts the entire structure of EMD algorithm (Fig. 1).

Morphological Filter Based EMD Algorithm

In the proposed morphological filter-based multi-channel bidirectional empirical mode decomposition, using morphological dilation as well as erosion filters which have the same window size for every channel, which retrieve the very same spatial extent from every channel image at the moment of decomposition, envelope edges for the inter image are produced. The lower (upper) envelope $D = (D_1, ..., D_n)$ $((U = (U_1, ..., U_n))$ for a multi-channel image $I = (I_1, ..., I_n)$ with window size

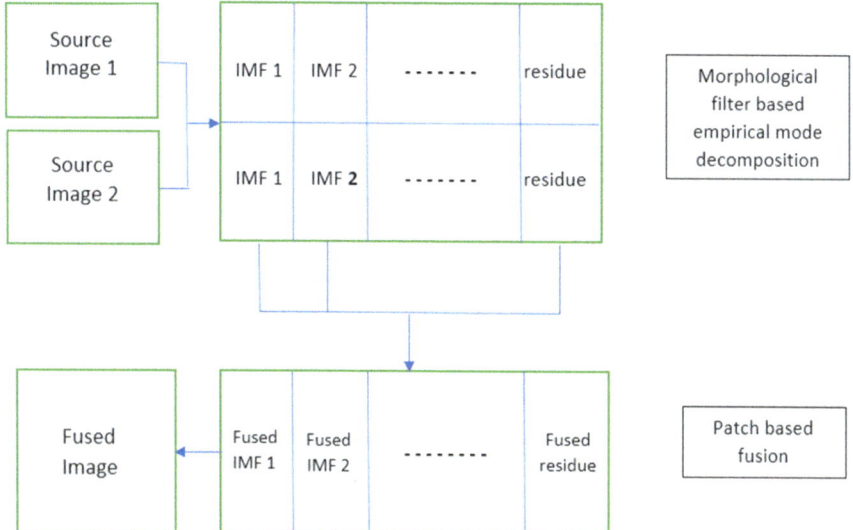

Fig. 1 Block diagram of morphology and EMD-based patch-wise image fusion

W × H can be generated by

$$U_k(x, y)|_{k=1,...,n} = (I_k \oplus b)(x, y) = \max_{(s,t) \in Z_{xy}} I_k(s, t)$$

$$D_k(x, y)|_{k=1,...,n} = (I_k \ominus b)(x, y) = \min_{(s,t) \in Z_{xy}} I_k(s, t) \tag{1}$$

where $\oplus$ represents the morphological dilation filter, $\ominus$ represents the morphological erosion filter, b characterises a binary group pointer function on Z_{xy}, and Z_{xy} denotes the group of pixels centred on the pixel (x, y) in the window w × w. To obtain significantly smoother envelopes, the average filter is utilised.

$$U'_k(x, y)|_{k=1,...,n} = \frac{1}{\omega \times \omega} \sum_{(s,t) \in Z_{xy}} U_k(s, t)$$

$$D'_k(x, y)|_{k=1,...,n} = \frac{1}{\omega \times \omega} \sum_{(s,t) \in Z_{xy}} D_k(s, t) \tag{2}$$

The window size w in Eqs. (1) and (2) is set to the smallest average extreme distance of all image channels in order to evaluate feature abstraction for all statistics channels of the source images.

$$\omega = min\{\omega 1, ..., \omega n\} \tag{3}$$

where average extreme space of kth channel picture I_k is represented by w_k ($k = 1, ..., n$) and is determined by

$$\omega_k = \sqrt{\frac{w \times h}{N_k}} \tag{4}$$

where N_k represents the average value of all I_k's local maxima and minima. In each iteration, this compares the values of each pixel and neighbourhood pixels in 3 × 3 window centred on it to locate all local maxima (minima) of I_k. This could iteratively extract IMFs of various scales using a sifting technique based on the envelope calculation technique described above, until residue is a monotonic function or a constant or the required number of IMFs is obtained.

EMD-Based Patch-Wise Image Fusion

The two source images I_1 and I_2 are combined to form a two-channel image $I = (I_1, I_2)$ that is decomposed by Algorithm 1 into K IMFs and a residue.

$$I = \sum_{I=1}^{K} F_i + R_k \tag{5}$$

where $F_i = (F_{i1}, F_{i2})(i = 1, ..., k)$ is the ith IMF and $R_k = (R_{K1}, R_{K2})$ is the associated residue. All intrinsic mode functions and residues are divided into several patches of size $M \times M$, with N overlapping columns/rows. This overlapping patch technique is developed to minimise the distortions that occur around partition boundary while using patch-based fusion techniques.

Fusion of IMFs

On measuring the energy levels of two related patches, the fused patch G_i^j is generated using a maximum selection rule based on energy levels, for the jth patch $F_i^j = (F_{i1}^j, F_{i2}^j)$ of ith IMF F_i

$$
G_i^j = \begin{cases} F_{i1}^j, & E\left(F_{i1}^j\right) \geq E\left(F_{i2}^j\right) \\ F_{i2}^j, & E\left(F_{i1}^j\right) < E\left(F_{i2}^j\right) \end{cases}
\tag{6}
$$

Equation (6) computes the energy of each patch by

$$
E(F_{ip}^j) = \sum_{(s,t)\in Z_j} F_{ip}^j(s, t)^2, \; p = 1, 2,
\tag{7}
$$

where Z_j stands for the pixels group in jth patch. The above formula is utilised to obtain the significant features from the source pictures.

Fusion of Residue

As for jth patch $R_K^j = (R_{K1}^j, R_{K2}^j)$ of residue R_K, two different methods are designed centred on the statistics collected by the intrinsic mode functions to extract fusion residue of patch H_k^j in accordance with the image types.

The first one combines multi-focus images using a maximum selection method that is energy-based. In the first IMF, the energy of two identical patches is compared as

$$
H_k^j = \begin{cases} R_{K1}^j, & E\left(F_{11}^j\right) \geq E\left(F_{12}^j\right) \\ R_{K2}^j, & E\left(F_{11}^j\right) < E\left(F_{12}^j\right) \end{cases}
\tag{8}
$$

where $E(F_{ip}^j)$ ($p = 1, 2$) represents the energy of initial intrinsic mode functions, and this could acquire the features up to finest scales. The above fusion method accurately describes the focused area of multi-focus pictures.

The second combines multi-modal images using an energy-based algorithm. The knowledge area retrieved by IMFs is used to merge the actually imply area of the residue patch, and the mean of a residue serves is helpful to fuse the illumination of each multi-modal image. The fusion equation can be obtained from

$$H_k^j = \sum_{p=1}^{2} a_p^j (R_{Kp}^j - \mu_{Kp}^j) + \sum_{p=1}^{2} b_p^j \mu_{Kp}^j$$

$$a_p^j = \left| \sum_{i=1}^{K} E\left(F_{ip}^j\right) \right|^l \bigg/ \left(\left| \sum_{i=1}^{K} E\left(F_{i1}^j\right) \right|^l + \left| \sum_{i=1}^{K} E\left(F_{i2}^j\right) \right|^l \right)$$

$$b_p^j = \frac{\left| \mu_{Kp}^j \right|^m}{\left| \mu_{K1}^j \right|^m + \left| \mu_{K2}^j \right|^m} \tag{9}$$

where μ_{kP}^j is mean of the jth patch of residue R_{kP}^j, The two non-negative exponent parameters for controlling feature guidance intensity as well as brightness fusion intensity, respectively, are $p = 1, 2, 1$, and m. If l and m are both set to zero, the result is just an average of the leftover information from two multi-modal images. The knowledge area retrieved by IMFs is used as the guidance to merge the brightness of each modal image if the virtues l and m have been set greater than 0, and the mean of the residue patch is used to fuse the mean divided section of the residue of each modal image if the results of l and m are set larger than 0. A higher value of l denotes that the merged result contains much stronger features, while a greater value of m indicates that more brilliant targets are included in the fusion result. Both the values of l and m are set to 6 in trials, which can produce successful outcomes.

Image Reconstruction

The value at each pixel (x, y) of the fused IMFs and residue is determined by averaging the values of the pixel (x, y) in all overlapping patches after all IMFs and residue patches have been fused.

$$G_i'(x, y) = \frac{1}{s(x, y)} \sum_j G_i^j(x, y)$$

$$H_K'(x, y) = \frac{1}{s(x, y)} \sum_j H_K^j(x, y) \tag{10}$$

where $S(x, y)$ represents the overlapping patch number at the pixel (x, y), and the resultant fused image I' is created by combining the fused IMFs and the fused residue.

$$I'(x, y) = \sum_{i=1}^{k} G_i'(x, y) + H_K'(x, y) \tag{11}$$

The proposed method is implemented by setting the initial value of fused IMFs at each pixel (x, y), the fused residue, then with the help of overlapping patch number by $G_i'(x, y)|_{i=1,\ldots,k} = 0$, $H_K'(x, y) = 0$, and $S(x, y) = 0$. Each time a patch is combined, and fused values are updated using

$$G'_i(x, y) \rightarrow G'_i(x, y) + G^j_i(x, y)$$

$$H'_K(x, y) \rightarrow H'_K(x, y) + H^j_K(x, y) \tag{12}$$

and at the jth patch, each pixel's overlapping patch value is modified by

$$S(x, y) \rightarrow S(x, y) + 1 \tag{13}$$

After fusing all the patches of K IMFs and residue, on every pixel, the resultant joined IMFs and residue are retrieved.

$$G'_i(x, y) \rightarrow \frac{1}{S(x, y)} G'_i(x, y)$$

$$H'_K(x, y) \rightarrow \frac{1}{S(x, y)} H'_K(x, y) \tag{14}$$

In pixel-based fusion, the noisy features of pixel-wise maps are reduced with the help of overlapping patch partition.

4 Experimental Investigations

4.1 Selection of Key Parameters

K Decomposition level. The most effective selection principle based on the energy thresholds is used to combine all IMFs and extract the most important information from the images. However, K is chosen over 2 for multi-modal images so because top two IMFs of method 2 are where the majority of the input images' information is concentrated in the tests performed multi-modal images.

Overlapping number N of rows/columns Most of the time, decreasing distortions while also increasing computing costs can be accomplished by increasing the number of rows and columns that overlap N in the patch split. The block sizes are $M = 2$ for multi-modal images and $M = 6$ for multi-focus images. More tests have shown that these decisions can generate desired fusion outcomes.

Block size M of the division. Combining all IMFs, the most efficient selection premise using the energy threshold values is used to get the most important information from the images. The very same decomposition threshold K of sample b is set to 1 in sequence for the initial IMF to signify that the focused province of multi-focus images is good. The top two IMFs of step 2 represent that the large number of input

Fig. 2 Multi-focus colour image fusion using proposed method

Fig. 3 Multi-focus greyscale image fusion using proposed method

tiny pictures' information is focused in the tests conducted with multi-modal image sets, so K is chosen over 2 in this case.

4.2 Results

The MATLAB R2017b software was used to create all the experimental results shown in this work on a laptop with an Intel Core i5 processor with Windows 11 operating system and RAM size is 16 GB.

Using morphology and EMD-based patch-wise image fusion, Figs. 2, 3, 4, and 5 show the fusion of multifocal (colour) photos, greyscale images, multi-modal (medical) images, and infrared images, respectively. Given that the maximum selection rule based on energy levels for the fusion of each IMF can extract more significant information, the patch-based fusion technique can enhance the fusion quality of each EMD method in visualisation while also reducing the distortions caused by pixel-wise fusion method. The structure of multi-modal images can also be better represented by the extracted IMFs' energy-based weighted averaging method, and the focused area of multi-focus images can be captured more effectively by the first IMF's activity level. It is clearly observed that the essential details present in the output image but absent in either of the source images.

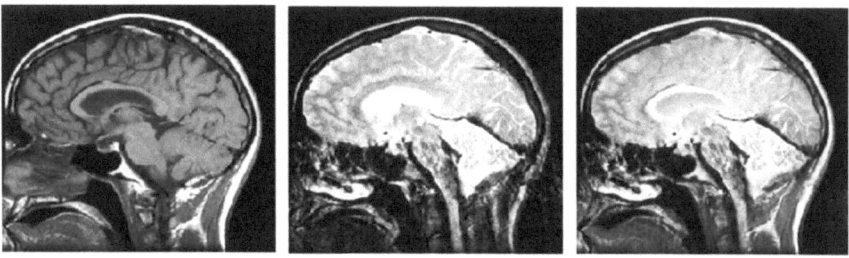

Fig. 4 Multi-modal medical image fusion using proposed method

Fig. 5 Multi-modal infrared image fusion using proposed method

5 Conclusion

To obtain good quality images, this paper describes a morphology and an EMD-based patch-wise image fusion. First of all, a morphological filter-based bidirectional EMD algorithm is developed for multi-channel images which uses dilation and erosion filters to calculate lower and upper envelopes of source images. This algorithm breaks down the input images into intrinsic mode functions of various sizes and a residue. This will gradually improve the computation efficiency. The IMFs and residue are then merged using a patch-based fusion method with overlapping partitions. With the aid of maximum selection rule based on energy levels, the IMFs are fused, and the residue is combined using the key information they have collected. The fused image is finally produced by combining the fused IMFs with all of the fused residues. The fused images for different sets of source images are displayed in Figs. 2, 3, 4, 5.

References

1. Karimullah S, Vishnuvardhan D, Basha SJ (2020) Floorplanning for placement of Modulesin VLSI physical design using harmony search technique. ICDSMLA 2019, Lecture notes in electrical engineering 601, Springer Nature Singapore Pte Ltd.
2. Jiang Y, Wang M (2014) Image fusion with morphological component analysis. Inf Fusion 18(7):107–118
3. Huang NE, Shen Z, Long SR, Wu MC, Shih HH, Zheng Q, Yen NC, Tung CC, Liu HH (1998) The empirical mode decom- position and the Hilbert spectrum for nonlinear and non-stationary time series analysis. Proc Math Phys Eng Sci 454(1971):903–995
4. Nunes JC, Bouaoune Y, Delechelle E, Niang O, Bunel P (2003) Image analysis by bidimensional empirical mode decomposition. Image Vis Comput 21(12):1019–1026
5. Wang X, Hu J, Guo L, Zhang D, Hong Q, Hao A (2018) Featurepreserving, mesh-free empirical mode decomposition for point clouds and its applications. Comput. Aided Geomet. Design 59:1–16
6. Trusiak M, Wielgus M, Patorski K (2014) Advanced processing of optical fringe patterns by automated selective reconstruction and enhanced fast empirical mode decomposition. Opt Lasers Eng 52:230–240
7. Liu Y, Chen X, Wang Z, Wang ZJ, Ward RK, Wang X (2018) Deep learning for pixel-level image fusion: recent advances and future prospects. Inf Fus 42:158–173
8. Yu L, Lei W, Juan C, Chang L, Xun C (2020) Multi-focus image fusion: a survey of the state of the art. Inf Fus 64:71–91
9. Qin X, Zheng J, Hu G, Wang J (2017) Multi-focus image fusion based on window empirical mode decomposition. Infrared Phys Technol 85:251–260
10. Reddy YPK, Vaishnavi A, Devi MS, Prasad MS, Reddy BS (2023) Multimodal medical image fusion approach using PCNN model and shearlet transforms via max flat FIR filter. In: Kumar A, Senatore S, Gunjan VK (eds) ICDSMLA 2021. Lecture Notes in Electrical Engineering, vol 947. Springer, Singapore. https://doi.org/10.1007/978-981-19-5936-3_73
11. Xia Y, Zhang B, Pei W, Mandic DP (2019) Bidimensional multivariate empirical mode decomposition with applications in multi-scale image fusion. IEEE Access 7:114261–114270
12. Zhu P, Liu L, Zhou X (2021) Infrared polarization and intensity image fusion based on bivariate BEMD and sparse representation. Multimed Tools Appl 80:4455–4471
13. Sufyan A, Imran M, Shah SA, Shahwani H, Wadood AA (2022) A novel multimodality anatomical image fusion method based on contrast and structure extraction. Int J Imag Syst Technol 32(1):324–342
14. Karimullah S, Vishnuvardhan D (2020) Experimental analysis of optimization techniques for placement and routing in Asic design. ICDSMLA 2019, Lecture notes in electrical engineering 601, Springer Nature Singapore Pte Ltd.
15. Shaik F, Sharma AK, Ahmed SM (2016) Hybrid model for analysis of abnormalities in diabetic cardiomyopathy and diabetic retinopathy related images. Springerplus 5:507. https://doi.org/ 10.1186/s40064-016-2152-2
16. Karimullah S, Vishnuvardhan D (2020) Iterative analysis of optimization algorithms for placement and routing in Asic design. ICDSMLA 2019, Lecture notes in electrical engineering 601, Springer Nature Singapore Pte Ltd.
17. Karimullah S, Vishnuvardhan D (2022) Pin density technique for congestion estimation and reduction of optimized design during placement and routing. Applied nanoscience
18. Nagaraju CH, Sharma AK, Subramanyam MV (2018) Reduction of PAPR In MIMO-OFDM using adaptive SLM And PTS technique. International Journal of Pure and Applied Mathematics, Special issue 118(17):355–373. ISSN: 1311-8080 (printed version); ISSN: 1314-3395
19. Karimullah S, Basha SJ, Guruvyshnavi P, Sathish Kumar Reddy K, Navyatha B (2020) A genetic algorithm with fixed open approach for placements and routings. ICCCE, Springer, pp 599–610

20. Karimullah S, Vishnuvardhan D, Arif M, Gunjan VK, Shaik F, Siddiquee KN (2022) An improved harmony search approach for block placement for VLSI design automation. Wireless Communications and Mobile Computing, vol 2022, Article ID 3016709, 10 pages. https://doi.org/10.1155/2022/3016709
21. Li H, Wang Y, Yang Z et al (2020) Discriminative dictionary learning-based multiplecomponent decomposition for detail-preserving noisy image fusion. IEEE Trans Instrum Meas 69(4):1082–1102

PSO-Based Evolutionary Image Segmentation System for Analysis of Fatty Liver Level Recognition

C. H. Nagaraju, S. Ramya Sree, P. Jameela, C. Kartheek, and B. Madhu Sudhan

Abstract One of the most frequent liver conditions nowadays is fatty liver, often known as liver hepatic glycogen. As our way of life pushes us toward this occurrence, clinical diagnosis of this ailment by human analysts is getting more common. Every clinic must have a quick and accurate expert method for fatty liver diagnosis that is the main reason that we wanted to create one detection technique. The suggested expert system uses several four markers used in image analysis algorithms and techniques to diagnose fatty liver and estimate its severity. The degree of disorder is assessed using four segmentation methods: Particle Swarm Optimization (PSO), Otsu, K-means, and watershed. The suggested system's precision will be assessed using the performance metrics such as IoU, F-score, and accuracy. Finally, the level of fatty liver is estimated utilizing several number of fat deposits incorporated inside the segmented image. Multiple high-resolution data samples with microscope zooming more than or equal to 200 are to be used for the experiments, which shall be performed in a systematic manner. When compared to conventional approaches, all performance indicators and comparisons produced results that were adequate. The proposed system could compare all samples to ground truth data.

Keywords Fatty liver · Image segmentation · Particle Swarm Optimization (PSO) · K-means clustering · Otsu · Watershed

1 Introduction

In many nations around the world, one of the leading causes of death is cancer [1]. The most dangerous medical problem that might endanger human life and health among them is liver disease. The second greatest cause of death in males and the sixth major cause of mortality in women, respectively, are liver tumors [2]. Around 841 new cases and 782 fatalities were recorded worldwide in 2018 [3]. Transplantation,

C. H. Nagaraju (✉) · S. Ramya Sree · P. Jameela · C. Kartheek · B. Madhu Sudhan
Department of Electronics and Communications Engineering, AITS, Rajampet, India
e-mail: chrajuaits@gmail.com

© The Author(s), under exclusive license to Springer Nature Singapore Pte Ltd. 2023
A. Kumar et al. (eds.), *Proceedings of the 4th International Conference on Data Science, Machine Learning and Applications*, Lecture Notes in Electrical Engineering 1038,
https://doi.org/10.1007/978-981-99-2058-7_15

resection, ablation, and radiation are the principal treatments used in routine clinical practice. However, there is a noticeable amount of risk involved with these treatment methods, which are also quite demanding and difficult [4]. When the fat content of the hepatocytes increases, the liver experiences fatty penetration. Patients with fatty liver frequently have no symptoms, and the condition is often discovered by accident. According to estimates, this illness affects 14–20% of people in the USA and Europe and is directly linked to obesity, diabetes, or alcoholism. The more reliable way to diagnose a fatty liver is with a liver biopsy. However, due to its invasive nature, it is only employed when all other non-invasive techniques have failed [imp]. The largest organ in your body is the liver. It aids in nutrient absorption, energy storage, and toxin removal in your body. When your liver becomes fatty, you have fatty liver disease. There are primarily two types:

Alcohol-unrelated fatty liver disease (NAFLD).

Alcoholic steatohepatitis, another name for alcoholic fatty liver disease.

NAFLD, a variety of fatty liver diseases, is not related to excessive alcohol consumption. Two categories are present:

Simple fatty liver is a condition in which there is little to no inflammation or liver cell death yet the liver is still fatty. Most of the time, a mild case of the fatty liver does not worsen or exacerbate the liver.

Non-alcoholic steatohepatitis (NASH) is characterized by inflammation, liver cell destruction, and liver fat. Scarring, or fibrosis, of the liver, can be brought on by inflammation and injury to liver cells. NASH can result in liver cancer or cirrhosis.

Heavy alcohol use is the cause of alcoholic fatty liver disease. The majority of the alcohol you consume is broken down by your liver, so it can be eliminated from your body. However, the breakdown process can produce dangerous byproducts. These chemicals can undermine your body's natural defenses, cause inflammation, and harm liver cells. Your liver deteriorates when you consume more alcohol. The initial stage of alcohol-related liver damage is known as alcoholic fatty liver disease. The following phases are cirrhosis and alcoholic hepatitis.

Diagnoses of liver illness are made using medical imaging techniques such as MRI, CT scans, and sonography. These tools have certain negative consequences and are expensive. As a result, researchers have suggested alternatives to imaging technologies for disease diagnosis [4].

A stochastic optimization technique called PSO is based on how swarms move and function. The idea of social interaction is utilized in PSO to resolve a conflict. It employs a swarm of particles (agents) that move about the search space in quest of the best answer. Each member of the swarm searches for the positional coordinates in the solution space that correspond to the best solution that the member has so far produced. It is referred to as a personal best or pbest. The PSO keeps track of the gbest, often known as the global best value. This is the highest value that has been determined by any nearby particle to date [5].

2 Literature Review

Researchers have proposed numerous image segmentation systems for fatty liver level recognition analysis. This survey presents several existing fatty liver techniques to illustrate the significance of the presented task.

A method for identifying liver disorders utilizing specific classification algorithms was put forth by Ramana et al. [6]. The classification techniques used are Support Vector Machines, C4.5, Naive Bayes classifier, and backpropagation neural networks. The efficiency of the technique is assessed in terms of accuracy, precision, sensitivity, etc. They replicated their work using WEKA. The method's accuracy for K-NN is 62.89%, for NBC it is 56.822%, for C4.5 it is 68.96%, for backpropagation it is 71.52%, and for SVM it is 58.22%. They presented the outcome in the paper by modifying the qualities.

In 2019, Liu et al. [7] confined that, to categorize, the backpropagation (BP) neural network and the C-means clustering technique are used, respectively. Two hundred patients' liver RF signals are used in the investigation. These RF signals, which each consist of 50 signals, represent four different types of the liver: severe fatty liver, moderately fatty liver, mildly fatty liver, and normal liver. A feature vector that is a weighted combination of PM, MSW, LWCM, and WMMM characteristics is used to categorize liver RF signals. There, the best feature vector that only contains PM, LWCM, and WMMM was discovered using the C-means clustering approach. Then, the BP neural network serves as the best feature vector. This study demonstrates that the ultrasonic RF signal is useful in determining the degree of liver fatty tissue, and this paper offers a new technique for fatty liver computer-aided diagnostics. A new computer-aided method based on an ultrasonic radiofrequency signal has been proposed in this research.

In 2020, the researchers Joloudar et al. [4], the researcher obtained mining methods that rely on an extraction, loading, transformation, and analysis (ELTA) strategy for a precise diagnosis. As a result, the ELTA method is used to analyze various data mining models, such as random forest, Support Vector Machine (SVM), Bayesian networks, Multi-Layer Perceptron (MLP) neural networks, and Particle Swarm Optimization (PSO)-SVM. The PSO-SVM model performs better than the others in terms of the criteria for specificity, sensitivity, accuracy, Area under the Curve (AUC), F-measure, precision, and False Positive Rate (FPR). Additionally, a tenfold cross-validation method is used to evaluate the models, enabling the models to be evaluated on data pertaining to liver disease. A methodical attempt was performed for Medical Data Mining of Liver disease on the UCI dataset. An accurate liver disease prediction that is delivered on time is crucial. In this study, the researcher examines five classification models: Random Forest, Particle Swarm Optimization (PSO), Support Vector Machine, MLP-Neural network, SVM Bayesian network, and PSO-SVM. The main objective is to determine which feature is most important for predicting liver illness using the ELTA technique. In the end, a PSO-SVM model was used to retrieve the seven attributes with the highest accuracy. In this study, the performance of the above models was also contrasted based on accuracy, sensitivity, specificity, AUC, F-measure, precision, and FPR standards.

In 2020, the researchers Wu et al. [8], this study aims to investigate the proposition that hepatocytes that have fat droplets deposited on them modify where their nuclei are located and how they are distributed spatially, changing the structural function (SF), a BSC factor. The distribution of nuclei determines the link between the fat fraction and the SF. To verify this hypothesis, digital images of 48 participants' liver histopathology slides were stained with hematoxylin and eosin. One to five sections from each participant's slide (453.6 m × 453.6 m) were selected, yielding a total of 218 images. For each image, hepatocyte nuclei and fat droplets were automatically identified. The SF in proportion to frequency was calculated using the nuclear distribution. The liver fat fraction was calculated using the fractional surface area of the fat droplets. The liver fat fraction and SF had a high correlation below 30 MHz (Pearson's r = 0.4, p = 10^{-4}). This study shows that the presence of fat droplets affects the distribution of hepatocyte nuclei, which influences the link between fat percentage and BSC.

In 2020, the researchers, Tamura et al. [9] show that this work proposed a compensating approach to extend the useful depth range of the HLSF based on the DND model through the use of HFU measurement. Twelve excised rat livers were studied for radio-frequency data; three of the livers were healthy (no lipid droplets were seen in the hepatocytes), and nine of the livers were fatty (10–70%). Based on the DND model parameter distribution that was acquired from healthy liver samples, each ROI was assigned a category by the healthy liver structure filter (HLSF). The functions of the depth-dependent Nakagami parameters were found by fitting the modified Gaussian distribution to the Nakagami parameters obtained from the three normal liver samples. HLSF(x) was developed using healthy liver datasets from focal depth −0.5 mm to focal depth + 3.5 mm at 1 mm intervals [10]. The filter is used to estimate DND characteristics at the same depth. This work improved the DND model and HLSF for lipid droplet quantification of steatosis liver. The suggested technique changed the depth dependency at B-mode data. The suggested strategy raised the depth range of dependable QUS (AUROC > 0.85) by 250% compared to half the depth of field, demonstrating that QUS could be used consistently with clinical HFU.

Ma et al. [11], the Dice coefficients are 92 and 90%, which are higher than the comparison segmentation networks. Furthermore, the experimental results suggest that the proposed method can lower computing consumption while maintaining higher segmentation accuracy, which is important for liver segmentation in practice and provides a good reference for doctors. To solve present problems in liver segmentation tasks, we used the VNet WGAN network in this study to automatically segment the liver. The algorithm fills the void. In the 2D segmentation network, three neighboring slices are used as input, and two convolution kernels are used to enhance the context information of the 3D data. As a result, our strategy significantly enhanced segmentation accuracy. The Dice coefficients are higher than the comparable segmentation networks at 92 and 90%. Additionally, according to the experimental findings, the suggested strategy can reduce processing requirements while keeping higher segmentation accuracy, which is crucial for liver segmentation in practice and serves as a useful guide for medical professionals. In this study, we employed the VNet WGAN network to automatically segment the liver in order to address current issues with liver segmentation tasks. The algorithm makes up the gap. Three adjacent slices are used as input in the 2D segmentation network, and two convolution kernels are employed to improve the context information of the 3D data. As a result, our approach greatly increased segmentation precision.

3 Methodology

3.1 Existing Method

K-Means

A popular vector quantization technique for cluster analysis in data mining is K-means clustering. It was first created for use in signal processing [12]. The cluster prototype is created by grouping observations into k clusters, with each observation being allocated to the cluster with the closest mean. As a result, the data space is divided into Voronoi cells. The most prevalent form of liver cancer is hepatocellular carcinoma or HCC for short [13]. Most occurrences of HCC are brought on by cirrhosis, which is most frequently brought on by alcohol consumption, or viral hepatitis infections (hepatitis B or C). It provides significant computational efficiency and backs multi-dimensional vectors as mentioned above clustering method.

This algorithm can be summarized as follows:

Step 1: By randomly picking data points from X, initialize the cluster centers $\{c1,..., cK.\}$.

Step 2: Distances between all data points and cluster centers are calculated.

Step 3: Label each data point with the clustered index (k) based on the shortest distance.

Step 4: Create new cluster centers by combining all member points in the same cluster.

Step 5: Do not alter until the cluster labels appear, repeat steps 2 through 4.

This research proposes a K-mean clustering-based optional segmentation technique. In CT liver pictures with high levels of image noise and some artefacts, it is intended to improve bone-segmented regions. There are two main requirements in the suggested method. In order to organize feature vectors and clustering regions, we first employed a hierarchical idea [14].

We developed a straightforward approach to determine the clustering region's kind and manage iteration. Therefore, before beginning the process, we are reluctant to reveal how often clusters there are altogether. In order to regulate the labels of clustered regions in the clustering results, we also developed a cluster indexing approach. By displaying the following indexes of background, soft tissue, and hard

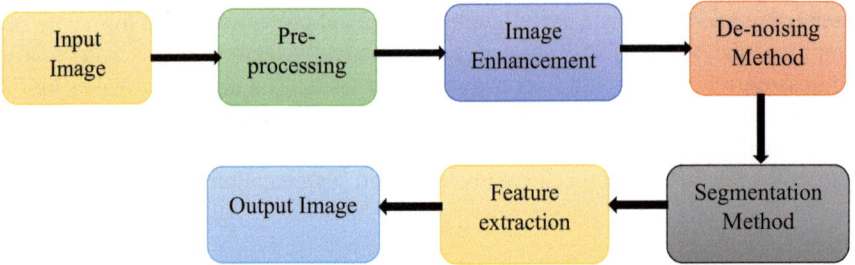

Fig. 1 Block diagram of Otsu algorithm

tissue sections, this indexing also connects to the different types of tissues or the CT standard numbers [15].

Otsu

In 1979, Scholar Otsu made the discovery of the Otsu technique. It is a thresholding technique that is widely utilized since it is both easy and efficient. This strategy employs a clustering-based approach [16]. It reduces the interclass variance and transforms a grayscale image into a binary image (Fig. 1).

Algorithm:

1. Use the threshold to divide the pixel into two groups.
2. Determine each cluster's mean.
3. Square the discrepancy in means.
4. Increase the number of pixels.

Watershed Segmentation Based on Markers

The region that indicates the presence of the necessary object is enhanced by the marker-controlled watershed segmentation method. The simplest technique to distinguish an image's objects from the backdrop is watershed categorization [17].

Algorithm:

1. Create a feature of segmentation in step one. The brown areas of this image serve as a representation of the things you're trying to separate.
2. Identify the symbols in the background. Each of the items contains these connected pixels in clumps.
3. Pick out foreground indicators. None of the objects contain these pixels.
4. Adjust the segmentation function so that the only locations where minima exist are at the foreground and background marker positions.
5. Calculate the watershed transforms of the revised segmentation function.

A contrasted image that has been absorbed is the end result of Otsu's augmentation. The watershed transformation technique will be used to segment the image. For a suitable and quick result, various preprocessing techniques such as image opening and closure have been carried out. In order to help doctors and radiologists pinpoint the precise location of the cancer cells in the liver, feature extraction is crucial in highlighting the cancer cells [18].

Watershed

The proposed model is a CAD model that uses the watershed Gaussian-based deep learning (WGDL) methodology to effectively identify the cancer tumor in CT scan images of the liver. The segmentation process's suggested workflow is shown. The Marker-controlled watershed segmentation approach was used to separate the liver from the other abdominal organs, and the Gaussian mixture model was used to segment the cancer tissue (GMM). To differentiate between the HEM, HCC, and MET types of liver tumors, the retrieved statistical, textural, and geometrical data were categorized using the DNN classifier.

Dataset

A total amount of 225 CT liver cancer images were acquired from Kaggle dataset which contain the standard images which are suitable for carrying out the proposed methodology.

Watershed Transform

The watershed transform is one method of region-based segmentation that is based on the concept of geography. This method treats the grayscale images as topographic relief, with a catchment basin-style regional minimum [19]. Flooding causes water to construct a wall and a watershed. The image was completely divided as a result of this strategy. The image structure is obtained via the morphological procedure. This procedure typically reduces the grayscale image's artifacts and system noise. On the gradient image, we applied the watershed transform to create a smooth boundary structure.

Model of a Gaussian Mixture

A matrix representing the image is shown, with each element standing for a single pixel. In the Gaussian mixture model (GMM), the image pixels are regarded as a random variable and denoted with the variable x, where x denotes a three-dimensional variable with RGB values.

The approach most frequently used to estimate the GMM parameter is maximum-likelihood estimation. The main objective of the estimation is to maximize the likelihood of the GMM dataset. The expectation maximization (EM) algorithm is used for estimation.

Feature Set

In order to effectively categorize the disease, a procedure known as feature extraction is used to extract useful information from an image. The statistical, geometrical, and texture features were retrieved from the segmented images and used for classification using the gray-level co-occurrence matrix (GLCM) method. The areas covered of an image are calculated: Cluster shadow, cluster prominence, energy, entropy, variance, sum average, homogeneity, sum entropy, sum variance, and inverse difference moment are all concepts used in the research.

Deep Neutral Networks

Numerous classification issues have been applied to the deep learning architecture, and the results have been excellent. In order to handle the problems of computer vision and machine learning, ANN is the conventional computational methodologies that are inspired by the network of biological neurons. One sort of artificial neural network that has more than three layers is a deep neural network (DNN), in which the outputs of the neurons are applied recursively to their own inputs. The classification is automatically carried out by the algorithm, and it also directly implements the decision-making pro.

Proposed Method

A swarm of potential solutions is how the PSO algorithm's fundamental variation operates. These particles are moved around the search space using a couple of simple equations. Both the best-known position of each individual particle and the movements of the particles are directed by the position of the cluster which is known to be the best. These will eventually start to direct the swarm's motions once better sites are found. It is hoped, but not guaranteed, that repeating the procedure will lead to the eventual discovery of a workable solution.

Image acquisition

The correctness of a machine vision model is one of the most important factors that may help your product flourish and be trusted in the market. For this, you need trustworthy image acquisition components that can create beautiful pictures. The machine vision system's speed is typically crucial to a company's manufacturing procedures for overall throughput and rapidity.

Histogram equalization

The contrast of images is improved using the computer method of image processing known as histogram equalization. This is obtained by efficient and effective image's intensity range and scattering the most frequent intensity levels. This tactic often increases the images' overall contrast when the useful info is compared with the expected values. As a response, regions with low local contrast can grab greater contrast.

RGB to Gray conversion

The RGB to gray function turns colored images into black and white images by reducing the brightness information but keeping the saturation and hue information.
Gray color and RBG color code have equal red, green, and blue values:

$$R = G = B$$

For each and every image with Red, Green, and Blue values can be as (R, G, and B):

$$R_1 = G_1 = B_1 = (R + G + B)/3 = 0.333R + 0.333G + 0.333B$$

This equation can be modified with different values for each R/G/B weight.

$$R_1 = G_1 = B_1 = 0.2126R + 0.7152G + 0.0722B$$

It is the same as

$$R_1 = G_1 = B_1 = 0.299R + 0.578G + 0.11B$$

Adjustment of image intensity values

Techniques for improving images are employed, with "improve" alluding to both objective and subjective criteria (e.g., increasing the signal-to-noise ratio) (e.g., making certain features easier to see by modifying the colors or intensities).

An image improvement technique called intensity adjustment translates the luminance values in an image to a new range. This picture uses a low-contrast image and its histogram as an example. Observe how all of the data congregate in the range's center in the histogram of the image.

Block Diagram

See Fig. 2.

PSO Block

The PSO is utilized to choose the best features. The features are selected based on their velocity factor, and the velocity factor will be used to determine the difference. Here, we discuss examining property conditions. When a character has a stronger impact on developing liver disease, the breakdown of the dependency factor follows, which also has a higher impact. Features are given a weight based on how likely they are to contribute to liver disease. With the right parameter alteration, another improvement should be achievable that facilitates molecule swarm streamlining to increase the viability of highlight choice. Every characteristic has a numerical weight that expresses the relationship between two features. Positive load indicators indicate fortification, whereas negative load indicators indicate constraint. For the purpose of determining its precise related point, the weight is standardized among speeds and loads. This provides a comprehensive solution for identifying liver issues early on.

PSO Algorithm

See Fig. 3.

The PSO algorithm follows the following procedure:

Step 1: Declare the parameters as well as the variables. Create a "population" of particles that are evenly distributed over X.

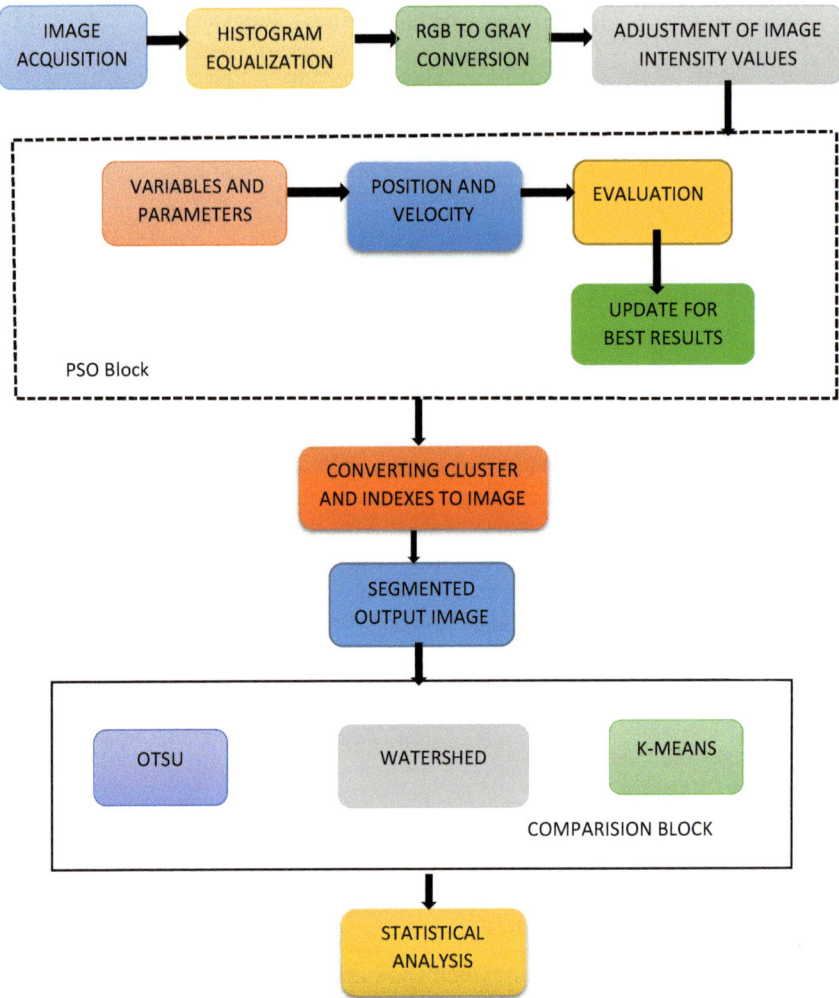

Fig. 2 Block diagram of the proposed methodology

Step 2: Examine each particle's position in relation to the objective function.

$$z = f(x, y) = \sin x^2 + \sin y^2 + \sin x \sin y \tag{1}$$

Step 3: Update a particle's position if it is currently at a better location than it was beforehand.

Step 4: Identify the best particle (based on the particle's most recent best locations).

Step 5: Particle velocities should be updated.

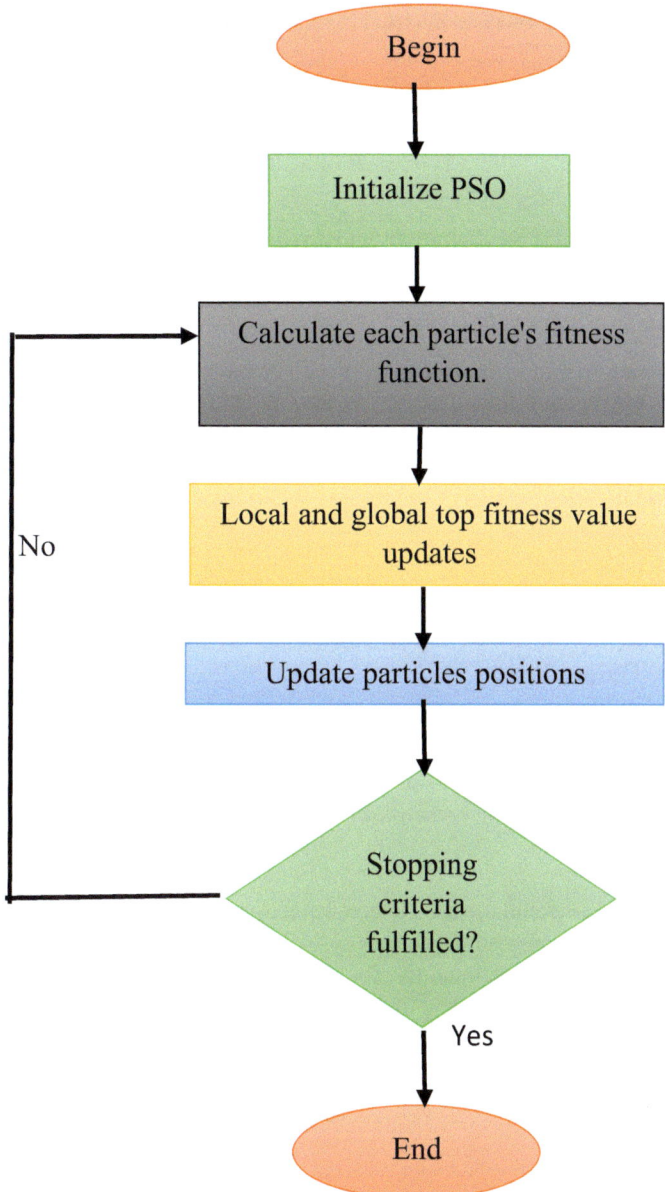

Fig. 3 Flowchart of the PSO algorithm

$$V_i^{t+1} = W.V_i^t + c_1 U_1^t (P_{b1}^t - P_i^t) + c_1 U_2^t (g_b^t - P_i^t) \dots \tag{2}$$

Step 6: Particles should be moved to their new locations.

$$P_i^{t+1} = P_i^t + v_i^{t+1} \dots \tag{3}$$

Step 7: Step 2 should be performed when the preceding requirements have been satisfied.

Converting and Indexes to Image

The segmentation of an image into various segments is known as image segmentation. The use of aggregation for image segmentation has been the subject of numerous studies. One of the most widely used techniques is the K-means clustering algorithm. But initially, the image is partially stretched improved to boost its quality before the K-means method is used. A data clustering technique called subtractive clustering creates the centroid based on the possible value of the output points. Therefore, the first centers are produced via subtractive clustering, and the K-means algorithm uses these initial centers to segment images. Finally, a medial filter is used to remove any undesirable areas from the segmented image.

Segmented Output Image

Image segmentation is a function that creates an output from input images. In the output, each pixel's object class or instance is indicated by a mask or matrix with numerous elements. For picture segmentation, a variety of relevant heuristics, or image quality attributes, can be helpful.

Otsu

Thresholding is the technique used to determine foreground pixels from surrounding pixels. The "Otsu's approach," put out by Nobuyuki Otsu, is one of the numerous methods for obtaining optimal thresholding. The criterion where the weighted volatility between the foreground and background pixels is the least is found using Otsu's variance-based method. The important thing is to measure the dispersion of background and foreground pixels by iterating through all of the threshold's potential values.

Watershed

Some sophisticated presently accessible the use of the watershed algorithm for segmentation since accurate results cannot be obtained with simple thresholding and outline detection.

The watershed algorithm is based on extracting particular background and foreground data, performing the watershed, and then utilizing markers to discover the exact borders. Generally speaking, this technique aids in the detection of touching and covering objects in photographs.

It can be user-defined for markers, such as by manually clicking and obtaining the dimensions for markers, or it can be done using pre-defined methods, like thresholding or any morphological processes. We are unable to directly employ watershed algorithms because of the presence of noise.

K-means

Since precise results are not possible with straightforward thresholding and outline detection, certain advanced segmentation techniques are already available that use the watershed method.

The watershed algorithm works by first extracting certain background and background data, running the watershed, and then using markers to identify the exact bounds. This method generally assists in the identification of contacting and concealing objects in photos.

It can be done using pre-defined methods, such as thresholding or any morphological process, or it can be user-defined for markers, such as by manually clicking and obtaining the dimensions for markers. Because of the noise, we are unable to use watershed algorithms directly.

Statistical Analysis

The collection of data for statistical analysis is performed to resolve trends and patterns. Here, after comparing with various methods such as Otsu, watershed, and K-means, statistical analysis generates an image as a result.

4 Simulation Results

The segmentation output accuracy can be increased by using image processing and segmentation. We can achieve a segmented result using the PSO approach. The PSO approach has improved the speed, accuracy, and detection of fatty liver. A human specialist can perfectly be replaced by this technique. When compared to the other Otsu, watershed, and K-means strategies, the PSO technique performs better.

- Figure 4 illustrates the original image that we took from the liver.
- Figure 5 illustrates the preprocessed image of the original image after applying the image acquisition.
- Figure 6 illustrate the segmented image of the original image after applying the histogram equalization.
- Figure 7 illustrates the ground truth image which is the input of the PSO Block.
- Figure 8 illustrates the output of the Otsu technique.
- Figure 9 illustrates the output of the watershed technique.
- Figure 10 illustrates the output of the K-means technique.
- Figure 11 illustrates the output of the PSO technique.

From the graph, we can conclude that the cost can be reduced and the iterations can be increased by using the proposed technique (Fig. 12, Table 1).

Fig. 4 Original image

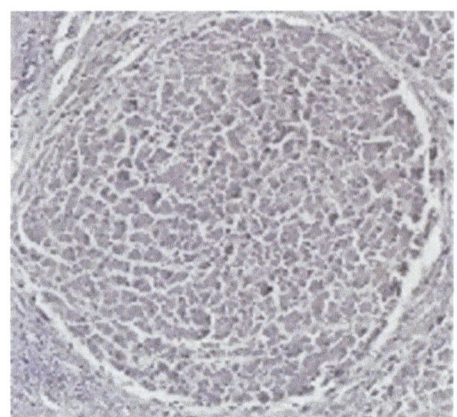

Fig. 5 Preprocessed image

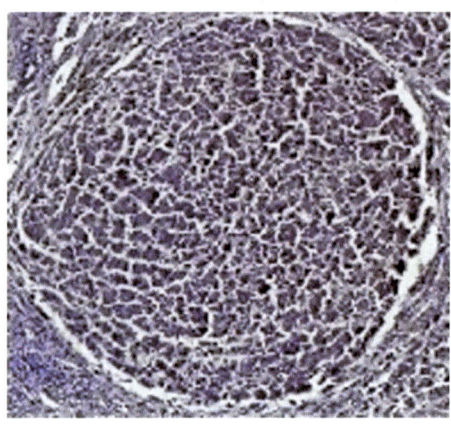

Fig. 6 Segmented image

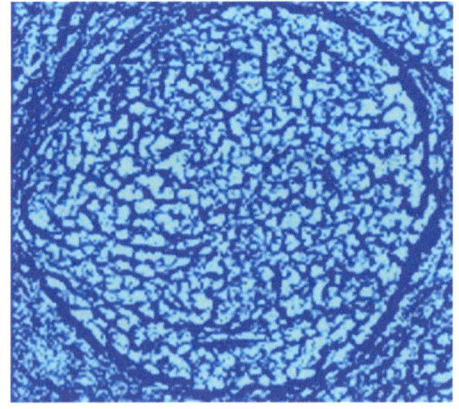

Fig. 7 Ground truth

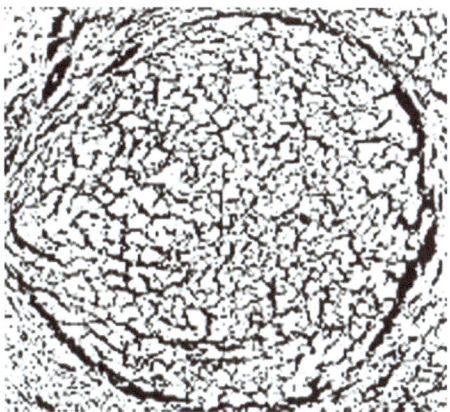

Fig. 8 Otsu image

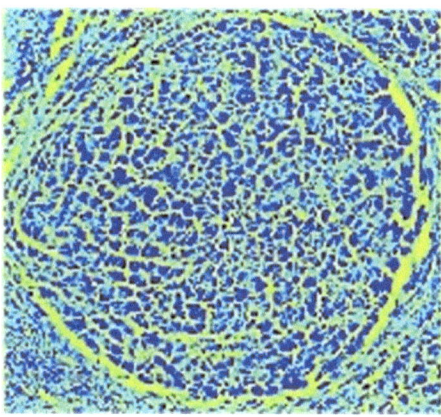

Fig. 9 Watershed image

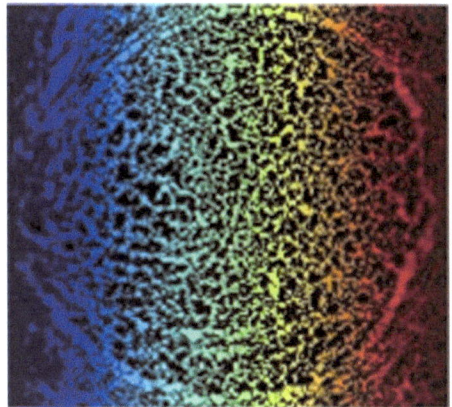

Fig. 10 K-means image

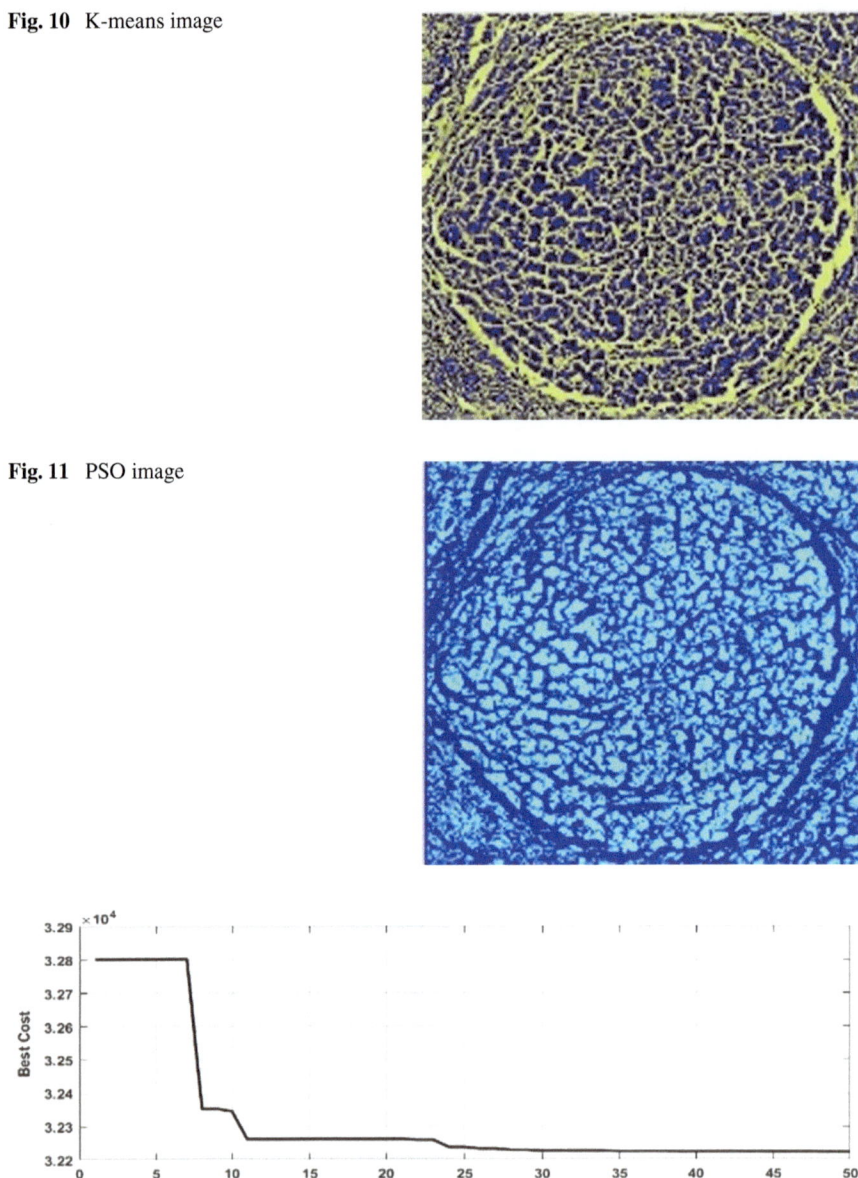

Fig. 11 PSO image

Fig. 12 Graph between the cost and the iteration of the different techniques

Table 1 Results

Methods	Accuracy (%)	F-scope (%)	IoU (%)
Otsu	78.2	65	78.9
Watershed	82.4	72.2	82.4
K-means	85.4	81.1	91.6
PSO	92.2	87.2	97.0

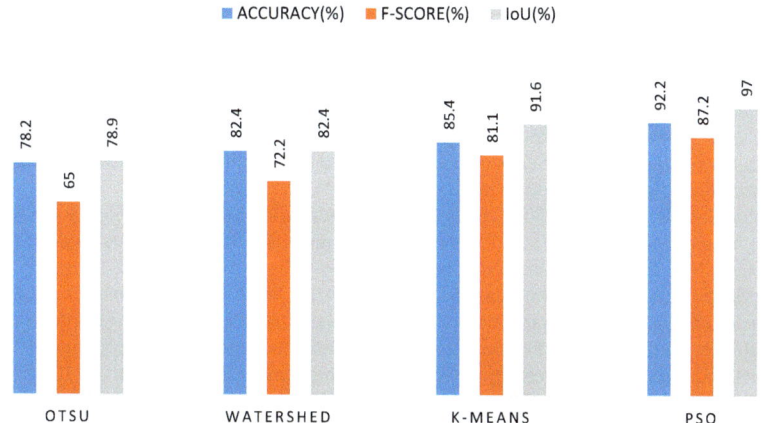

Chart 1 Accuracy, F-scope, IoU values for OTCS, watershed, K-means, and PSO

The image shows the original image, preprocessed image, segmented image, and ground truth image which we get the image as the final output. And the graph shows the cost comparisons of Otsu, watershed, K-means, and the PSO techniques (Chart 1).

From the figure, we can see the compared segmented image of the different techniques such as Otsu, watershed, K-means, and PSO.

5   Conclusion

In conclusion, both ALFD and NAFLD disorders are treated using the PSO technique. In the future, we can employ segmentation methods like the genetic algorithm, which is nature-inspired. It is advised to apply the given strategy to various datasets with different microscopic zoom levels.

And the accuracy, F-scope, and the IoU values are increased in the proposed method.

References

1. Massoptier L, Casciaro S (2007) Fully automatic liver segmentation through graph-cut technique. Article in conference proceedings: annual international conference of the IEEE engineering in medicine and biology society. IEEE Engineering in Medicine and Biology Society. Conference Feb 2007
2. Ahmad M, Qadri SF, Qadri S, Saeed IA, Zareen SS, Iqbal Z, Mizanur Rahman SM (2022) A lightweight convolutional neural network model for liver segmentation in medical diagnosis. Hindawi Computational Intelligence and Neuroscience, vol 2022, Article ID 7954333, 16 pages. https://doi.org/10.1155/2022/7954333
3. He K, Liu X, Shahzad R, Reimer R, Thiele F, Niehoff J (2021) Advanced deep learning approach to automatically segment malignant tumors and ablation zone in the liver with contrast-enhanced CT. This article was submitted to cancer imaging and image-directed interventions, a section of the journal Frontiers in Oncology Published on 15 July 2021
4. Joloudari JH, Saadatfar H, Dehzangi A, Shamshirband S (2019) Computer-aided decision-making for predicting liver disease using PSO-based optimized SVM with feature selection. Contents lists available at ScienceDirect Informatics in Medicine Unlocked journal homepage: http://www.elsevier.com/locate/imu
5. Ribeiroand R, Sanches J (2009) Fatty liver characterization and classification by ultrasounds. In: Araujo H et al (eds) Partially supported by FCT, under ISR/IST Plurilingua funding: IbPRIA 2009, LNCS 5524. Springer, Berlin, pp 354–361
6. Rezwanul Haque M, Milon Islam M, Kamrul Hasan M (2018) Performance evaluation of random forests and artificial neural networks for the classification of liver disorder, February 2018. In: Conference: international conference on computer, communication, chemical, materials and electronic engineering at: Rajshahi, Bangladesh. https://doi.org/10.1109/IC4ME2.2018.8465658
7. Liu Z (2019) Study on diagnosis of fatty liver based on ultrasounding RF signal. In: 2019 3rd international conference on imaging, signal processing and communication
8. Wu Y, Lopez L, Andr MP (2021) Liver fat droplet dependency on ultrasound backscatter coefficient in nonalcoholic fatty liver. 978-1-7281-5448-0/20/$31.00 ©2020 IEEE Authorized licensed use limited to: Univ of Calif Santa Barbara. Downloaded on May 21, 2021 at 15:11:13 UTC from IEEE Xplore
9. Tamura K, Mamou J, Hachiya H (2021) Effective depth expansion for reliable fatty liver assessment using a double Nakagami distribution model. 978-1-7281-5448-0/20/$31.00 ©2020 IEEE Authorized licensed use limited to: Carleton University. Downloaded on May 28, 2021 at 02:14:03 UTC from IEEE Xplore
10. Peng J, Wang Y, Kong D (2014) Liver segmentation with constrained convex variational mode. Contents lists available at ScienceDirect Pattern Recognition Letters journal homepage: www.elsevier.com/locate/patrec
11. Ma J, Deng Y, Ma Z, Mao K, Chen Y (2021) A liver segmentation method based on the fusion of VNet and WGAN. Correspondence should be addressed to Jinlin Ma; 624160@163.com. Received 26 Feb 2021; Accepted 21 Sept 2021; Published 8 Oct 2021 Academic Editor: Cristiana Corsi
12. Karimullah S, Vishnuvardhan D, Arif M, Gunjan VK, Shaik F, Siddiquee KN (2022) An improved harmony search approach for block placement for VLSI design automation. Wireless Communications and Mobile Computing, vol 2022, Article ID 3016709, 10 pages. https://doi.org/10.1155/2022/3016709
13. Albu A, Precup R-E, Teban T-A (2019) Results and challenges of artificial neural networks used for decision-making and control in medical applications. Facta Univ Ser Mech Eng 17(3):285–308
14. Karimullah S, Vishnuvardhan D (2022) Pin density technique for congestion estimation and reduction of optimized design during placement and routing. Applied Nanoscience

15. Shaik F, Sharma AK, Ahmed SM (2016) Hybrid model for analysis of abnormalities in diabetic cardiomyopathy and diabetic retinopathy related images. Springerplus 5:507. https://doi.org/10.1186/s40064-016-2152-2
16. Karimullah S, Basha SJ, Guruvyshnavi P, Sathish Kumar Reddy K, Navyatha B (2020) A genetic algorithm with fixed open approach for placements and routings. ICCCE, Springer, pp 599–610
17. Karimullah S, Vishnuvardhan D (2020) Experimental analysis of optimization techniques for placement and routing in Asic design. ICDSMLA 2019, Lecture notes in electrical engineering 601, Springer Nature Singapore Pte Ltd.
18. Karimullah S, Vishnuvardhan D, Basha SJ (2020) Floorplanning for placement of Modulesin VLSI physical design using harmony search technique. ICDSMLA 2019, Lecture notes in electrical engineering 601, Springer Nature Singapore Pte Ltd.
19. Nagaraju CH, Sharma AK, Subramanyam MV (2018) Reduction of PAPR in MIMO-OFDM using adaptive SLM And PTS technique. International Journal of Pure and Applied Mathematics, Special issue, 118(17):355–373, ISSN: 1311-8080 (printed version); ISSN: 1314-3395

Blake–Zisserman Model of Segmentation Method for Low-Contrast and Piecewise Smooth Image

B. Lakshmi Devi, E. Haripriya, D. Jaya, P. Anjali, G. R. Venkatesh Prasad Gowd, and K. Hemanth Kumar

Abstract Although extensively investigated, image segmentation is still a challenging topic, particularly for new and developing imaging modalities with different levels of extremely strong noise. Edge-based models and region-based models are two types of variational segmentation methods that typically offer more robust solutions for complicated images. The primary challenge of segmenting low-contrast images with interesting visual features having piecewise smooth intensities is addressed in this project study. Whites and blacks are replaced by a plethora of grey tones in low-contrast images that have little to no tonal contrast. The difficulties posed by the two factors low contrast and piecewise smooth features are in fact well-known difficulties. Due of its potential to enhance the MS model, the Blake–Zisserman model is taken into account in this research. Similar to the MS, the model has never been explicitly solved, making it impossible to apply it exactly and directly. A convex-relaxed game formulation of the less well-known Blake–Zisserman model is applied in this study to segment images with poor contrast and heavy noise. Programming in MATLAB with a version of R2017b or higher will be used for this project. This paper suggests using segmentation to detect weak leaps and finer edges without being vulnerable to noise. By segmenting using a CRCV model, it can successfully improve photos that have undergone piecewise smoothing and low-contrast images. With the alternate direction approach of multipliers, try to solve the coupled issue using the split-Bregman variation (ADMM).

Keywords Alternating Direction Method of Multipliers (ADMM) · Mumford–Shah (MS) · Blake–Zisserman · Image segmentation

B. Lakshmi Devi (✉) · E. Haripriya · D. Jaya · P. Anjali · G. R. Venkatesh Prasad Gowd · K. Hemanth Kumar
Department of ECE, AITS, Rajampet, Andra Pradesh, India
e-mail: bodagalalakshmi.devi@gmail.com

© The Author(s), under exclusive license to Springer Nature Singapore Pte Ltd. 2023   155
A. Kumar et al. (eds.), *Proceedings of the 4th International Conference on Data Science, Machine Learning and Applications*, Lecture Notes in Electrical Engineering 1038,
https://doi.org/10.1007/978-981-99-2058-7_16

1 Introduction

Despite the fact that image segmentation has been studied for a while, a theoretical foundation is still needed. While image restoration involves removing picture degradations like noise, blur, or occlusions in order to create a better image than the original image, image segmentation involves grouping portions of an image that have similar properties together. By segmenting an image, you can process just the crucial parts of it as opposed to the complete thing. Image segmentation is a significant and challenging issue. By segmenting low-contrast and piecewise pictures, this segmentation is used in the Blake–Zisserman model to tackle higher-order discontinuities. Segmented images have a high contrast, no unnecessary noise, and no blur. The constants in pieces two of the most well-known variational models in the fields of image segmentation and restoration study are Mumford–Shah and Rudin–Osher–Fatemi. This model can help with image segmentation through image restoration and threshold, which in turn can help with image manipulation. Due to the observer-dependent nature of the image segmentation problem, it is more challenging. In a variety of domains, including computer vision, medical image processing, and others, image segmentation is frequently employed. The topic of image segmentation in the field of medical image processing is now open for discussion. The processing and analysis of medical images must first segment the images for analysis. Its objective is to represent as faithfully as possible the anatomical structure of interest or a particular tissue area. Many scientists are looking into novel medical picture segmentation techniques because traditional machine learning-based approaches for doing so have some drawbacks.

In recent years, deep learning has attracted a lot of attention in various artificial intelligence applications, and digital image processing has been applied in a number of applications. In order to make individual organs or lesions easier to access and understand, medical image segmentation involves separating them from background pictures like CT or MRI scans. There has not been a thorough analysis of deep learning-based medical image segmentation. Classification, detection, and registration are only a few of the additional elements of medical image analysis that are covered. The vast treatment of networks, capabilities, and flaws in this article leaves out some important information. A computational segmentation model based on a full convolution network showed promise in the segmentation of medical images. The development of U-Net and its application of three-dimensional convolution were studied by Milletari et al. There is a deconvolution layer for every convolution layer. As a result, medical image segmentation is carried out automatically. In medical image processing, image segmentation is utilised to isolate the trouble spot by applying various models and techniques to various regions of the body. The three sections of an article are methodologies (network structures), training methods, and problems. The most popular network structures used for picture segmentation are introduced in the network structures section along with their key benefits and drawbacks.

2 Literature Survey

Rapidly rising in importance as a preferred method for assessing medical image segmentation is deep learning techniques. This study examines various contributions made to the field of deep learning in medicine, including the most prevalent problems raised in recent publications learning research can be used for a variety of tasks, including object detection, segmentation, registration, and image classification. First, the fundamental concepts behind deep learning techniques, programmes, and frameworks are presented. We provide a brief explanation of the optimum deep learning strategies [1].

The process of selective segmentation is crucial to image processing. There are numerous uses for being able to accurately segment an object in an image, especially in medical imaging. Clinicians can use robust approaches to help in diagnosis, surgical planning, etc. When there is little contrast between two things and it is difficult to see an edge, it is still challenging. In this situation, relying only one-edge limitations will not work. In order to create a segmentation model that is robustly capable of segmenting regions in an image even in the presence of low contrast, we intend to apply area restrictions in addition to edge information [2].

For the purpose of resolving picture restoration issues, introduce and investigate a multi-scale approximation of a nonlinear elliptic functional first proposed by Blake–Zisserman. The functional is a fourth-order PDE variational model that seeks to identify geometric characteristics in the image that are first- and second-order singularities. The approach entails the introduction of a set of discrete linear energies that approximate the convergence of the BZ functional. These functional describe the behaviour of the associated high-order diffusion operators on picture restoration while maintaining corners and contours and are constructed using an adaptive finite element method [3].

Deep convolutional networks' convolution filters exhibit rotation-variant behaviour. This research demonstrates that current techniques for leveraging rotation-variant characteristics can be enhanced by proposing a grid-based graph convolutional network, even though learned invariant behaviour can be partially attained. We draw the conclusion that the inherit nature of spectral graph convolutions is capable of learning invariant behaviour [4].

3 Existing Method

Mumford–Shah is the existing method. Mumford–Shah is never directly solved and cannot be implemented exactly. However, for images with weak edges that may be without unwanted noise, and without blurriness this method can be used. Mumford–Shah type models limitation is the stair casing effect for gradients. Compared to Mumford–Shah model (MS), Blake–Zisserman model is very useful for difficult images.

4 Methodology

To effectively deal with low images, particularly for treating higher-order discontinuities by MS model, this model is implementable. This work proposes enhancements for treating higher-order discontinuities, particularly in low-contrast images.

4.1 Block Diagram for Extracting Image

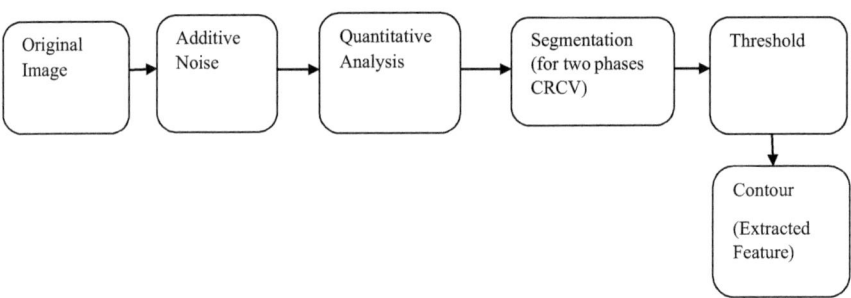

The block diagram is related to extracting the feature "image" without discontinuities. The above block diagram shows the original image, additive noise, qualitative analysis, segmentation, threshold, and contour image. The original image is a clean image that was used for analysis and it is considered as input. The output is connected to additive noise. Generally, we know noise is unwanted material which is almost present in images during processing steps of image processing. By the use of filters, noise is removed in the images. Here, the Gaussian noise is added. The output of additive noise is connected to qualitative analysis. A set of computational tools, including binarization, simplified expanding, simple image statistic threshold, were used to improve the qualitative analysis of scanning electron microscope images. Segmentation is the process of separating a digital image into various segments also known as image objects in digital image processing and computer vision. For analysing segmentation result, we are using minimiser from stage from stage one later threshold by the use of its parameters. We get two-phase results when we use CRCV; otherwise, we get multiphase results. The output is connected to threshold for further process. The output is connected to the thresholding. It is the process of conversion of a colour or greyscale image into a binary/digital image, which is simply as black and white. The threshold method is the most basic method of image segmentation. To convert a greyscale image to a binary image, this method uses a clip level (or a threshold value). The main part of this method is the threshold. First method is related to k-means algorithm. Next is to define threshold values manually, which produces useful results in most cases.

These are classified into three types based on the threshold operator.

1. The global threshold operator is to find out by the pixel's grey values.
2. The local threshold operator is determined by the local properties.
3. The dynamic threshold operator is to find out by the local properties position, pixel's grey values.

After thresholding process, output is connected to contour for extracting enhanced image.

Contour: the curve connects all continuous points (along the boundary) of the same colour or intensity. Contours are an effective tool for shape analysis as well as object detection and recognition. Use binary images for greater precision. The extracted image was extracted with higher picture quality.

5   Algorithm

K-means algorithm the automated method is being used. Later, k clusters of input $(k > 2)$ are performed, a threshold is set, and then results are manually entered, allowing for a more easily selected range of values. Mumford–Shah and extremely loud noises are possible. Natural image approximation is using the proposed algorithm for minimising the piecewise smooth Mumford–Shah function. Our method yields high-quality solutions that are independent of initialisation.

Drawback:

The MS model is limited in its ability to treat higher-order discontinuities in low-contrast and piecewise images.

Applications:

In terms of medical imaging (X-rays, scanning, and other applications).

6   Proposed Method

To deal with low-contrast images more effectively, especially when treating higher-order discontinuities, the Blake–Zisserman model was proposed rather than the Mumford–Shah model. The Blake–Zisserman model, which can improve on the MS model, cannot be implemented directly and precisely, as the MS model was never solved directly. Based on the Blake–Zisserman model, we now propose a solvable model. The existence of image is established as the solution to game reformulation. It depicts the numerical algorithm solving the coupled problems using ADMM. ADMM is an algorithm that solves convex optimization problems by breaking them down into smaller pieces, each of which is then easier to handle. The block diagram represents the process of extracting an image without discontinuities.

6.1 Block Diagram for Extracting Image for Proposed Method

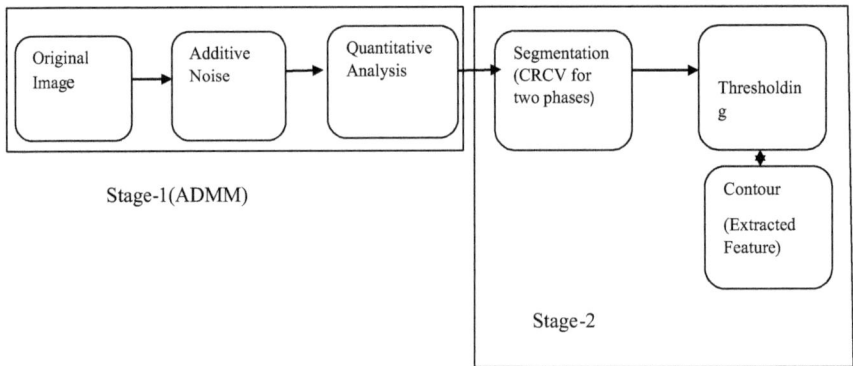

The above block diagram depicts Stage 1 and Stage 2. Stage 1 is related to ADMM solver. The ADMM (solver) process uses the original image, additive noise, and quantitative analysis to treat regularizations. In the second stage, the feature "Image" is extracted.

The Alternating Direction Method of Multipliers (ADMM) is an algorithm. This algorithm solves difficult problems by splitting into tiny pieces to make easier to maintain. This improves the image and it treats higher-order discontinuities particularly for low-contrast and piecewise images. The regularizations are treated, where it is used to fix or solve the problem of image restoration. The next stage is related to segmentation. Later, CRCV is used for two phase results. Next, thresholding is performed. It is about conversion of greyscale into digital image. Finally, the feature image is extracted which enhances the image.

Advantages:

Image improvement is collected here. Without unwanted noise, the image is extracted and improved. The B–Z model is used to extract contours with enhancement features and treat higher-order discontinuities.

Applications:

Scanning, X-rays, and other forms of medical imaging.

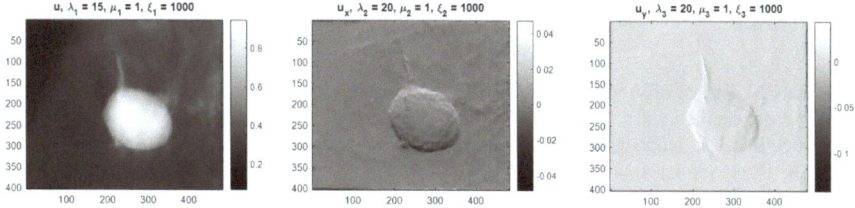

Fig. 1 Output of our model

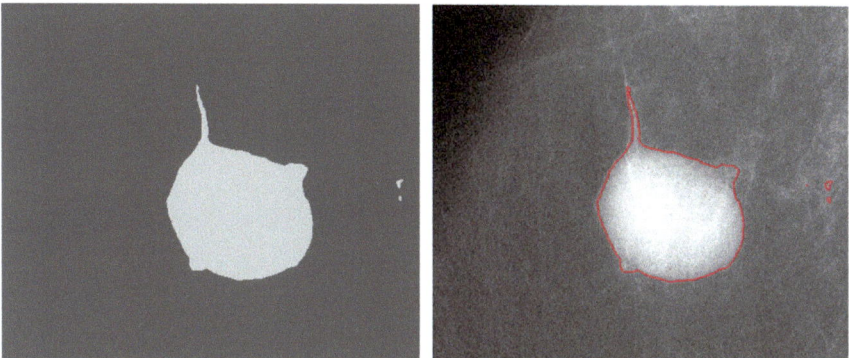

Fig. 2 Feature extraction (contour image) and threshold output image

7 Results and Analysis

The clear image without noise is analysed by this model. The quantitative results are related to six methods. Those are CRCV, CCZ, CNCS, TROF, DL, Ours. From this analysis, we can justify the image is clear, and compared to original image, the image is improved (Figs. 1 and 2).

8 Conclusion

This model enhances a feature image; it has some features like without noise, without blurriness. Especially treating of higher-order discontinuities, this model was implemented. It can improve the image with advancement. By the use of ADMM solver, it was implemented. For two-phase results, we are using CRCV model and it is only for two-phase segmentation.

References

1. Zanetti M, Ruggiero V, Miranda M (2016) Numerical minimization of a second-order functional for image segmentation. Commun Nonlinear Sci Numer Simul 36:528–548
2. Roy S, Borzì A, Habbal A (2017) Pedestrian motion modelled by Fokker-Planck Nash games. R Soc Open Sci 4:170648
3. Hesamian MH, Jia WJ, He XJ, Kennedy P (2019) Deep learning techniques for medical image segmentation: achievements and challenges. J Digit Imaging 32:582–596
4. Burrows L, Chen K, Torella F (2020) On new convolutional neural network based algorithms for selective segmentation of images. In: Proceedings of the MIUA 2020 Proceedings, Oxford, UK, 15 July 2020; Communications in computer and information science book series (CCIS), vol 1048. Springer, Berlin/Heidelberg, Germany, pp 93–104

Performance Enhancement of Wireless Systems Using Hybrid RIS Technique

**Shaik Karimullah, M. Lalitha, Mounika, S. Mahesh Reddy,
S. Mahammad Waseem, and Shaik Mohammad Irfan**

Abstract In this, we must characterize statistically and examine the performance of MRCIS—multiple reconfigurable intelligent surface-supported systems. We must first take into account that RIS through varied physical shapes are ordered distributively, the RISs' channels are supposed to be independent yet distributed non-identically, and the system experiences a fading environment. Using the two DMRIS-aided wireless methods, ERA and ORIS-aided, one can calculate the approximations for the OP-outage probability along with EC-ergodic capacity (ORA). In particular, we present a paradigm for putting into practice statistical end-to-end channel characterization. These schemes' end-to-end channel magnitudes can have either a LND—Log-Normal distribution or a GD—Gamma distribution. With these discoveries, we assess the tight approximation formulas for the OP and EC.

Keywords Reconfigurable intelligent surface · Hybrid RIS · Gamma distribution · Ergodic probability · Outage probability

1 Introduction

Reconfigurable intelligent surfaces (RISs) have been envisioned as a new wireless technology capable of dynamic, main objective control of radio signals between such as the transmitter and the receiver, effectively transforming the network resource into a service [1]. The reconfigurable intelligent surface (RIS) is a feasible method for creating future spectral and energy efficient, yet cost-effective wireless networks by cleverly changing the wireless propagation environment, which is typically thought to be random and uncontrollable. IRS can perform a variety of useful operations, such as three-dimensional (3D) passive beamforming, spatial interference nulling as

---

S. Karimullah (✉) · M. Lalitha · Mounika · S. Mahesh Reddy · S. Mahammad Waseem ·
S. M. Irfan
Annamacharya Institute of Technology and Sciences, Rajampet, Andhra Pradesh, India
e-mail: munnu483@gmail.com

well as cancelation, as well as other useful operations, by controlling signal reflection via a large number of low-cost passive elements [2].

The researchers propose a RIS selection method that selects the RIS with the maximum SNR to aid communication in order to achieve low-complexity transmission. The researchers suggest multi-RIS-aided systems both for inside and outside communications in situations in which a shortest link of both destination and source is not available [3]. Though small-scale diminishing was disregarded, effectiveness of the RIS choice technique remained not investigated. Authors investigated the three-dimensional throughput of multi-cell and single systems once multiple RISs remain used for the support of numerous users while attempting to argue with and interference or inter-cell–intra-cell [4]. Future wireless communication systems have been thought to benefit from the upcoming technology known as reconfigurable intelligent surface (RIS). These man-made surfaces have been decided to make of reconfigurable electromagnetic substances that can be exploited and programmed by interconnected electronic devices, potentially improving spectrum and energy efficiencies [5]. When compared to conventional relaying technology, RISs do not require as much expensive gear or overhead to operate. Additionally, RISs have the ability to modify the wireless environment by the utilization of almost reflecting components, giving system inventers complete control over the electromagnetic reaction of the interrupting signals to the surrounding objects [6].

Because of RISs, the definition "smart radio environments" seems to be currently under development. A smart radio environment is a wireless system in which the surroundings are transformed into a smart transportable space that actively engages in information transfer and processing. This is in contrast to today's wireless networks, in which the environment is beyond the operators' control. Smart radio environments essentially expand the concept of software networks. Future wireless networks, in particular, are rapidly evolving more toward a software dependent but also reconfigurable platform in which every element of the network can respond to environmental changes [7]. However, in this optimization process, the environment itself continues to be an unpredictable component because it is blind to the communication process taking place within it.

As a result of their distinct qualities, RISs become one of the fundamental technologies to achieve the futuristic vision of a smart radio environment. Reconfigurable intelligent surface (RIS), which is based on recent meta-materials' developments, is seen as a promising technology for wireless networks in the future, not only for its significant improvements in spectral and energy efficiency but also for its ultimate goal of smart propagation environment control. Particularly, an IRS is made up of inexpensive, essentially passive reflecting components that can change how electromagnetic waves interact with one another [8].

2 Literature Review

According to a review of the technical literature, RIS and RIS-assisted wireless systems have received a lot of attention in terms of design, demonstration, optimization, and analysis. As an illustration, the authors presented a RIS for indoor applications that uses 102 MSs and works at 2.47 GHz [9]. Similar to this, research writers described a RMS-reconfigurable MS with scattering, variable polarization, and focusing controls, and they provided intelligent walls that had frequency-selective MSs [10].

A MS that can rotate an electromagnetic wave that is linearly polarized by 90° has also been described, and a microwave-band MS that is extremely thin and based on phase discontinuities has been proposed [11]. Furthermore, a RIS design that has used varactor-tuned ring resonator to achieve configurable PS by vastly differing the slanted high voltage to the variable resistor was delivered and demonstrated its effectiveness. The final RIS element was RIS elements with PIN diodes attempting to control their EM response. An exponential growth uplink dynamical system making an analysis of RIS-assisted system applications was decided to carry out under Rician fading, but also basic optimization structures for maximizing received RF power throughout RIS-enabled wireless communication schemes were presented underneath the presence of optimization [12].

Additionally, designs for the best precoder which is linear, power distribution, and RIS phase matrix that made use of comprehensive statistics channel awareness and sought to maximize the least SINR present with base station stayed described [13]. Similar to this, the problem of maximizing the weighted sum rate of all users is formulated and solved by jointly optimizing the active precoding matrices at the base stations and the PSs in RIS-assisted multi-user wireless networks, while the joint intensification of the sum rate besides the energy efficiency remained examined for a multi-user downlink scenario [14]. In RIS-assisted systems, the simultaneous wireless information and power transfer optimization problem were also researched. Meanwhile, in a downlink multi-user situation, array antenna base station that can conduct digital beamforming interacts with a variety of users via a finite size [15].

3 Existing Method

The ORA and ERA schemes also utilize multiple RIS, were put out. The statistical characterization of the two methods was our main concern. In order to achieve this, we put up an accurate context to ascertain the circulation with e2e fading medium in each scheme. We go into further detail on the distinctions between the G—Gamma and LN—Log-Normal distributions. When the OP estimated using the G—Gamma and LN—Log-Normal circulations are compared and a slight discrepancy that may be insignificant. The difference at the left tail (the lower tail) of the two distributions, rigorously speaking, is the cause of this minor discrepancy in OP. Let $N(0, 1)$ stands

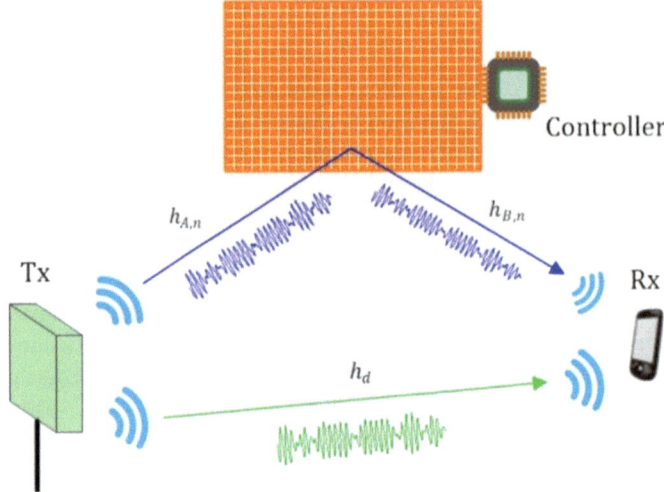

Fig. 1 RIS wireless systems

for a typical RV with a Gaussian distribution. A RV Y with PDF in has a mean given by $E[Y]$ and a variance VAR$[Y]$ of, $Y \sim \alpha \beta 1 + N(0, 1) \sqrt{\alpha}, \alpha \to \infty$.

Both the OP and EC are significantly impacted by the quantity of passive elements deployed at RISs. The reason for this is that under fading medium, each RIS is susceptible to a variable level of fading severity, meaning that the shape and spread parameters of the Nakagami-m distributions between various RISs vary. The ORA procedure is more responsive than ERA system to get element in settings when analyzing performance variations between various element situations. For instance, the transition power of ERA process is required to adjust for the change from element setting L4 to L2, but the ORA scheme needs to compensate for 6.3 dBm to achieve the same OP (Fig. 1).

Because there are not many reflecting elements in total, the Gamma distribution approximation has lower DKL and DKS than LG—Log-Normal circulation approximation, suggesting that it gives greater precision than the LG—Log-Normal circulation. But when the sum of the reflecting elements is large enough, the Gamma and Log-Normal distributions offer equivalent approximation accuracy. The CDF and PDF of the Log-Normal Distribution are given as,

$$FW(w; v, \zeta) = 1wp2\pi\zeta 2e - (\ln w - v)2\, 2\zeta 2$$

$$FW(w; v, \zeta) = 12 + 12\, \mathrm{erf}\, \ln w - vp2\zeta\, 2!$$

The CDF and PDF of the Gamma Distribution are given as,

$$FY(y; \alpha, \beta) = \beta\alpha\Gamma(\alpha)y\alpha - 1e - \beta y, \quad y \geq 0$$

$$FY(y; \alpha, \beta) = \gamma(\alpha, \beta y)\Gamma(\alpha), \ y \geq 0.$$

4 Proposed Method

We provide hybrid RIS systems, which may assess the theoretical foundation and enactment analysis of both ORA and ERA methods of e2e fading medium estimation with a rough distribution. The e2e channel estimation of RISA systems in both structures in relations of OP—outage probability and EC—ergodic capacity are also provided in this work. We may assess the mathematical foundation of the e2e fading channel and performance analysis by merging these two approaches. Ergodic capacity can be determined, and we can assess the effectiveness of these two systems in terms of outage probability. All RISs in the ERA scheme aid in transmission between the source and the destination. When compared to the ERA method, the ORA scheme can handle less reflecting signals.

We demonstrate that true circulation of the size of the e2e medium constant of the ERA structure can alternatively be approximated by the Log-Normal spreading, providing further insights into the fading model of the ERA system. We examine the derived approximation distribution's correctness. The justification between simulated genuine circulation and matched circulation is specifically used to assess the accurateness of the developed arithmetical model (Fig. 2).

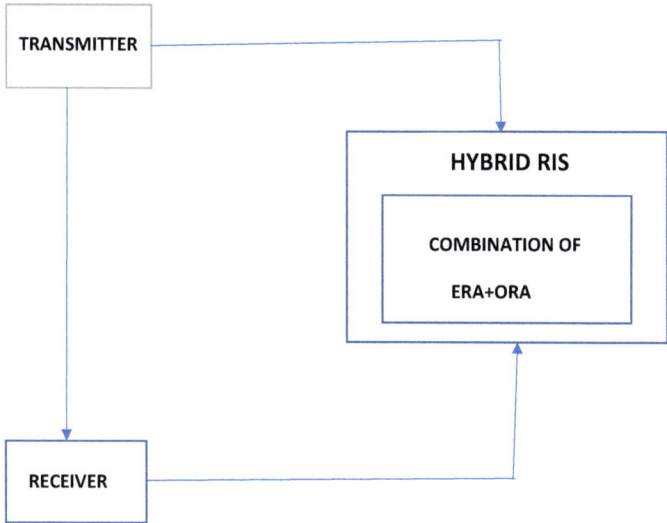

Fig. 2 Block diagram of hybrid RIS technique

4.1 Implementation

The entire analyses is carried out in PC with RAM 8 GB. The operating system used is Windows 10 with an Intel(R) Core i5-8250U CPU @ 1.60 GHz incorporated in it. It is 64 bit operating system. Entire analysis is carried out in MATLAB, which is latest version with more advanced tools. In this MATLAB, tools like communication tools are used.

5 Results and Analysis

See Fig. 3, Tables 1 and 2.

If we observe the above graphs, we can analyze variations of ERA and ORA schemes of transmit power. We can observe that ORA scheme curve started earlier than ERA scheme curve and attains the maximum value earlier than ERA scheme for different RIS values. We can observe that there is an increment of transmit power values with respect to different RIS values. The higher the transmit power, the farther a signal can travel, and the more obstructions it can effectively penetrate (Fig. 4).

If we observe the above graphs, we can analyze that variations of ERA and ORA schemes of energy efficiency. The ORA values start at higher rate when compared to ERA scheme and also ORA scheme attains maximum energy efficiency. It uses less energy to perform the ERA and ORA schemes. The ORA scheme has more efficiency when compared with the Era schemes. By using different RIS values, we analyze that the ERA scheme has less efficiency (Tables 3 and 4).

6 Conclusion

In this study, we proposed the ERA and ORA schemes of Hybrid RIS-aided systems. We concentrated on performance enhance of the two approaches. To this purpose, we put forth a mathematical methodology for figuring out how the e2e fading channel is distributed over both approaches. The framework specifically found that the G—Gamma/LN—Log-Normal distributions can roughly approach valid distribution of e2e medium coefficient for the ERA method. We discovered estimated equations for PDF and CDF of e2e channel's enormousness for the ORA scheme. The context also found that a LNG distribution is aided to simulate fading channel in the ORA method. We assessed the fading models produced as a result.

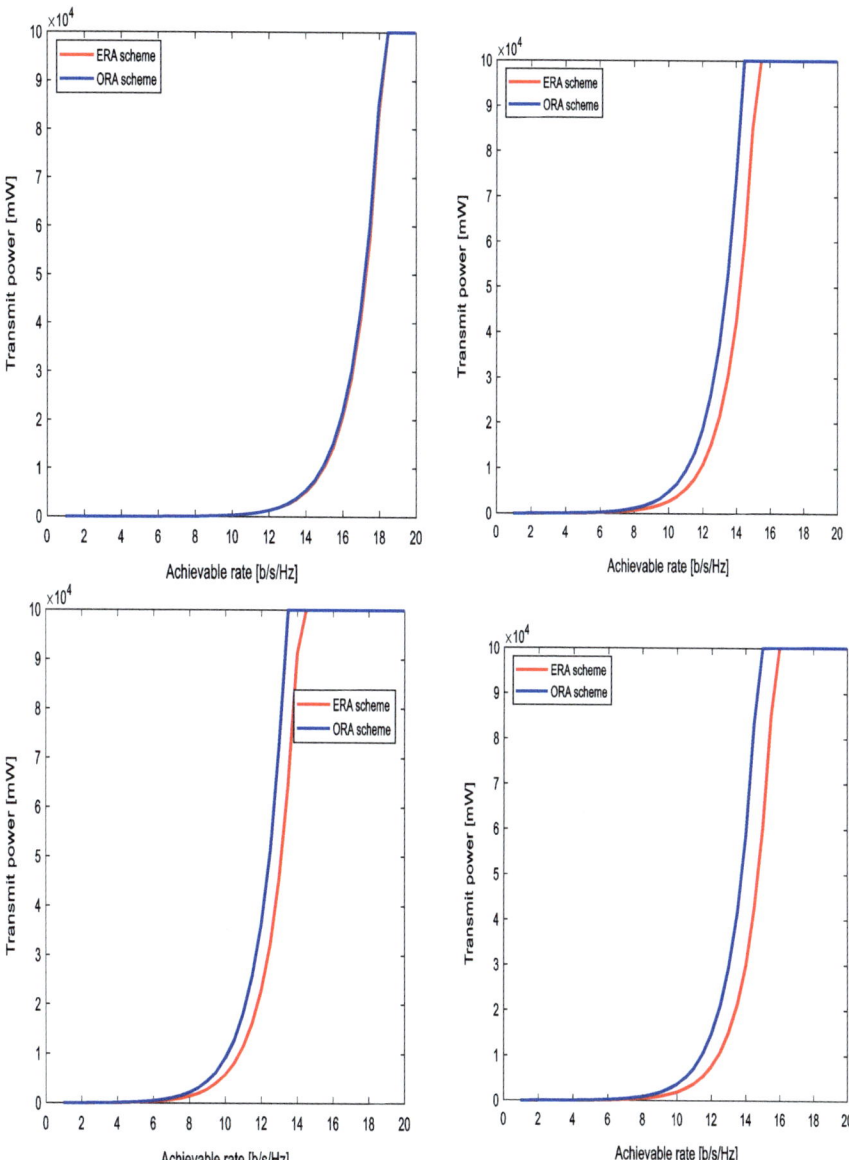

Fig. 3 Transmit power of ERA and ORA schemes

Table 1 Transmit power of ORA scheme

Achievable rate	Transmit power			
RIS	2	3	4	5
2	0	0	0	0
4	0	0	0	0
6	0	0	0	0
8	0	0.00002	0.00003	0.00001
10	0	0.0001	0.00012	0.00005
14	0.00005	0.0008	0.0009	0.0003
18	0.001	0.001	0.001	0.001

Table 2 Transmit power of ERA scheme

Achievable rate	Transmit power			
RIS	2	3	4	5
2	0	0	0	0
4	0	0	0	0
6	0	0	0	0
8	0	0.00001	0.00002	0
10	0	0.00003	0.0001	0.00003
14	0.00005	0.0004	0.001	0.0004
18	0.001	0.001	0.001	0.001

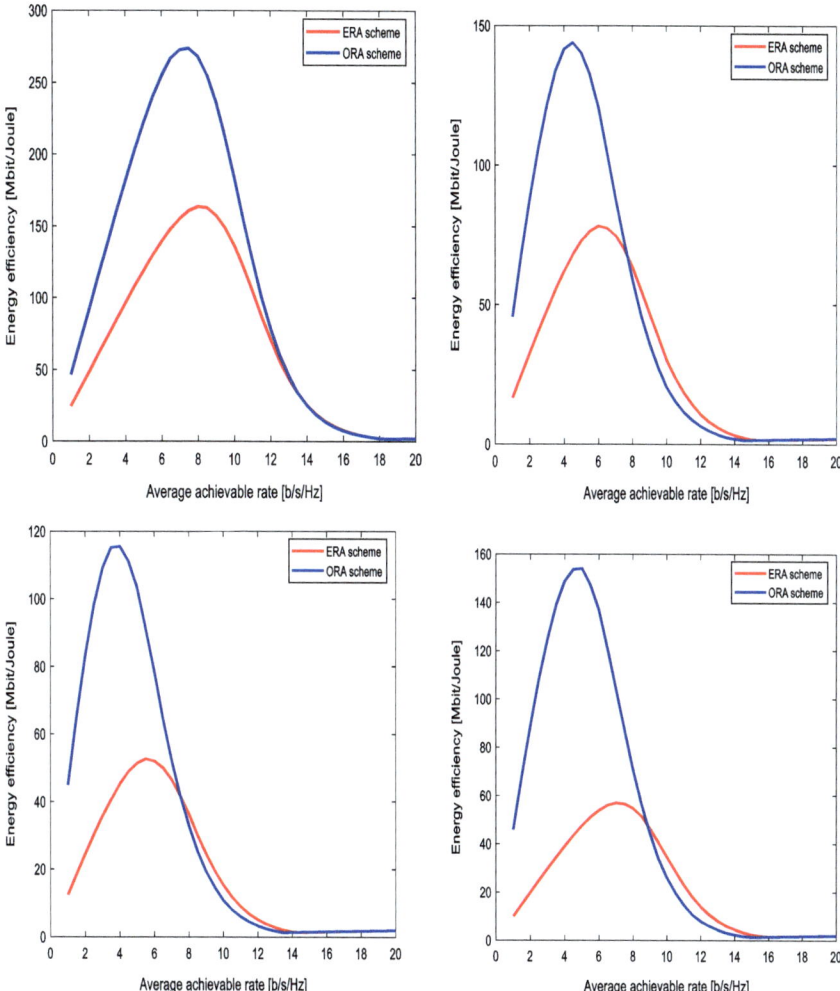

Fig. 4 Energy efficiency simulation of ORA and ERA schemes

Table 3 Energy efficiency of ORA scheme

Average achievable rate	Energy efficiency			
RIS	2	3	4	5
2	75	60	58	50
4	150	110	100	90
6	250	135	90	130
8	220	70	30	60
10	130	30	20	20
12	80	10	10	9
14	30	5	4	3
18	0	0	0	0

Table 4 Energy efficiency of ERA scheme

Average achievable rate	Energy efficiency			
RIS	2	3	4	5
2	40	25	20	15
4	75	60	45	35
6	125	80	56	50
8	170	70	45	42
10	140	35	30	20
12	75	20	7	15
14	30	10	3	5
18	0	0	0	0

References

1. Basar E (2019) Transmission through large intelligent surfaces: A new frontier in wireless communications. In: Proceedings of European Conference on Networks and Communications (EuCNC), Valencia, Spain, pp 1–6. https://arxiv.org/pdf/1902.08463.pdf
2. Karimullah S, Vishnu Vardhan D (2022) Pin density technique for congestion estimation and reduction of optimized design during placement and routing. Appl Nanosci
3. El Faouzi N-E, Leung H, Kurian A (2011) Data fusion in intelligent transportation systems: progress and challenges—a survey. Elsevier Inform Fusion 12(1):4–10
4. Karimullah S, Vishnuvardhan D (2018) A review paper on optimization of placement and routing techniques at NC'e-TIMES # 1.0 IN2018, IJET ISSN: 2395-1303
5. Karagiannis G, Altintas O, Ekici E, Heijenk G, Jarupan B, Lin K, Weil T (2011) Vehicular networking: a survey and tutorial on requirements, architectures, challenges, standards and solutions. IEEE Commun Surv Tutor 13(4):584–616
6. Karimullah S, Vishnuvardhan D (2020) Experimental analysis of optimization techniques for placement and routing in Asic design. In: ICDSMLA 2019, Lecture Notes in Electrical Engineering 601. Springer Nature Singapore Pte Ltd
7. Sommer C, Dressler F (2014) Vehicular networking. Cambridge University Press

8. Karimullah S, Vishnu Vardhan D, Javeed Basha S (2020) Floorplanning for placement of modulesin VLSI physical design using harmonysearch technique. In: ICDSMLA 2019, Lecture Notes in Electrical Engineering 601. Springer Nature Singapore Pte Ltd
9. ETSI (2013) Intelligent Transport Systems (ITS); Vehicular Communications; GeoNetworking; Part 4: Geographical addressing and forwarding for point-to-point and point-to-multipoint communications; Sub-part 2: Media-dependent functionalities for ITS-G5. ETSI, TS 102 636-4-2 V1.1.1
10. Karimullah S, Vishnu Vardhan D (2020) Iterative analysis of optimization algorithms for placement and routing in Asic design. In: ICDSMLA 2019, Lecture Notes in Electrical Engineering 601. Springer Nature Singapore Pte Ltd
11. Chen S, Hu J, Shi Y, Peng Y, Fang J, Zhao R, Zhao L (2017) Vehicleto-everything (V2X) services supported by LTE-based Systems and 5G. IEEE Commun Stand Mag 1(2):70–76
12. Karimullah S, Vishnuvardhan D, Arif M, Gunjan VK, Shaik F, Noor-e-alam Siddiquee K (2022) An improved harmony search approach for block placement for VLSI design automation. Wirel Commun Mob Comput 3016709:10 pages. https://doi.org/10.1155/2022/3016709
13. Simsek M, Aijaz A, Dohler M, Sachs J, Fettweis GP (2016) 5G-enabled tactile internet. IEEE J Sel Areas Commun (JSAC) 34(3):460–473
14. Karimullah S, Basha SJ, Guruvyshnavi P, Sathish Kumar Reddy K, Navyatha B (2020) A genetic algorithm with fixed open approach for placements and routings. In: ICCCE, pp 599–610. Springer
15. Khaledian S, Farzami F, Erricolo D, Smida B (2017) A full-duplex bidirectional amplifier with low DC power consumption using tunnel diodes. IEEE Microw Wirel Compon Lett 27(12):1125–1127

Creation of a Platform for Artisans to Promote Their Product Using Blockchain as NFT

Bipin Kumar Rai⬡, Tanisha, Sushant, Gautam Kumar, and Rashmi Pathak

Abstract Before blockchain technologies, the idea of securing digital asset is remained susceptible to tampering but after few many years of studies and improvements in blockchain brought about the development of Non-Fungible Tokens (NFTs), that are tokens which might be virtual property and have evidence of possession embedded. The Blockchain Platform for Artisans (BPFA) is an internet platform in which customers/artists can buy or promote non-fungible paintings portions/products and earn cryptocurrencies. Therefore, by this technology, promoting art using NFT and creating revenue by earing cryptocurrencies are easy and authorized creating and selling NFTs. In this paper, we are presenting an idea for developing a platform for artisans to promote their art using blockchain and with security assurance also how they can make NFT as an investment option. Blockchain Platform For Artisans (BPFA) uses interactive UI so that local artisans can also use platform easily and promote their art like image in the form of NFT and blockchain provides security as it has immutable property. Moreover, our paper will also help to know the key demanding situations and challenges in edition of blockchain from the attitude of privateness, safety, and selling.

Keywords Non-fungible tokens (NFTs) · Cryptocurrencies · Blockchain

B. K. Rai (✉) · Tanisha · Sushant
ABES Institute of Technology, Ghaziabad, India
e-mail: bipinkrai@gmail.com

G. Kumar
CMR Engineering College, Hyderabad 501401, India

R. Pathak
National Institute of Technology, Silchar, Assam, India

1 Introduction

In order to establish a precise reputation and promote their work at competitive costs, artist must cultivate robust relationships with paintings' sellers and galleries. As such, the market is often characterized by low transparency and high network activity. The big offers with over thousands and thousands or even billions USD had been made in the contemporary high-stop art marketplace and artwork festivals. A lot of high-end art sales were made like "Girl with balloon" by "Banksy", "Infinity Net" painting by "Yayoi Kusama", and "Graffiti-inspired" drawings by "Keith Haring" which are some examples of high art. Publics lack information of the way the artwork market works and are primarily interested in the impressive sales. Moreover, the art market lacks transparency and has unreported rules. In order to be successful as an artist, you must have top connections with different artists and institutions (galleries, museums). Also, studies show that people from middle-class backgrounds face greater challenges as artists. In recent years, the art market has undergone some changes due to NFT's trend. Due to the emergence of NFT, artwork gala's and galleries have started the sale of NFT artwork as well [1].

NFT, shorten for "non-fungible token", is virtual belongings which are consultant of bodily or virtual innovative paintings along with tune, virtual artwork, games, gifs, videos, and more. In approach of NFT every token isn't replaceable with any different token, creating every token a completely specific entity that represents a single precise item [2]. Those tokens encompass virtual facts inside the shape of media (tune, video, photograph) the cost of which may be recorded in phrases of cryptocurrency. The distinctiveness and non-exchangeability of NFTs, if not absolutely eradicates, the problem of authenticity and counterfeits to a huge volume with the aid of a virtual identity of the proprietor covered in each token such that an asset is without difficulty identifiable to its proprietor. NFT is a manufactured from blockchain technology, and the idea of blockchain is not always new and is presented by Stuart Haber and W. Scott Stornetta in 1991. The blockchain technology helps cryptocurrencies like Bitcoin and Ethereum.

NFTs have commenced a new international for digital artists. The convenience with which a user can replica an picture online makes selling digital art an assignment. However, the NFT procedure gives artists a way to receive credit for it and sell their art, and different buyers see NFTs as an funding opportunity. The marketplace for NFT development is greater than doubled from $142 million in 2019 to $338 million closing year. As of December 2021, non-fungible tokens (NFTs) have generated over $22 billion in income, and numerous organizations like OpenSea, Sorare, and Sky Maven have secured billion-dollar valuations. There have been diverse incidents of safety breach on the NFT market-location portals as an example—OpenSea is valued approximately. $13 billion with about 1.5 million customers, in step with Dune Analytics. The data breach ought to impact approximately 1.8 million e-newsletter subscribers and customers, consistent with the Verizon 2021 facts Breach Investigations record, insider threats account for almost 1/4 (22%) of every records breaches [3]. There are numerous already existing NFT platforms used worldwide are shown in Fig. 1.

Fig. 1 NFT existing platforms

Blockchain technology lets in the transaction of art work to grow to be more digital and greater measurable in today's art market. The exciting traits of blockchain are distributed ledger and robust safety, while non-repudiation is the significant belongings of statistics protection in blockchain. Blockchain helps in various other areas also such as health care, supply chain, trade finance, real estate, and so on. There are various other research papers present in such fields. As healthcare information is extremely sensitive, it poses a danger of invading people' privacy if stored or exported without right security measures [4] Blockchain technology is used as a allotted ledger, decentralized gadget, and interoperability property that makes it greater unique and facilitates in relaxed transactions and trace our asset [5]. Some of the advantages of blockchain are:

Distributed database: Blockchain provides a new way for comparing the identification and traceability of artistic endeavors. The emerging blockchain generation in current years is a dispensed database evolved by means of the underlying technology of Bitcoin.

Transparency: Blockchain is decentralized, that means any community member can affirm facts stored into the blockchain. Consequently, the general public can agree with the community.

Traceability: Blockchain creates an permanent audit trail, permitting clean tracing of adjustments at the network.

Immutability: As quickly as all transactions are recorded on the blockchain gadget no person is capable of alternate, delete or alter information. All transactions at the blockchain are time-stamped and date-stamped, so it is miles a permanent shop [6].

2 Research Objective

(a) To develop a blockchain-based platform for promoting art.
(b) Using blockchain technology creating a platform to promote art through buying and selling of art as NFT, while ensuring data availability.
(c) Provide an easily accessible framework.

3 Related Work

The preceding research indicates the opportunity of applying the era of blockchain within the subject of art (Non-Fungible Tokens). Blockchain has the capability to come to be an effective disruptive pressure. A survey of 800 executives indicates fifty-eight percent considering that up to ten percentage of worldwide GDP will be saved the use of blockchain technology [7]. Digital artwork is a brand new area and selling an art digitally includes various issues than physical art. Over 1 billion customers get entry to YouTube each day in 2016 document, and in 2014 over 1.9 billion pics have been reported to be shared online daily. Copyright infringement has additionally been normalized. Research inside the song industry has shown that at least 28% of the populace participate in unlawful record sharing. Whether it is miles obtained legally or illegally, much of this content material can be ate up at the displays of laptops or smartphones, by no means taking bodily shape. There are additional demanding situations with regards establishing the provenance of virtual artwork. The problem of virtual artwork possession has been known as "the elephant within the room" of the artwork world [8]. We found out the interrelationships between NFT income and price of Bitcoin and Ether. According every day statistics among 2018 and 2021, determined that Bitcoin fee marvel triggers a boom in NFT sale, additionally. The outcomes recommend that (huge) cryptocurrency affect the increase, improvement of the NFT marketplace; however, there may be no impact within less than half of a 12 months (by means of 2021), and loads of hundreds of NFTs properly valued over $800 million had been traded. Maximum those stated art, track, in-activity product. Besides the not unusual infrastructure, NFT platform which includes OpenSea and Rarible moreover use cryptocurrencies, maximum normally Ether, price of buying and selling alternative. If customers commonly required cryptocurrencies to shop for NFT, it is low-cost to expect that the cryptocurrency marketplace has an impact on the smaller NFT marketplace [9]. As of December 2021 report, we found that non-fungible tokens (NFTs) have generated over $22 billion in sales, and several companies like OpenSea, Sorare, and Sky Maven have secured billion-dollar valuations. Market intelligence firms have begun providing tools that summarize sales and price data for different NFT collections and present this information to market participants. These statistics are often cited in the media to compare different collections [10]. Despite the developing popularity of NFTs, extant research on this rising phenomenon is scant, commonly focusing at the technical elements of blockchain-based totally protocols and marketplace alternate networks [11], including digital works of art to the blockchain manner to file their existence along with info regarding their provenance, cost, exhibition records, and so on, [12]. Another term crypto art-Crypto art is digital artwork that uses blockchain technology to confirm ownership. Crypto artwork gives the entire availability of art work records (images and metadata), transaction facts (bids and income), and social records (likes and perspectives) and consequently will permit researchers to look at the mechanics of achievement in paintings and modern industries, in all likelihood in fantastic detail [13].

4 Preliminaries

Blockchain-decentralized database shops all transactional record within the shape of a series, and the blocks are linked to each other. The Genesis block is the first block in the blockchain. The block header and a list of records are saved in each block. Blockchain database shops all transactional records within the sequence. Blockchain is the technology that can be the answer of security and data security [14].

(A) Consensus Protocol:

The blockchain consensus protocol includes some particular goals inclusive of coming to an consent, cooperation, alliance, and obligatory participation of every node within the consensus technique. Thus, a consensus set of rules ambitions at locating a not unusual concurrence that may be a win for the entire network. Now, we will talk diverse consensus algorithms and the way they work:

- Proof of work.
- Proof of stake.
- Practical Byzantine Fault Tolerance.
- Proof of burn.
- Proof of elapsed time.

Consensus protocols shape the spine of blockchain with the aid of helping all of the nodes inside the network verify the transactions. Bitcoin makes use of proof of work (PoW) as its consensus protocol, that is strength and time-in depth.

Blocks are join to every other referring to the preceding block within the chain. Every block is diagnosed through hash algorithm.

Figure 2 explains the whole structure of blocks, where the first block is called genesis block. All blocks are linearly connected and secured. Every block header consists of previous block hash, current block hash, Merkle root, nonce with other details. For adding a new block in a chain, blockchain uses consensus mechanism such as POW, POS. Every block also consists of set of transactions, metadata, and other mining informations.

(B) Solidity:

It is a contract-oriented, high-stage programming language for enforcing smart contracts. Solidity is enormously stimulated by other programming languages. This language is statistical types and it supports complex features for programming like inheritance and user-defined data types. The little view of solidity and format used for BPFA is shown in Fig. 3.

Figure 3 Solidity is a programming language for writing smart contracts through code. IDE such as Remix, VScode, Atom, Embark is used for writing smart contracts. These programs are stored in blockchain and run when predetermined conditions are met. They are used to automate the execution of an agreement.

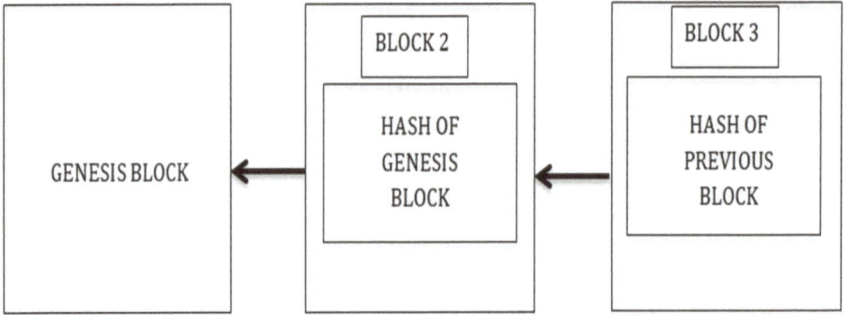

Fig. 2 Simple blockchain structure representing a chain of blocks

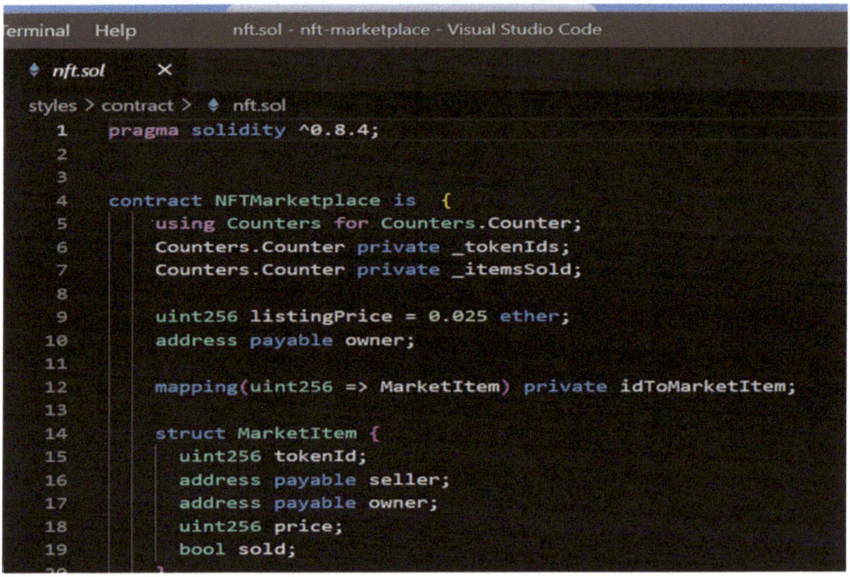

```
Terminal   Help              nft.sol - nft-marketplace - Visual Studio Code

 ◆ nft.sol       ✕

styles > contract > ◆ nft.sol
    1      pragma solidity ^0.8.4;
    2
    3
    4      contract NFTMarketplace is  {
    5          using Counters for Counters.Counter;
    6          Counters.Counter private _tokenIds;
    7          Counters.Counter private _itemsSold;
    8
    9          uint256 listingPrice = 0.025 ether;
   10          address payable owner;
   11
   12          mapping(uint256 => MarketItem) private idToMarketItem;
   13
   14          struct MarketItem {
   15              uint256 tokenId;
   16              address payable seller;
   17              address payable owner;
   18              uint256 price;
   19              bool sold;
```

Fig. 3 Solidity programming used in BPFA

5 Proposed Work

For the purpose of discussion about NFT, the major role and the solution are based on blockchain technology. The blockchain is distributed across a multitude of co-present instantiations, for each of which is computationally proven each time new entries are delivered. This consistent synchronization among co-existing copies of the records saved on the blockchain evidently calls for the statistics to be available to all individuals. This transparency, which renders the blockchain a totally public ledger, is a key safekeeping mechanism. This functioning is essential for providing authenticity and proof of ownership to artisans. Blockchain generation permits the transaction of art work to come to be more virtual and more measurable in nowadays art marketplace. The exciting features of blockchain are decentralized ledger and sturdy safety, while non-repudiation is the vital property of facts safety in blockchain.

5.1 BPFA Architecture

Figure 4 represents the complete structure of BPFA where all the entities playing their element one by one and attaining the goal. Entities includes: vendor, customer, MetaMask wallet, database which is mongoDB, smart contract, hardhat, and ganache.

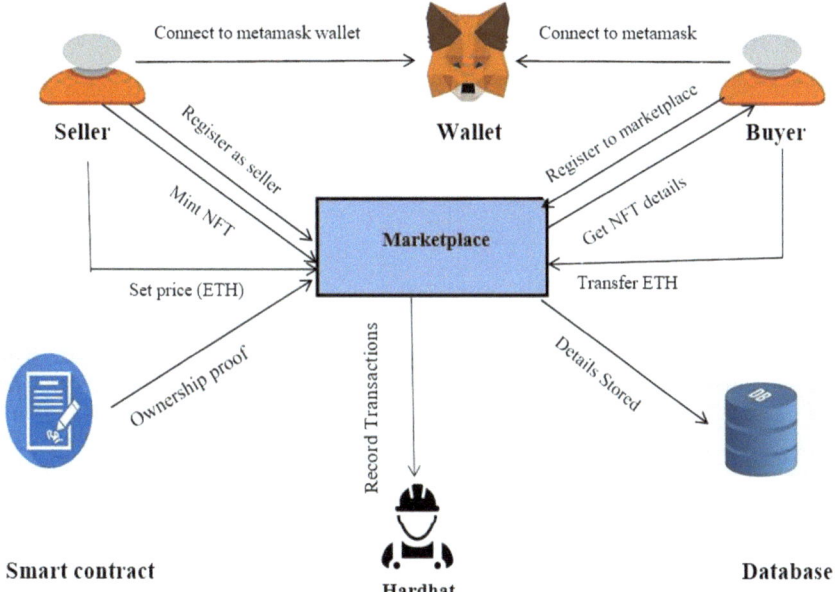

Fig. 4 Design and architecture of BPFA

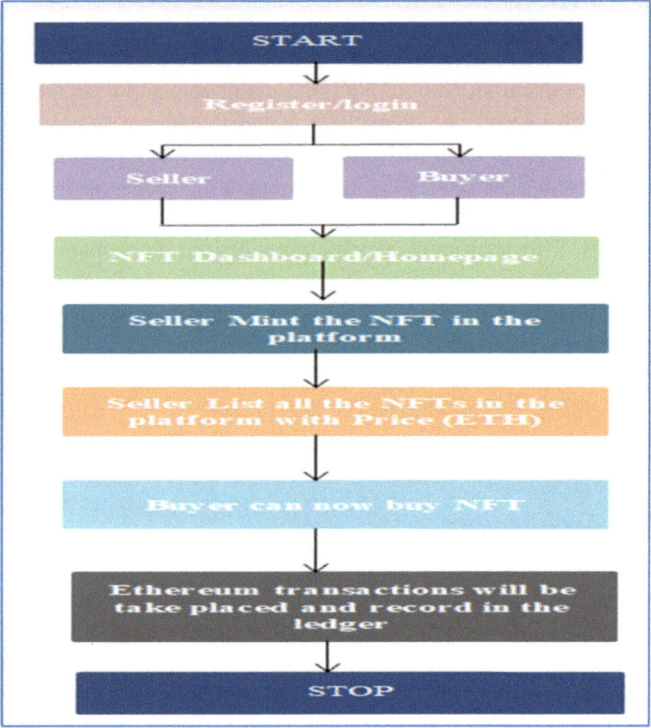

Fig. 5 Flowchart of BPFA

They all are linked to the principle BPFA and then all of the modules integrate to perform simultaneously.

5.2 Flowchart

Figure 5, in this chart, all the steps are listed one by one from initial register to final transaction, and all the steps taken by seller and buyer are mentioned above. There are total seven steps by which seller can mint and buyer can buy the NFTs.

5.3 Sequence Diagram

Figure 6 shows collection of interactions taking place between seller, client, and all the other entities worried in the machine, i.e., MetaMask wallet, hardhat, login interfaces, and so on, for getting and selling the NFTs.

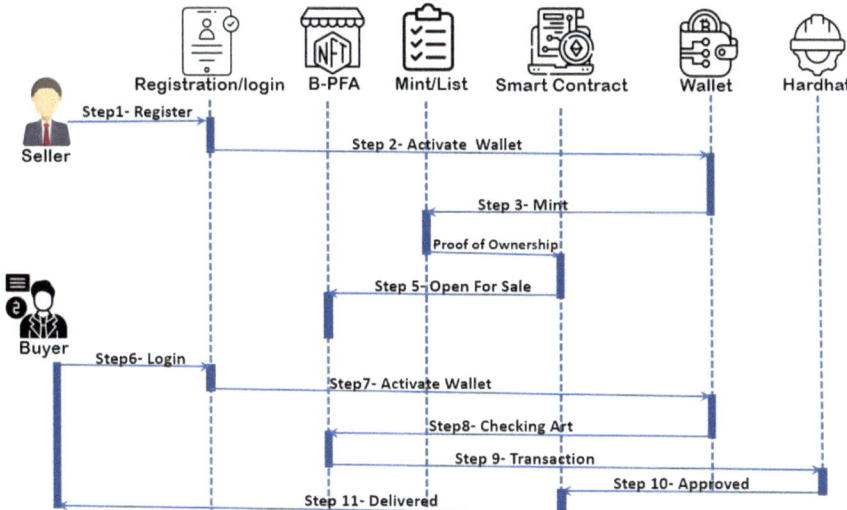

Fig. 6 Sequence diagram of BPFA

According to the survey that we have studied so far we came through various loopholes present in some existing NFT platforms that they are less secure, some of the platforms are only for the popular artist, and similarly, there are various other drawbacks present in existing solutions. Hence, in our paper, we are focusing on such major issues and creating an easy framework so that it will also be helpful for local artisans. Here is the comparison between OpenSea, Nifty with our solution BPFA (Table 1).

Table 1 Comparison between BPFA with other existing platforms

	OpenSea [17]	Nifty Gateway [18]	BPFA
Low fees	✗	✗	✓
Local art promote	✓	✗	✓
Fixed time auction	✓	✗	✗
Wallet needed	✓	✗	✓
Exclusive art	✗	✓	✗
Security	✗	✓	✓
Pay using credit card & debit card	✗	✓	✗
Royalties (Artist's choice)	✓	✗	✓

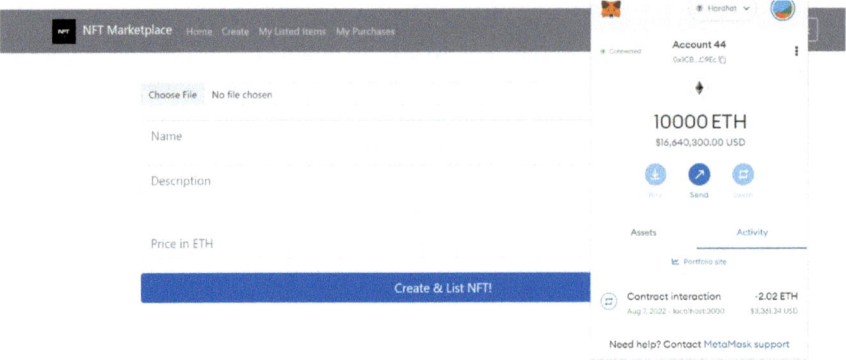

Fig. 7 Frontend of BPFA

6 Process of Creating and Listing NFT

Figures 7 and 8 both depict the interface and interaction for promoting/purchasing the NFTs. During the registration in the platform, it is essential to provide the complete details of the product that one wants to sell and then list the product in the platform also setting the amount of product in the form of ether as blockchain deals with cryptocurrencies. All the data filled by the seller are beneficial in future for both the seller and buyer. The smart contract running in the system takes all the details of the product and make an agreement between seller and purchaser. Once the seller lists the NFT in the platform, then it will be visible to all the user and whoever wants to purchase they make an agreement with user with the help of smart contract and make a transaction which is highly secured and records in the ledger. The whole process is simple and easy and takes less effort in listing the product as NFT as well as for future investment.

Fig. 8 Transaction records

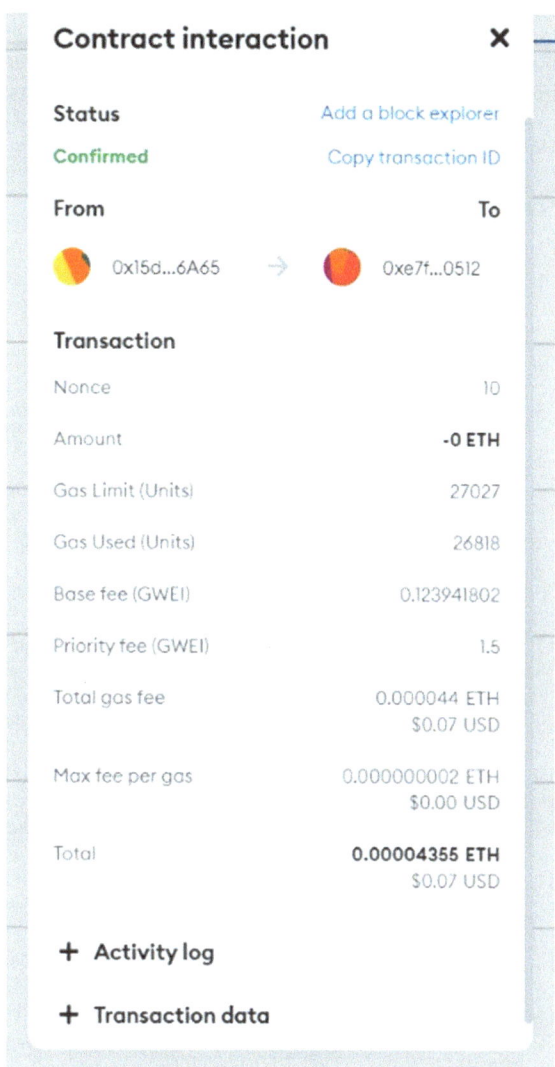

7 Conclusion

In this paper, we discussed the acceptance of blockchain in art market for promoting arts of local artisans and how it benefits the local artist. Non-fungible tokens (NFTs) are making waves within the worlds of cryptocurrencies and virtual artwork. Hence, in our research, we found out the jaw-dropping trading prizes and growth where the peoples are using this trend for long-term investment as well as for promoting their arts. Using blockchain, we resolve various issues like privacy and information security. After all the researches, we conclude that blockchain is a remarkable technology for the privacy and security of information, and in art market, making art as non-fungible token is a huge benefit for all the artists.

References

1. Ling Ng W, Inge Zwart S (2022) A new era of the visual art market? A platform analysis of the biggest NFT market OpenSea on diversity and equality. https://www.diva-portal.org/smash/get/diva2:1692740/FULLTEXT02.pdf. Accessed 31 Dec 2022
2. Abdou D, Elnasr F (2021) Prospects of investment in digital art: case of ethereum and non-fungible token (NFT). https://doi.org/10.37708/el.swu.v3i2.3
3. Jeong SYE (2022) Value of NFTs in the digital art sector and its market research. https://digitalcommons.sia.edu/stu_theses/101/. Accessed 31 Dec 2022
4. Prasad PS, Sunitha Devi B, Janga Reddy M, Gunjan VK (2019) A survey of fingerprint recognition systems and their applications. Lect Notes Electr Eng 500:513–520
5. Rai BK (2022) PcBEHR: patient-controlled blockchain enabled electronic health records for healthcare 4.0. Health Serv Outcomes Res Method. https://doi.org/10.1007/s10742-022-00279-7
6. Franceschet M et al (2021) Crypto art: a decentralized view. Leonardo 54(4). https://doi.org/10.1162/leon_a_02003
7. Jeong SY (Esther) (2002) Value of NFTs in the digital art sector and its market research
8. Kireyev P (2022) NFT marketplace design and market intelligence. https://papers.ssrn.com/sol3/papers.cfm?abstract_id=4002303. Accessed 31 Dec 2022
9. Mazur M (2021) Non-fungible tokens (NFT). The analysis of risk and return. https://ssrn.com/abstract=3953535
10. Michalska J (2016) Blockchain: how the revolutionary technology could change the art world. Art Newspaper 26(283):38–39. https://www.theartnewspaper.com/2016/09/01/blockchain-how-the-revolutionary-technology-behind-bitcoin-could-change-the-the-art-market. Accessed 31 Dec 2022
11. Monrat AA, Schelén O, Andersson K (2019) A survey of blockchain from the perspectives of applications, challenges, and opportunities. IEEE Access 7:117134–117151. https://doi.org/10.1109/ACCESS.2019.2936094
12. Pichuhina Y (2021) Blockchain: new economic and management solutions in art industry. Municipal Econ Cities 7:167. https://doi.org/10.33042/2522-1809-2021-7-167-18-22
13. Rehman W, e Zainab H, Imran J, Bawany NZ (2021) NFTs: applications and challenges. In: 2021 22nd international arab conference on information technology, ACIT 2021. https://doi.org/10.1109/ACIT53391.2021.9677260
14. Rai BK (2022) BBTCD: blockchain based traceability of counterfeited drugs. Health Serv Outcomes Res Method. https://doi.org/10.1007/s10742-022-00292-w
15. Rai BK, Kumar G, Balyan V (eds) AI and blockchain in healthcare. In: Advanced technologies and societal change. Springer, Singapore. https://doi.org/10.1007/978-981-99-0377-1
16. Rai BK, Sharma S, Kumar G, Kishor K (2022) Recognition of different bird category using image processing. Int J Online Biomed Eng 18(7)
17. Rai, BK, Srivastava S, Arora S (2023) Blockchain-based traceability of counterfeited drugs. Int J Reliable Qual E-Healthc (IJRQEH) 12(2):1–12. https://doi.org/10.4018/IJRQEH.318129
18. Rai BK, Srivastava AK (2017) Prototype implementation of patient controlled pseudonym-based mechanism for electronic health record (PcPbEHR). Int J Res Eng IT Soc Sci 07(07):22–27. ISSN 2250-0588

Face Recognition Based Home Security System to Detect Usual/Unusual person Using IoT

Syed Musthak Ahmed, Jooluri Aditi, Arshiya Afreen, Guduru Tharun, and Vinit Kumar Gunjan

Abstract Security and safety are top concerns in our digital age, whether it is the safety of our own home or the security of the data. Doors serve as barriers that keep intruders out. In order to improve home security, a door security system is absolutely necessary. Face recognition-based door unlock system is used to protect guarded areas. The ability to identify visitors entering the house is a crucial part of any home security system. In this project, face recognition module is incorporated. Here, the images of authorized people are captured and stored in the database. Once the person arrives at the door, his/her image is compared with images stored in database. The door opens only when the person's image matches with the database. When an unauthorized person arrives, due to mismatch of image the door remains locked and gives signal by glowing red LED. Alternately, in case of emergency, the visitor can open the door by entering the manual pin provided by the admin. The developed project offers double factor authentication that is leading to high secure door locking contributing a smart home concept. These features are obtained by integrating Arduino UNO, ESP32 CAM, L298N motor driver, GSM sms900a and 4×4 matrix keypad. This module can be implemented in apartments, commercial buildings and security required places like bank and corporate sectors, etc., to keep unauthorized persons like thieves and other sort of dangers at bay.

Keywords Home security · Face recognition · Arduino UNO

S. M. Ahmed (✉) · J. Aditi · A. Afreen · G. Tharun
Department of ECE, SR Engineering College, Warangal, Telangana, India
e-mail: syedmusthak_gce@rediffmail.com

V. K. Gunjan
Department of CSE, CMR Institute of Technology, Hyderabad, India

© The Author(s), under exclusive license to Springer Nature Singapore Pte Ltd. 2023
A. Kumar et al. (eds.), *Proceedings of the 4th International Conference on Data Science, Machine Learning and Applications*, Lecture Notes in Electrical Engineering 1038,
https://doi.org/10.1007/978-981-99-2058-7_19

1 Introduction

Security is a major problem for everyone in daily life. The major three areas where security is require include management security, operational security and physical security. Home security is one which is very important for residence to look in for unauthorized persons such as thieves, robbers and foes. Thefts from homes and banks are a frequent and steadily growing problem in today's world. Home security systems are being developed gradually in several nations. Earlier, traditional methods such as lock and keys, ID card, tags were used to access the door. However, this is not a totally dependable security system. The advancement of technology has given a path for the invention of several measures for home security. Biometric takes over conventional existing system to deliver promising security. Biometric technology includes the ability to identify people by their fingerprints, faces, irises, voices, retinas, etc. The trends moved from fingerprint to face recognition. Therefore, we prefer a face recognition system for unlocking the door. Facial recognition is widely used in many different corporate sectors and industries. Facial recognition is a quick, efficient and simplest method, and it only requires a person to gaze at the camera and not put any physical effort to unlock the door. The solenoid lock on the door opens if the image is present in the database. A security alert message in the form of SMS is sent to the owner of the house in the case of an intruder.

2 Literature Review

There are several existing methods proposed and published in different journals. Every paper that we mentioned in this part approaches the challenge in a different way. A few of the existing methods are discussed below.

In [1–5] authors proposed face recognition-based door unlocking system using Raspberry Pi. In this, the system is composed by the combination of face recognition security system, password security system and the alert system through GSM module. If the person is detected, the door opens, else it sends an alert text to the admin. As an alternative, there is a manual password entering system.

In [6] author proposed Door Unlock by Face Recognition (DUFR). DUFR before implementing the application is trained on datasets that comprise the image containing face feature of the people. On successful match of face the system triggers the door to unlock but on a false match system notifies to mobile device where the user can access door on their personal choice.

In [7] author demonstrated a Raspberry Pi-based face recognition smart door lock system. They have stated that images will be identified from a live video feed. If the image is authorized, door will be unlocked and door gets locked automatically after a given time.

In [8] author proposed Securing an IOT based Home using Digital Image Processing and an Android Application. In this application if the image is recognized, the door will be unlocked automatically, else the picture is forwarded to the owner's Android application. From there, the owner can take action via buttons like accept, reject and buzzer.

In [9] author illustrated Smart Door System using Face Recognition Based on Raspberry Pi. The systems' face recognition algorithm will interact with the webcam and solenoid lock using the Raspberry Pi to operate the door.

In [10] authors demonstrated design of face recognition-based embedded home security system. The proposed system is based on 'Remote Embedded Control System' (RECS). The WIU module sends the notification regarding the visitor via mail/SMS. All the communications are done between the owner and the microcontroller.

In [11] author discussed as IoT is becoming more common in homes, threats to security and privacy are increasing. But they conducted a semi-structured interview with 15 smart home residents, they discover gaps in threat models due to a lack of technical knowledge and awareness about smart home. They derived suggestions for future research and smart home technology designers from these and other findings.

In [12] author discussed smart home and security and gave a review of the tools related to smart home security.

In [13] author developed face recognition technology for an automatic door access system. The door will open automatically for a known face using the command of the microcontroller. On the other hand, for an unidentified person, an alarm will sound.

In [14, 15] author proposes an efficient and comfortable lock system design for doors that can be opened or locked without a handle. It is operated with special features using Bluetooth technology. The system is compatible and it employs low-power Bluetooth technology.

In [16] author developed a password-based home security system by making use of electronic technology at a reasonable cost. A four-digit password input by the user will appear as "****" on the LCD. Therefore, nobody else can see what the user enters. If the password is right, welcome text is displayed on the LDC, else if it displays password is incorrect.

In [17] author utilized the fingerprint-based remote monitoring concept, the technology can recognize a vivid fingerprint with accuracy and provide unlock ID information to the owner.

In [18] author explores the idea of a door locking and unlocking system that operates over GPRS. The user must register before his or her information is saved in a database. The door will be moved by the Dc Motor, either locking or unlocking it. If there is an intruder, the IR sensor will detect the activity and send the registered user a message of warning.

In [19] the author used fingerprint recognition that uses image from the user and grant access to the door. The process checks the data from finger print sensor and sends an alert message to the admin user.

In [20] author utilized three main components, viz. image capture, faces detection and recognition and email notification. Facial images for numerous people can be saved in the database. Using the Telegram Android app, the door lock can be accessed remotely from anywhere in the world.

3 Proposed System

The suggested systems block diagram is displayed in Fig. 1. The proposed approach is a smart door unlock system that uses face recognition. When an authorized person visits, the door opens immediately. Here, the image compared with the image of the person whose faces are already stored in the database. When the image matches action will be initiated for door operation, this process is done by ESP32 CAM, which identifies and detects an individual. The L298N motor receives a high signal from the ESP32 CAM module when a visitor is recognized. The solenoid lock is unlocked by the driver using the power supply. There is a backup technique to unlock the door using the keypad in emergency cases when the face is not recognized. A green LED will illuminate to show that the door has opened if the user is successful in entering the proper PIN. The GSM module generates a text message and sends it to the house owner when the door is open. In case, if the persons face is not recognized, a red LED is illuminated, and if wrong pin is entered, buzzer starts beeping and red LED starts glowing. Eventually, the GSM module will notify the owner through alert notice about the attempt made by the unknown person.

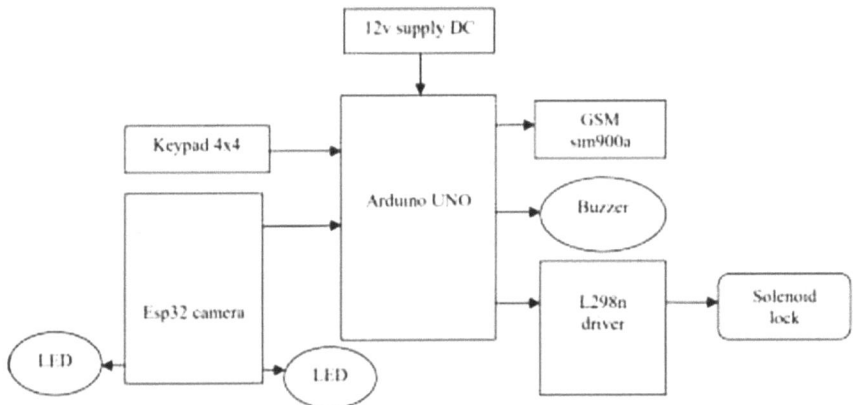

Fig. 1 Block diagram of the proposed method

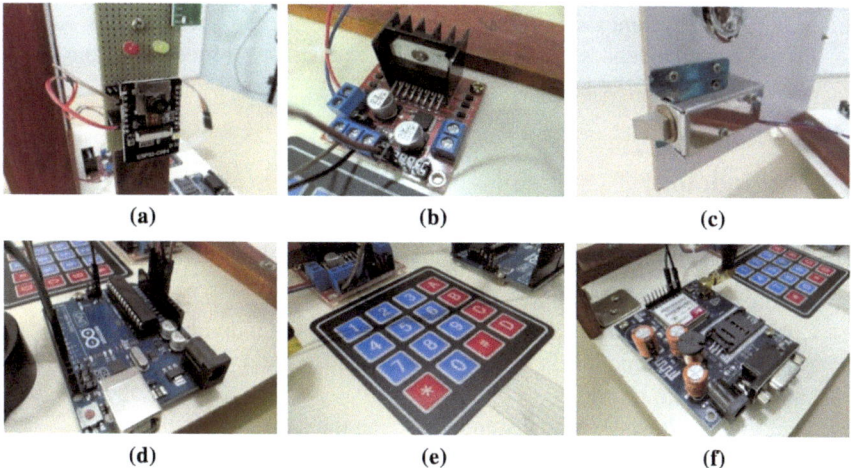

Fig. 2 **a** ESP32 CAM, **b** L298N motor driver, **c** solenoid lock, **d** Arduino Uno, **e** 4 × 4 matrix keypad and **f** SIM900A GSM

4 Hardware Description

The various hardware devices and components used in developing the module are briefly described in Figs. 2a–f.

ESP32 CAM (Fig. 2a) is a compact, low-power camera module based on the ESP32. It is perfect for Internet of Things (IOT) devices that need cameras with advanced features like image tracking and recognition. Figure 2b is a motor driver which controls electromagnetic lock. It can simultaneously control the speed and direction of two DC motors. Figure 2c is a solenoid lock which refers to a latch used for electrical locking and unlocking. Upon receiving the signal from the ESP32 camera, the L298N motor driver uses a 12 V power supply to unlock the solenoid lock. Figure 2d is Arduino UNO, and it is an open-source microcontroller board that can be interfaced with various electronic components. Figure 2e is a 4 × 4 matrix keypad used to provide input values. It consists of 16 keys. These keys are arranged in a matrix of columns and rows. Figure 2f is SIM900A GSM, and almost everything a typical cell phone can do is done with this shield, including sending and receiving SMS text messages and making or receiving phone calls.

5 Objectives

The aim and objective behind the product development is to

1. Replace the traditional door lock system with face recognition door operating system.

2. Make use of current state of the art technology to make smarter homes.
3. Protect ones residences from theft/unauthorized persons from entering the house.
4. Develop a concept of home security system.

6 Methodology

The procedure adopted to meet the objectives is listed below.

1. Atomizing the door lock mechanism.
2. Controlling the door by identifying and detecting the individual.
3. An ESP32 CAM module is fixed to the entrance of the door to capture the image and identify the visitor.
4. The captured image with the stored image in the database is evaluated.
5. The door opens and an SMS text message is delivered to the house owner if the photographs match.
6. The door can also be unlocked by entering the password on the keypad in the event that a match cannot be made.
7. The door automatically opens if the password is correct.
8. Otherwise, the password is wrong an SMS will be automatically generated to the authorized user and the door will continue to be locked.
9. A buzzer is triggered if the authentication attempts are unsuccessful while entering password.

7 Flowchart of the Implementation

The experimental setup for the proposed system is shown in Fig. 3. The ESP32 CAM module records an image of the person approaching the door and compares it to the database of images that have already been stored. If the comparison is successful, the individual is granted access to the door. In case if face is not detected, there is an alternate way to unlock the door by entering the PIN on the keypad. The door is opened and successful match of face and password is indicated by a green LED. An SMS text is automatically generated and sent to the admin. A red LED and blown buzzer are signs if the PIN code does not match. The concern persons will be informed about the intruder detection via SMS.

8 Implementation and Working of the System

Implementation of smart door unlock system using face recognition is shown in Fig. 4a, while the schematic of circuit developed is shown in Fig. 4b. The components used in the project are ESP32 camera, L298N motor driver, SIM900A GSM, buzzer, keypad which are interfaced to Arduino UNO.

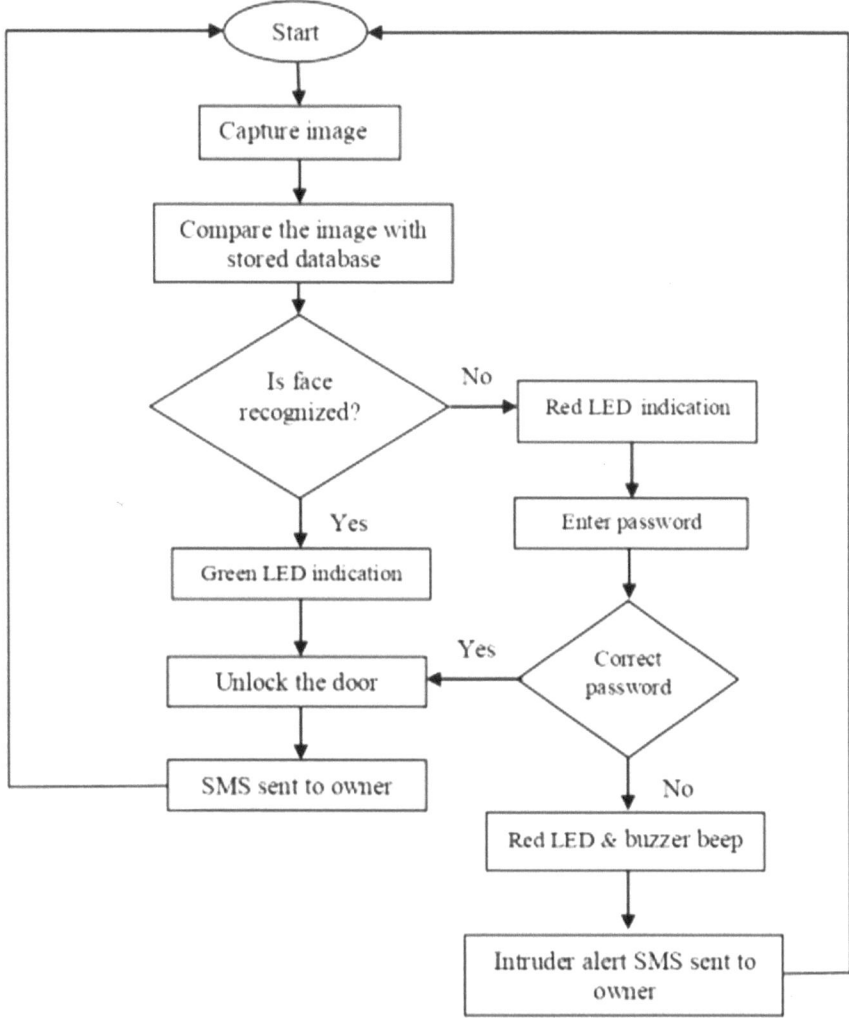

Fig. 3 Overall flow of smart door unlock system

The working of the module is explained as follows:

(i) **Initialization**: The system is built by integrating the part of face recognition circuit with IoT. Face recognition is operated in five steps, viz. collecting images, creating database, preprocessing, training and testing images. Initially, the authorized person's images are collected by capturing pictures through camera module. These images are used for training the system for authorization of selective person. In this model, a total of five sample pictures of individuals are taken from different positions. These images are stored in the database and a portal is created to enroll data as shown in Figs. 5a and b.

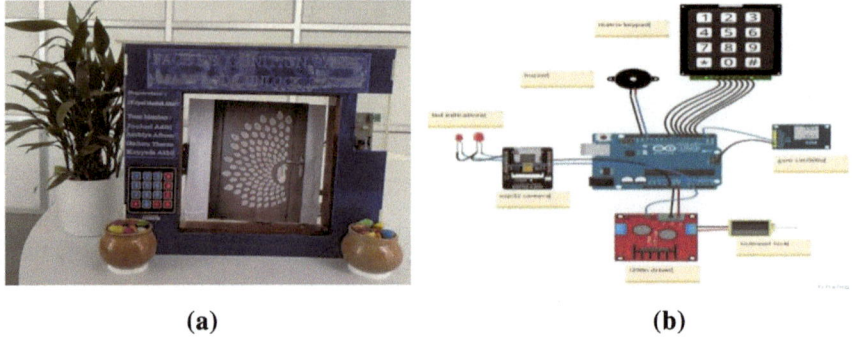

Fig. 4 **a** Model of face recognition-based door unlock system and **b** schematic design showing the inter-connection of model

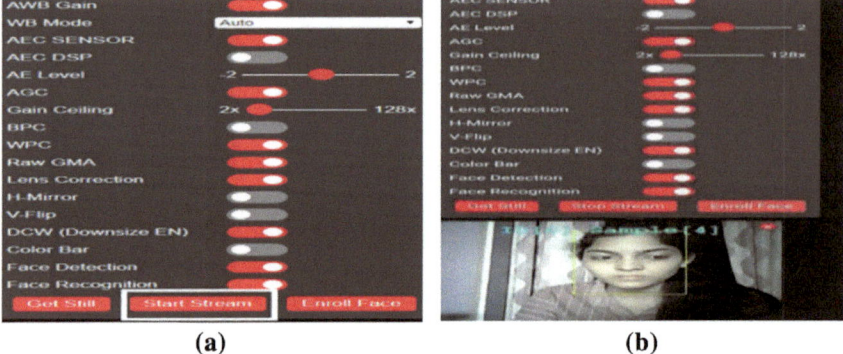

Fig. 5 **a** Portal to enroll images into the database and **b** capturing multiple samples' images of an individual

(ii) **Authentication—door opening/closing**: The face recognition system verifies for the correlation between displayed image on the camera with that of the stored images in the database. This operation can be configured in three stages, viz. presence of authentic person, presence of unauthentic person and opening by keypad due to non-recognition of authorized/unauthorized person. This is illustrated in the following sections.

(a) **Authentic person visit**—The door is closed initially (Fig. 6a). When an authorized person approaches the door as shown in Fig. 6b, the camera module captures the image and compares it with the enrolled images. As the images are already registered in the database, the circuit takes action and opens the door as shown in Fig. 6c. An SMS text is also sent to the admin as shown in Fig. 6d.

(b) **Unauthentic person visit**—In this condition also primarily door is closed (Fig. 7a). When an unauthorized person approaches near the door, the

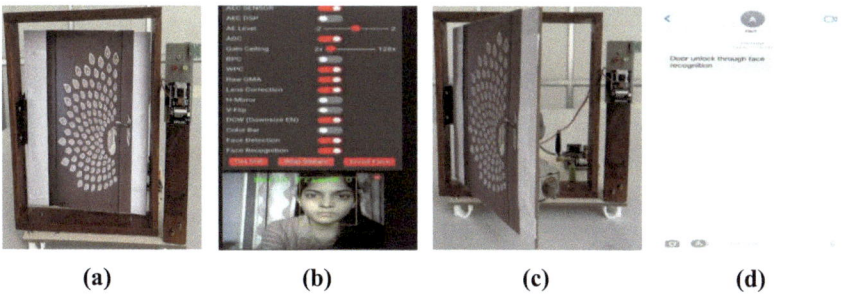

(a)   (b)   (c)   (d)

Fig. 6 **a** Door closed, **b** authorized person is detected, **c** door opened and **d** SMS is sent to the admin

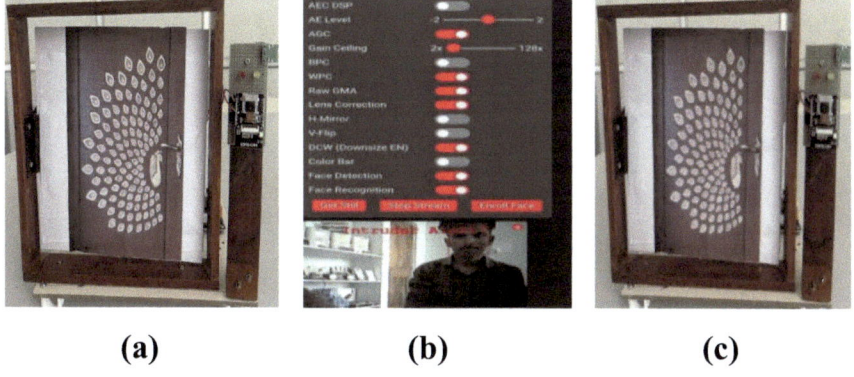

(a)   (b)   (c)

Fig. 7 **a** Door closed, **b** unauthorized person detection and **c** door remains closed

      captured image does not correlate with image stored in database (Fig. 7b). Hence, the door remains closed as shown in Fig. 7c.

(c) **Door unlocking by keypad**: The door is shut in this instance as well. When a person approaches the door whether it is authorized/unauthorized, one has to enter the PIN through the keypad as shown in Fig. 8a. If the right PIN is entered, the door opens and an SMS text is sent to admin as shown in Fig. 8b. If the wrong PIN is entered, the door remains closed and SMS alert is sent to the admin stating that an attempt is made by unknown to open the door as shown in Fig. 8c.

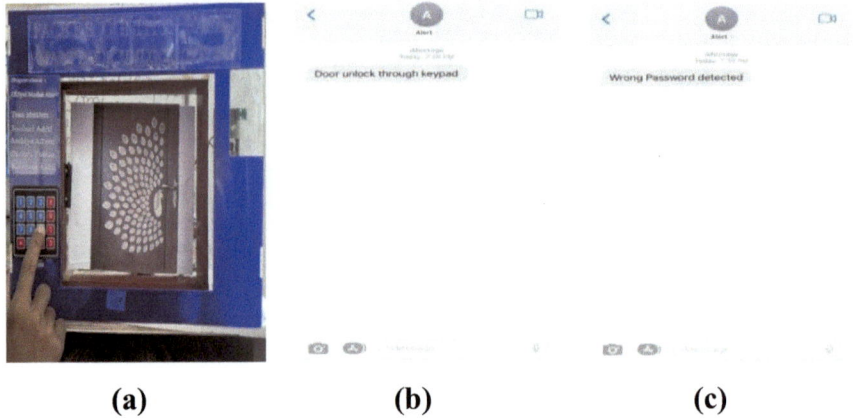

(a)   (b)   (c)

Fig. 8 **a** Entering manual PIN, **b** SMS generated when door is open using keypad and **c** SMS generated when entered wrong PIN

9 Conclusion

The door access control is implemented by using face recognition, which grants access to only authorized people to enter. Here, the images are stored in the database. The door automatically opens when the authorized person approaches it. The lock is closed when an unauthorized person approaches the door. An alternate method is implemented to enter four digits passkey in case of face recognition mechanism fails. The door is unlocked with either ways of successful match. The unauthenticated attempts are indicated with a buzzer sound and a red LED. An SMS alert is sent to the house owner about the attempt. This face recognition module can be installed in residential houses, banks, malls and in other public locations, where authentication of unauthorized person is essential. The present system is implemented with a combination of ESP32CAM, solenoid lock, GSM SIM900A, L298N motor driver, Arduino UNO and keypad. In this work an attempt is made to develop a home security system which is accessible, affordable and yet effective to make a smart home.

Acknowledgements The authors acknowledge Mr. Sridhar Head, center of design and Mr. Mettupelly Kumaraswamy, center of design, S R Engineering college for their support in preparing the model of the project. The authors also extend their gratitude to Principal and Management of S R Engineering college for helping us in carrying out this project work in the campus.

References

1. Vamsi TK, Sai KC, Vijayalakshmi M (2019) Face recognition based door unlocking system using Raspberry Pi. Int J Adv Res Ideas Innov Technol 5(2):1320–1324
2. Misal P, Karule M, Birdawade D, Deshmukh A, Pathak M (2014) Door locking/unlocking system using SMS technology with GSM/GPRS services. Int J Electr Commun Comput Eng 5(4):192–294
3. Sharma Y, Sijariya R, Gupta P (2023) How deep learning can help in regulating the subscription economy to ensure sustainable consumption and production patterns (12th Goal of SDGs). In: Deep learning technologies for the sustainable development goals: issues and solutions in the post-COVID era. Springer Nature, Singapore, pp 1–20
4. Siddiquee KNEA, Islam MS, Singh N, Gunjan VK, Yong WH, Huda MN, Naik DB (2022) Development of algorithms for an iot-based smart agriculture monitoring system. Wirel Commun Mob Comput 2022:1–16
5. Mani MR, Srikanth T, Satyanarayana C (2022) An integrated approach for medical image classification using potential shape signature and neural network. In: Machine learning and internet of things for societal issues. Springer Nature, Singapore, pp 109–115
6. Bouazzi I, Zaidi M, Usman M, Shamim MZM, Gunjan VK, Singh N (2022) Future trends for healthcare monitoring system in smart cities using LoRaWAN-based WBAN. Mob Inf Syst
7. Pathak R, Gupta SS (2020) A study on natural computing: a review. In: ICDSMLA 2019: Proceedings of the 1st international conference on data science, machine learning and applications. Springer, Singapore, pp 1975–1983
8. Singh N, Gunjan VK, Nasralla MM (2022) A parametrized comparative analysis of performance between proposed adaptive and personalized tutoring system "seis tutor" with existing online tutoring system. IEEE Access 10:39376–39386
9. Gaddam DKR, Ansari MD, Vuppala S, Gunjan VK, Sati MM (2022) Human facial emotion detection using deep learning. In: ICDSMLA 2020: Proceedings of the 2nd international conference on data science, machine learning and applications. Springer, Singapore, pp 1417–1427
10. Belmon AP, Auxillia J (2022) IoT-based continuous glucose monitoring system for diabetic patients using sensor technology. In: Machine learning and internet of things for societal issues. Springer Nature, Singapore, pp 35–41
11. Mokhlesabadifarahani B, Gunjan VK (2015) EMG signals characterization in three states of contraction by fuzzy network and feature extraction. Springer
12. Kotha MK, Pavan KK (2022) Deep learning for object detection: a survey. In: Proceedings of the international conference on computer vision, high performance computing, smart devices and networks: CHSN-2020. Springer Nature, Singapore, pp 61–84
13. Gunjan VK, Singh N, Shaik F, Roy S (2022) Detection of lung cancer in CT scans using grey wolf optimization algorithm and recurrent neural network. Heal Technol 12(6):1197–1210
14. Pathak R, Soni B, Muppalaneni NB (2023) Role of blockchain in health care: a comprehensive study. In: Gunjan VK, Zurada JM (eds) Proceedings of 3rd international conference on recent trends in machine learning, IoT, smart cities and applications. Lecture Notes in Networks and Systems, vol 540. Springer, Singapore. https://doi.org/10.1007/978-981-19-6088-8_13
15. Prabhu Das S, Jagadesh BN, Prabhakara Rao B (2022) Performance evaluation of segmentation algorithms in non contrast and contrast MRI images for region of interest. In: Proceedings of the international conference on computer vision, high performance computing, smart devices and networks: CHSN-2020. Springer Nature, Singapore, pp 95–111
16. Ahmed M, Ansari MD, Singh N, Gunjan VK, BV SK, Khan M (2022) Rating-based recommender system based on textual reviews using iot smart devices. Mob Inf Syst
17. Gunjan VK, Prasad PS, Pathak R, Kumar A (2020) Machine learning methods for extraction and classification for biometric authentication. In: ICDSMLA 2019: proceedings of the 1st international conference on data science, machine learning and applications. Springer, Singapore, pp 1984–1988

18. Usman M, Wajid M, Shamim MZ, Ansari MD, Gunjan VK (2021) Threshold detection scheme based on parametric distribution fitting for optical fiber channels. Recent Adv Comput Sci Commun (formerly: recent patents on computer science) 14(2):409–415
19. Kumar S, Gunjan VK, Ansari MD, Pathak R (2022) Credit card fraud detection using support vector machine. In: Proceedings of the 2nd international conference on recent trends in machine learning, IoT, smart cities and applications: ICMISC 2021. Springer, Singapore, pp 27–37
20. Murru GS, Kakollu C, Kenguva AK, Surya Chandra P (2020) Door unlocking system using fingerprint sensor for home automation. Int J Sci Res Eng Trends 6(2):1071–1073

Automatic Detection and Cleaning of Manhole Blockages Using IoT

Farheen Sultana, Syed Musthak Ahmed, and Vinit Kumar Gunjan

Abstract Manhole and sewage system is most common in rural and urban areas. If the system is left undetected which causes blockages, overflow of drainage water, bad odour, and suffocation leads to several diseases. Monitoring and cleaning of sewage system are very essential to maintain cleanliness in both urban and rural areas. Even the workers who are involved in the process of cleaning are prone to accidents by getting inside the manhole due to inhaling of poisonous gases. Cleaning of manholes manually leads to disasters due to ignorance. The proposed system developed is an automatic cleaning process avoiding human intervention. This saves time, cost, and manual labour for immediate attention. The developed system makes use of ESP32 WROOM, gas sensor, ultrasonic sensor for blockage detection, while communication circuit along with IoT module is used for information system. A servo robotic arm is employed to remove the blockages, and exhaust fan with associated circuit is used to suck out the poisonous gases. Upon detection of any blockages, the communication system gives information to the municipal corporation authorities about the location where the blockage is, so that the module developed can be deployed to the place of blockage for cleaning purpose.

Keyword Clogs · Manhole · Robotic arm · Servo motors · WiFi

F. Sultana
Department of ECE, M.Tech (ES), SR University, Warangal (Urban), Telangana, India

S. M. Ahmed (✉)
Department of ECE, SR University, Warangal (Urban), Telangana, India
e-mail: syedmusthak_gce@rediffmail.com

V. K. Gunjan
Department of CSE, CMR Institute of Technology, Hyderabad, Telangana, India

© The Author(s), under exclusive license to Springer Nature Singapore Pte Ltd. 2023
A. Kumar et al. (eds.), *Proceedings of the 4th International Conference on Data Science, Machine Learning and Applications*, Lecture Notes in Electrical Engineering 1038,
https://doi.org/10.1007/978-981-99-2058-7_20

1 Introduction

Sewage is the wastewater generated by several means such as domestic houses by cooking, cleaning, bathing, etc., industrial wastages, commercial places, clinics, and so on. This wastewater contains many chemicals which are used in our daily life. Some of the chemicals which arise from the industries and domestic usage are also harmful for living beings. Due to this wastewater, people are exposure to sewage by ingestion (or) contact, leads to illness, and sometimes to death. In appropriate sewage maintenance can contaminate ground water and leads to dangerous diseases. During rainy season, drainage becomes clogged, disrupting people's daily lives. Hence, it is very essential to clean the sewages in time to reduce pollution and keep the environment clean. The manual manhole cleaning system is shown in Fig. 1. A manual manhole cover is a detachable plate that functions as a lid to cover to prevent any person or objects from falling in. Unwanted manhole openings have recently posed a serious threat to society, resulting in health problems, pedestrian accidents, and even fatalities.

The central pollution control board of India identified around 351 rivers polluted due to sewage. The statistical data of various states polluted due to sewage are shown in Fig. 2 and are tabulated in Table 1.

2 Literature Survey

Himanshu [1], in his work, presented about the blockages that occur in the sewer system and the risk associated for the workers and the pollution caused due to sewage blockages. To avoid this, he employed an IoT system to warn the municipal officials about the problem via E-mail or mobile app. The developed system helps to clear the clog, besides identifying foul gas, and temperature. He developed a simple and low-cost, time-saving, human-friendly device.

| (a) | (b) | (c) |

Fig. 1 **a** Manual manhole lid, **b** manhole with clog, **c** manual manhole cleaning system

Fig. 2 Various states polluted due to sewage

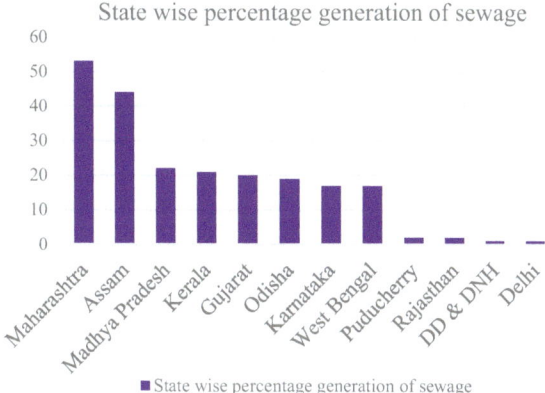

State wise percentage generation of sewage

■ State wise percentage generation of sewage

Table 1 Indian states and rivers polluted due to sewage

S. No.	State	Identified polluted rivers due to sewage
1	Maharashtra	53
2	Assam	44
3	Madhya Pradesh	22
4	Kerala	21
5	Gujarat	20
6	Odisha	19
7	Karnataka	17
8	West Bengal	17
9	Puducherry	2
10	Rajasthan	2
11	DD and DNH	1
12	Delhi	1

Karthikeyan et al. [2], in their work, they proposed the system using wireless sensor networks, in which they designed with the assistance of sensor nodes. This system is less expensive and requires little maintenance. This system is monitored using an IoT-based monitoring system. This system reduces human-caused accidents and manual labour.

Rohit shende et al. [3], in their work, they designed a drainage monitoring system in this paper. This drainage is monitored using an Arduino, a gas sensor, a flow sensor, and an NRF. These systems monitor the blockage that has occurred between two manholes and detect harmful gases that are hazardous to workers. This system will keep track of everything.

Ruheena et al. [4], in their work, they created the system with the aid of sensor nodes and advocated the use of wireless sensor networks. This technology is less expensive and requires little maintenance. This system is monitored using an IoT-based monitoring system. This technology reduces manual labour, saves time, and prevents human errors.

Chandhini et al. [5], in their work, they created a network system to monitor poisonous gases in the sewage system.

Gaurang sonawane et al. [6], in their work, employed a microcontroller module with a gas sensor and level indicator interface to it, and an NRF is installed in the drainage manholes. The system will monitor if a blockage occurs between two manholes as well as sense various gases that are harmful to humans and a water level monitoring system, and it will then trigger an alarm from which the appropriate action will be taken.

In [7, 8], the authors developed a manhole monitoring and steering system to avoid accidents.

Chang et al. [9], in their work, employed an RFID tag to detect the underground manhole solution for searching, authenticating, and even positioning the manhole. According to their findings, this creates a level surface that is also safe for the general public.

Nataraja et al. [10], they have portrayed an alerting system in which a buzzer alerts the surrounding area and uses GSM to transmit the sensed data to the controlling authorities. And this warning is based on identifying manholes that have opened owing to sewage water overflow, due to an increase in pressure or due to a rise in temperature that could cause the covers of the manholes to crack. This could cause the society a lot of issues. They came to the conclusion that adopting this tactic will assist society in avoiding such occurrences.

In [11, 12], in his work have sewer inlets for easy clearance in all roads, residences, workplaces, and commercial sectors. Every city necessitates a dependable sewage system. To avoid or reduce obstruction in sewer systems, various strategies have been developed. We present an easy-to-use yet effective Intelligent Clogged Sewer Control System that eliminates this problem by notifying the appropriate governing authorities when the sewer is likely to become completely blocked by garbage or debris. They concluded that it removes the majority of contaminants from wastewater before releasing it into the natural water stream.

In [13–16], the authors developed an efficient accident-avoidance manhole system. The ultimate aim was to prevent risk of death for manual scavengers.

Girisrinivaas et al. [17], they suggested system will test the gas and water levels in the sewage system, store the results in the cloud for analysis, and then send an SMS with the sewage system's status to a GSM module located close to the corporation headquarters, which concludes the avoid the health conditions of the people.

In [18, 19], in their work, developed a sewer inspection robot which works wirelessly. Their design was compact and adaptable to the sewage system.

Yves Abou Rjeily et al. [20], in their work, they used sensor monitoring for both storm water and waste water.

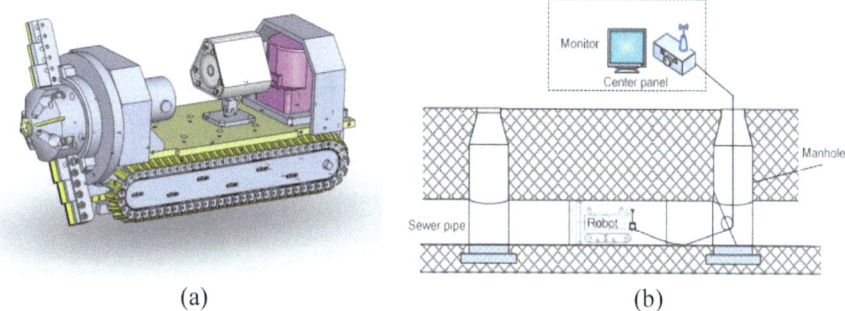

Fig. 3 **a** Model of pipe-cleaning and inspection robot in sewage system, **b** view of robot cleaning the sewage system

3 Existing System

The system discussed here is the issues confronting Vietnam's public sewage system, such as leaks, soil erosion, and contamination from chemicals and heavy metals. The wastewater contains toxic substances from both industrial and domestic sources, which can harm the environment and human health [21]. They developed and designed a sewer robot capable of travelling through sewers autonomously. A cable controls the sewer robot, which is outfitted with a camera and a light [22–24]. They describe the use of various competencies in the mechanical design of the robot along with motor control and robot control systems. The cleaning and inspection procedures include a camera and a light to create a video record of the sewer's condition. Using an active vision system, the robot was designed to maintain orientation within the sewer pipe as it moves. The robot is controlled remotely via a cable. They tested the robot both in the lab and in the field. The tests were designed to assess the robot's performance and effectiveness in real-world situations.

Figure 3a and b show the model developed used for pipe-cleaning and inspection purpose.

4 Proposed System

The block diagram of proposed system is shown in Fig. 4. It consists of Esp32 Wroom, ultrasonic sensor, gas sensor, Blynk server, Arduino Uno, exhaust fan, L298n driver, buzzer, servo motors, channel relay, and linear actuator. The ultrasonic sensor, gas sensor are fed into the Esp32. The Blynk IoT platform is used to receive notifications. The ESP32 and Arduino communicate via a single wire. The servo motor is controlled by Arduino. There are two servo motors, one MG996R and the other is MG995. The MG996R servo gripper is used to remove the clog from the manhole system. The MG995 servo lid is used to open and close the manhole lid that rotates at an angle

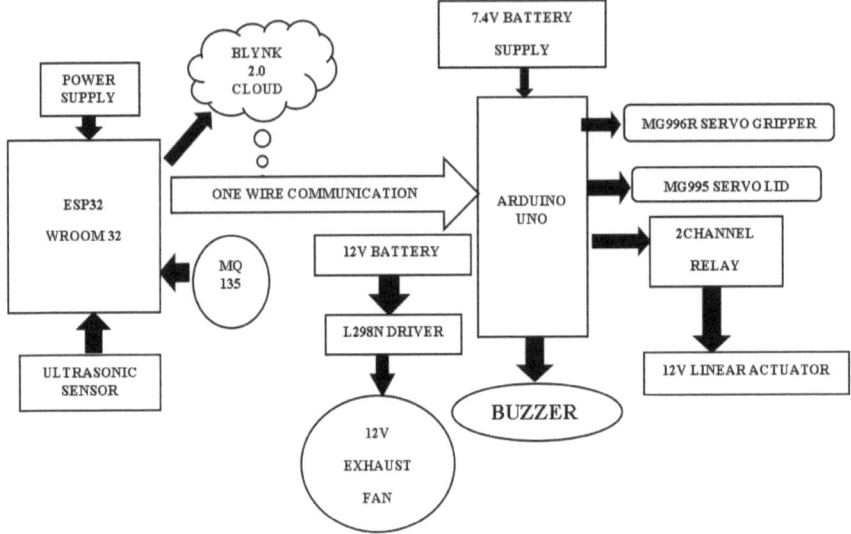

Fig. 4 Block diagram of the proposed sewage monitoring system

of 0–120°. Two-channel relays are used to control the polarities of the actuator. A linear actuator along with the gripper is used to remove the clog [25]. A fan is fixed above the gripper used to remove harmful gases. The L298N driver turns the exhaust fan on and off.

5 Objectives

This module of automatic detection and cleaning of manhole is developed to reach the following objectives.

- Avoid manual cleaning of manholes to prevent human intervention and save human life.
- Identifying the presence of clog in the sewage/drainage system.
- Find the level of drainage water due to blockage.
- Locate the position of blockage.
- Prevent odour due to clog and create a clean environment.
- Save the government from revenue cost, time, manual dependence, and intervention.

6 Methodology

The step-by-step procedure adopted in developing this module is:

- Identifying the required sensors to resolve the set problem.
- A gas sensor (MQ135) is selected to sense the harmful gases in the sewage system (Ex: CO_2, methane…).
- The ultrasonic sensor (HCSR04) to determine the distance of the presence of clog in the manhole.
- Two servo motors are deployed, one servo motor MG996R is used for gripper operation, and the other MG995 is used for opening and closing of manhole lid.
- The actuator with a gripper function as a robotic arm. The actuator is used for up and down operation.
- An exhaust fan fixed above the gripper is used to suck out the harmful gases from the manhole.
- The Blynk IoT platform is employed to get the notifications of the blockage and harmful gases and also for sending the E-mail alerts.

7 Hardware and Software Requirements

The hardware components and devices used in developing the automatic detection and cleaning of manhole blockage using IoT are shown in Figs. 5a–h. The brief discussion of these components is described. (i) Figure 5a shows an Esp32 WROOM, and it belongs to a microprocessor family with built-in Wi-Fi and Bluetooth. It operates at 160–240 MHz with the performance capacity of 600 DMIPS. It has 320 KB built-in RAM and 448 KB ROM. (ii) Figure 5b shows the MQ135 gas sensor. It operates at 5 V DC. It is highly sensitive, has fast response, good life time, and is stable [26–28]. It can detect gases like benzene, smoke, CO_2, and other pollutants in air. (iii) Figure 5c shows HCSR04 ultrasonic sensor. It is used to detect the blockage inside the manhole and measures the distance of the clog. As there is a blockage, then it sends the web notification to the user and also sends signal to the Arduino. It uses ultrasonic waves for blockage detection. (iv) Figure 5d shows Arduino Uno R3. It is set up to control robotic ARM movement as well as manhole plate opening and closing. It is powered by a lithium-ion battery with a 3.7 V voltage. ESP32 sends signals to the Arduino during the detection of clog blockage [29]. The Arduino will then open the manhole system's gate, complete the drilling and cleaning operation, and close the manhole. Until the manhole is opened during the process, the buzzer will sound to warn those nearby. (v) Figure 5e shows linear actuator. It converts the rotatory motion into the linear motion. It consists of 12 V by cascading batteries (3 lithium ions batteries) are used. Two-channel relay is used to control the polarities of the actuator [30, 31]. This relay is controlled by the Arduino. (vi) Figure 5f shows L298n motor driver. The purpose of it is to control the direction and speed of DC motor. (vii) Figure 5g shows the servo motors MG996R, MG995. The MG996R is

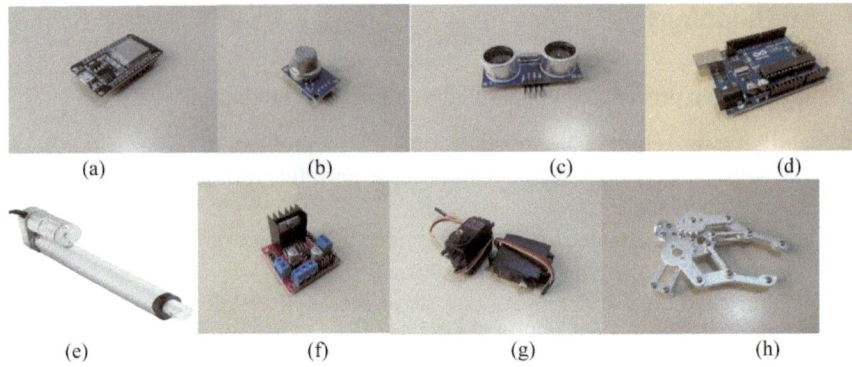

Fig. 5 The hardware components and devices used in developing the automatic detection and cleaning of manhole blockage using IoT

used for gripper operation whose angle of rotation is 0–180°. The other servo motor MG995 is used for opening and closing of manhole lid. (viii) Figure 5h shows the metallic gripper. Its function in our project is to hold the clog from the sewage system [32].

The module is developed using the following software.

(i)   The Arduino Integrated Development Environment (IDE) or Arduino Software (IDE): It has a text editor for writing code, a message area, a text console, a toolbar with buttons for common functions, and a series of menus. It controls the hardware remotely. It stores sensor data and also displays it to upload and communicate with programmes. Here, a BLYNK IoT PLATFORM is used.

8   Implementation and Working of the System

Figure 6a is the module developed for prototype testing of automatic detection and cleaning of manhole. The schematic circuits interconnected to Arduino with other accessories such as sensors, fan, buzzer, actuator, relay are shown in Fig. 6b.

Figure 7 shows the working of the module.The operation is explained in following steps:

(i)   Opening of the lid.
(ii) Automatically inserting the robotic arm into the manhole to remove the clog.
(iii) Sucking off of poisonous gases.
(iv) Closing of the lid.

If the blockage occurs inside the manhole, the authorities receive a notification about the location of the blockage. This operation of notification is carried out by the Blynk IoT App. Once the message is received, the authorities then takes the initiation of cleaning the clog. The operation to remove the clog with the existing

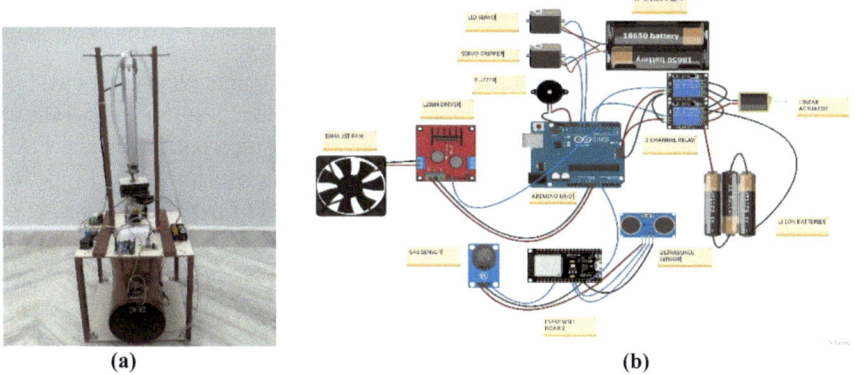

Fig. 6 **a** Schematic diagram showing the inter-connection of devices of the manhole system and **b** model of automatic detection and cleaning of manhole system

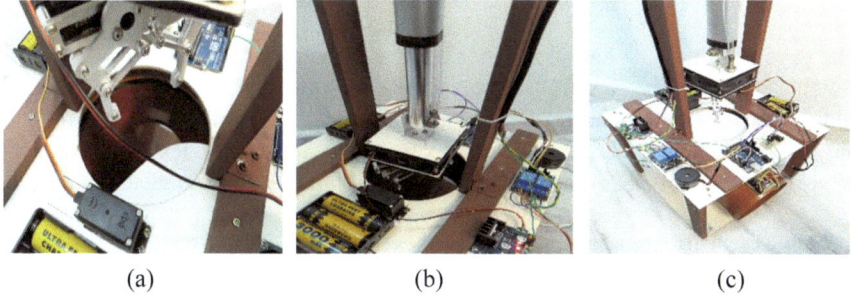

Fig. 7 **a** Opening of the manhole lid, **b** inserting of the robotic arm to remove the harmful gases, **c** closing of the manhole lid

module is first carried out by removing the lid [33]. Here, the servo motor MG995 is used for the opening of the lid. It has 180° freedom but is programmed to rotate the lid by 120°. The next step is to insert the actuator into the manhole for cleaning of the clog and removing the poisonous gases. To perform this operation, the servo motor MG996R is employed for the gripper operation. The gripper attached to the actuator removes the clog. Simultaneously, the exhaust fan attached above the gripper sucks of the harmful gases. The two operations are performed simultaneously. During this process, a buzzer makes a sound to alert people who are nearby the manhole to prevent them from the danger of approaching towards the manhole and/or cautioning them due to accident falling into the manhole [34]. The clog is removed by linear pull-out of the actuator. Once the clog is removed, the lid is automatically closed by the servo motor MG995, thus completing the operation of clog removal and sucking of poisonous gases.

9 Conclusion

The present work is designed to develop and automatic drainage/manhole cleaning system avoiding human intervention. This developed module saves time, cost, besides labour dependence. Here, a vertical axis robotic arm is incorporated which has a linear top–bottom moment. A gripper is attached to the actuator whose function is to remove the clog. Two servo motors are used, one for opening and closing of manhole lid, while the other supports to perform gripper operation. An exhaust is fixed above the gripper which helps in sucking out the poisonous gases stored inside the manhole. Notification operation is performed by Blynk IoT App which sends an E-mail alerts to the authorities about the blockage. The complete operation is performed automatically without any human intervention.

10 Future Scope

The developed module has an actuator that has only up and down vertical moment which can just lift the clog from the manhole. The work can be further extended by adding few features like horizontal moment of the actuator. This feature will not just only lift the clog by the gripper but also takes out the clog and put it into the bin. This needs an additional electromechanical circuitry to make it the complete system.

Acknowledgements The authors acknowledge the centre of IoT, SR University and MUFA Tech solutions, Warangal, for their technical support in developing the prototype model of the project. The authors also acknowledge the principal and management of SR University in carrying out the work smoothly in campus.

References

1. Himanshu S, Bharani Kumar J, Shashank K, Rama Swamy T (2022) IoT based manhole detection and monitoring system. Int J Res Appl Sci Eng Technol 1–14
2. Karthikeyan VM (2022) Manhole monitoring system implemented in smart city. Int J Eng Res Technol 10:43–47
3. Shende R (2020) Case study of smart real time drainage monitoring system. Int J Res Eng Sci Manag 2:325–326
4. Ruheena M (2021) Manhole detection and monitoring system. Int J Eng Res Technol (IJERT) 9:1–6
5. Chandhini G (2020) IoT based underground drainage monitoring system. Int J Recent Technol Eng 9:247–249
6. Sonawane G (2018) Smart real time drainage monitoring system using IoT. Iconic Res Eng J 1:1–6
7. Mankotia A, Shukla AK (2022) IOT based manhole detection and monitoring system using Arduino. ASET 57(5):2195–2198
8. Patel R (2007) IoT based waste water spillage detection system. J Phys Conf Ser 1–10

9. Chang AY, Yu C-S, Lin S-C, Chang Y-Y, Ho P-C (2009) Search, identification and positioning of the underground manhole with RFID ground tag. In: Fifth international joint conference, pp 1899–1903

10. Nataraja N, Amruthavarshini R, Chaitra NL, Jyothi K, Krupaa N, Saqquaf SM (2018) Secure manhole monitoring system employing sensors and GSM techniques. In: International conference on recent trends in electronics, information & communication technology, pp.2078–2082

11. Pullan P, Niranjan V (2019) Intelligent clogged sewer control system. In: 2019 IEEE international WIE conference on electrical and computer engineering, pp. 1–4

12. Sakthipriya D, Logeswari V, Nishanthi K, Reethika B (2018) Manhole monitoring system implemented in smart city. In: International conference on emerging trends in IoT & machine learning, pp 1–9

13. Ganpat MA, Ganpat O, Bansoe, Shivaji S (2022) Smart manhole detection. Int J Eng Appl Sci Technol 6:139–141

14. Menakadevi T, Akash M, Dilip Kumar B (2021) IoT based automated manhole detection. Int J Eng Res Technol 8:1–4

15. Aarthi M, Bhubaneshwar A (2021) IoT based drainage and waste management monitoring and alert system for smart city. Ann RSCB 25:6641–6651

16. Aly HH, Soliman AH, Mouniri M (2015) Towards a fully automated monitoring system for Manhole cover: smart cities and IOT applications. In: 2015 IEEE first international smart cities conference (ISC2), pp 1–7

17. Girisrinivaas R, Parthipan V (2017) Drainage overflow monitoring system using IoT. In: 2017 IEEE international conference on power, control, signals and instrumentation engineering, pp 2133–2137

18. Scholl K-U, Kepplin V, Berns K, Dillmann R (1999) An articulated service robot for autonomous sewer inspection tasks. In: 1999 IEEE/RSJ international conference on intelligent robots and systems, pp 10–14

19. Kawaguchi Y, Yoshida I, Kurumatani H, Kikuta T, Yamada Y (1995) Internal pipe inspection robot. In: IEEE international conference on robotics and automation, pp 857–862

20. Abou Rjeily Y, Sadek M, Hage Chehade F, Abbas O, Shahrour I (2017) Smart system for urban sewage: Feedback on the use of smartsensors. In: Institute of electrical and electronics engineers, pp 1–4

21. Prasad PS, Sunitha Devi B, Janga Reddy M, Gunjan VK (2019) A survey of fingerprint recognition systems and their applications. Lect Notes Electr Eng 500:513–520

22. Verma N, Jain A (2023) Intelligent self-tuning control design for wastewater treatment plant based on PID and model predictive methods. In: Deep learning technologies for the sustainable development goals: issues and solutions in the post-COVID era. Springer Nature, Singapore, pp 69–82

23. Rudra Kumar M, Pathak R, Gunjan VK (2022) Diagnosis and medicine prediction for COVID-19 using machine learning approach. In: Computational intelligence in machine learning: select proceedings of ICCIML 2021. Springer Nature, Singapore, pp 123–133

24. Das N (2023) Digital education as an integral part of a smart and intelligent city: a short review. In: Digital learning based education: transcending physical barriers, pp 81–96

25. Rashid E, Ansari MD, Gunjan VK, Ahmed M (2020) Improvement in extended object tracking with the vision-based algorithm. In: Modern approaches in machine learning and cognitive science: a walkthrough: latest trends in AI, pp 237–245

26. Pradhan AK, Swain S, Kumar Rout J (2022) Role of machine learning and cloud-driven platform in IoT-based smart farming. In: Machine learning and internet of things for societal issues. Springer Nature, Singapore, pp 43–54

27. Lakshmanna K, Shaik F, Gunjan VK, Singh N, Kumar G, Shafi RM (2022) Perimeter degree technique for the reduction of routing congestion during placement in physical design of VLSI circuits. Complexity 2022:1–11

28. Singh A (2023) Transportation management using IoT: deep learning to predict various traffic states. In: Deep learning technologies for the sustainable development goals: issues and solutions in the post-COVID era. Springer Nature, Singapore, pp 203–226

29. Gaddam DKR, Ansari MD, Vuppala S, Gunjan VK, Sati MM (2022) A performance comparison of optimization algorithms on a generated dataset. In: ICDSMLA 2020: proceedings of the 2nd international conference on data science, machine learning and applications. Springer, Singapore, pp 1407–1415

30. Kumar Raja DR, Hemanth Kumar G, Lakshmi Sagar P (2022) Data mining approach for prediction of various risk factors in supply chain management. In: Proceedings of the international conference on computer vision, high performance computing, smart devices and networks: CHSN-2020. Springer Nature, Singapore, pp 173–180

31. Usman M, Wajid M, Shamim MZ, Ansari MD, Gunjan VK (2021) Threshold detection scheme based on parametric distribution fitting for optical fiber channels. Recent Adv Comput Sci Commun (formerly: recent patents on computer science) 14(2):409–415

32. Rama Santosh Naidu P, Lavanaya Devi G, Kondapalli VR, Neelapu R (2022) A novel and self adapting machine learning approach of ECG signal classification in association with cardiac arrhythmia. In: Proceedings of the international conference on computer vision, high performance computing, smart devices and networks: CHSN-2020. Springer Nature, Singapore, pp 195–207

33. Bouazzi I, Zaidi M, Usman M, Shamim MZM, Gunjan VK, Singh N (2022) Future trends for healthcare monitoring system in smart cities using LoRaWAN-based WBAN. Mob Inf Syst

34. Pathak R, Soni B, Muppalaneni NB (2023) Role of blockchain in health care: a comprehensive study. In: Gunjan VK, Zurada JM (eds) Proceedings of 3rd international conference on recent trends in machine learning, IoT, smart cities and applications. Lecture Notes in Networks and Systems, vol 540. Springer, Singapore. https://doi.org/10.1007/978-981-19-6088-8_13

Design and Implementation of a Smart Door Locking System with Automatic Appliance Switching

Syed Musthak Ahmed, Abbidi Shivani Reddy, Nagineni Sahithi, Beryl, and Vinit Kumar Gunjan

Abstract The implementation of a smart door for residences is discussed in this paper, which allows them to operate household devices as they leave the house. In the present days, electricity tariff is quite high. The majority of people forget to turn off their appliances, which results in higher electricity bills. In order to save electricity, it is necessary to automatically control the devices. One such technique is adopted here using smart door locking system. As residents leave the house, the devices turn off automatically while nobody is inside. As they arrive, to unlock the door, a password-entering system is introduced at the entrance. Once the door is unlocked, based on person's status appliances are either put to off or on. Arduino Uno is incorporated to control the door operation. The operation is performed in two ways: one is manually by using a keypad, and the other is digitally using a Bluetooth module. The system depends upon the pre-decided password concept with proper security. This developed system has strict access control with security features for controlling the door. Thus, the implementation is effective in terms of security, cost and safety.

Keywords Arduino Uno · Electricity tariff · Smart door · Bluetooth module · Security

1 Introduction

Electricity is very important for every one of us in our daily life, starting from domestic needs, agriculture, industrial purpose, commercial and so on. Unnecessary wastage of electricity leads to increase in tariff and energy consumption. Most of the time people forget to switch off the appliances before they exit. Due to this the

S. M. Ahmed (✉) · A. S. Reddy · N. Sahithi · Beryl
SR Engineering College, Warangal, Telangana, India
e-mail: syedmusthak_gce@rediffmail.com

V. K. Gunjan
CMR Institute of Technology, Hyderabad, Telangana, India

A. Kumar et al. (eds.), *Proceedings of the 4th International Conference on Data Science, Machine Learning and Applications*, Lecture Notes in Electrical Engineering 1038, https://doi.org/10.1007/978-981-99-2058-7_21

appliances consume power. Domestic appliances are one such where people forget to turn off the appliances. To make home smarter, several methods have been developed by incorporating the current technology. Smart home system makes use of automatic connection to control devices with environment via media such as internet and phone, thereby making life easier and smarter.

In Indian scenario the tariff varies from the units of consumption. Most of the times the tariff is recorded high in spite of non-utilization of appliances due to several reasons. Hence, there is a need to save or reduce the usage of electricity by providing the necessary security for houses called "home automation". Smart door automation is one such technique where appliances in the home are going to switch off automatically except devices like refrigerator, Wi-Fi, washing machine, etc. Several methods have been developed to make home smarter as discussed in literature survey.

2 Literature Review

Several automated door locking system has been developed to make residence smarter, by incorporating the advances in technology. Smart home is one such concept being proposed by several scholars. The main aim behind is to save electricity.

In [1–4], the authors developed a smart door locking system using a bio-metric controlled by Arduino Uno. The library of finger prints of all authorized persons is created and stored. Whenever the authorized person places his finger print, door automatically opens. Upon checking the matched finger print, a conformation OTP is sent to the authorized user. In the event of an unauthorized person while there is a mismatch in the fingerprint, buzzer is blown indicating the stranger.

In [5–8], Bluetooth is controlled by Arduino Uno to open door using fingerprint. It sends SMS to the user in case of intrusion, gas leakage or fire accidents, with the help of Android phone application and also if in case the fingerprint is failed to match. Additionally, a buzzer is employed to inform the nearby residents. A One Time Password (OTP is also delivered by GSM to the authorized person's cellphone.

In [9–12], the authors presented smart door locking system using face recognition. In their work they incorporated Internet of Things which uses Wi-Fi module. The resident's facial features are stored in the library and are checked every time while trying to open the door. If at all the face do not match with already stored data then the user gets alerted with either email or SMS. By integrating this technology, the user need not carry any sort of keys or RFID cards. Without any physical contact to the door, identification and unlocking the door is made faster.

In [13, 14], the authors in their work developed a door locking system which is opened using RFID. In their work, the barcode on ID card is scanned. It then checks for the authorized user from the library where the data of the user's ID card is stored. If the ID card is matched the door opens and will intimate the person who is inside with a message that the door is opening.

In [15–17], the authors in their work developed a door locking system which is operated using Internet of Things connecting it with Android application using Wi-Fi application. The password is entered in the Android phone to open the door. Opening and closing the door is controlled by the user using the smart phone rather than to open it manually with help of a key.

Kadhim et al. [18] in their work implemented a smart home system to detect intruders using IoT. They incorporated a door locking system which is opened using MATLAB. To unlock, the user's picture which is already stored is compared every time while trying to open the door, and the intruder's picture is sent to user through email.

In [19, 20], the authors developed a voice-controlled door locking system which is opened using voice recognition with the help of Bluetooth module. By incorporating this technology, the door is unlocked without any physical contact. This can be mostly helpful for the users such as elderly, disabled and kids.

3 Existing System

Earlier implementation was carried out using Android Development Tools (ADT), Java Development Kit (JDK), Android SDK (Software Development Kit), etc. Figure 1 depicts the basic block diagram of the current system implemented using Arduino software with Android smart phone. Here the Arduino serves as the central processing unit. Bluetooth is incorporated in transmitting and receiving data with microprocessor. The system consists of two parts—the input and the output. The input system constitutes of Android smart phone controlled by Bluetooth which is connected to Arduino Uno, operated by a 12 V DC power supply. The output system has LED, relay and solenoid [21]. The LED and associate circuit indicate the presence of electric current, while the solenoid is controlled by a relay.

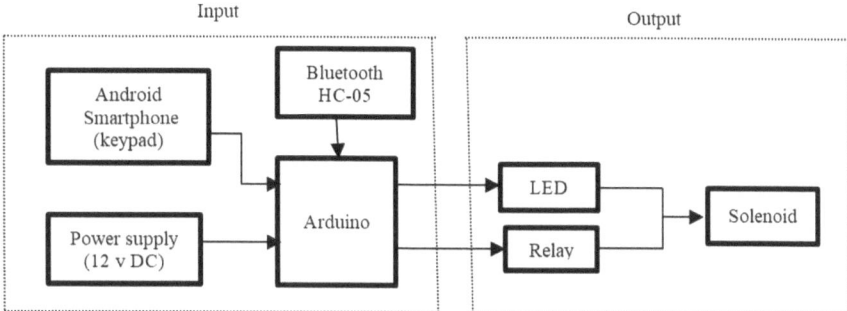

Fig. 1 Basic block of existing system

4   Proposed System

Figure 2 shows the proposed system block diagram. The model consists of Arduino Uno that controls the switching of domestic appliances through sensors. In the proposed project the axial fan and bulb are used as domestic appliances for prototype testing [22]. The axial fan is interfaced to Arduino Uno through L298N motor driver, while the bulb is interfaced through a single-channel relay. A keypad is directly interfaced to the Arduino Uno to enter the password for opening the door [23]. Under situations where there is a need to open the door, upon information to the authorized person the door can be opened by the facility created on the mobile. Upon information the authorized user gets a display of a button on his phone. Once the button is pressed on his phone the door gets unlocked [24, 25]. This feature is achieved by incorporating Android Bluetooth module.

A solenoid lock is used to unlock the door. When the entered password is matched with prerecorded password then electric current from Arduino Uno is passed to solenoid lock and high magnetic field is generated. It attracts the slug, and the door is opened using motor driver. In case the wrong password is entered, the door remains closed and buzzer blows [26]. The IR sensor is connected to Arduino Uno, to check the person condition. If it detects the person, the appliances are turned on automatically. The LDR sensor is used to check the light intensity; if the intensity detected is low, the bulb is turned on automatically, and if it is high, the bulb remains off [27, 28]. While exiting the house, the IR sensor checks the person's presence and controls the appliances. If in case the person is inside, the appliances remain in on condition, else they get turned off [29].

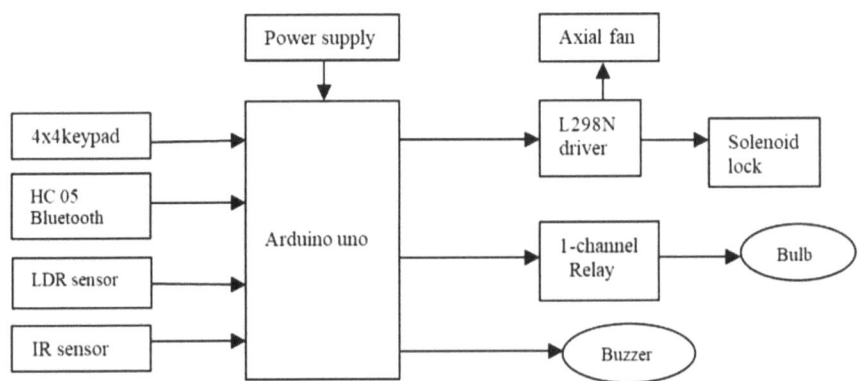

Fig. 2 Block schematic illustrating the suggested approach

5 Hardware Description

The various hardware devices and components are used in developing the module, and each of these components is illustrated and described briefly from Fig. 3a–i.

Figure 3a is the Arduino Uno used in the system which operates at 5 V DC, and it is basically used to control the sensors and actuators. Figure 3b is a 4 × 4 keypad matrix used to enter the password. There are 4 rows and 4 columns in the keypad and total of 8 keys. Figure 3c is IR sensor used to detect the person and operates at 50 mA and 5 V. Figure 3d is a motor driver, connected to fan and solenoid lock to operate because they both have windings and AC load. It operates at 150 mA and 5 V. Figure 3f is a buzzer which is used to alarm and alert and notify the user, and it has a built-in oscillator circuit as the sound frequency is fixed. Figure 3g is a Bluetooth module used so that the user can unlock the door [30].

Fig. 3 **a** Arduino Uno, **b** 4 × 4 matrix keypad, **c** IR sensor, **d** L298N motor driver, **e** 1-channel relay, **f** buzzer, **g** Bluetooth module, **h** solenoid lock, **i** LDR sensor

It permits simultaneous direction and speed control of two DC motors. With a peak current of up to 2 A, it operates between 5 and 35 V of voltage. Figure 3e is a relay that can be used to control high voltage, high current load such as motor and using Android app. Figure 3h is a solenoid lock also known as electromagnetic lock which works on Faraday's law that has a slug with a slanted cut. It is basically an electronic lock and is designed for safety of the door. Figure 3i is LDR sensor used to check the intensity of light. The resistance of LDR changes depending upon the light intensity. This makes LDR to be used in light sensing devices.

6 Objectives

The aim and objective behind the product developed is to

1. Incorporate the existing technologies to make smart home.
2. Save electricity tariff while appliances are not in use and at the same time in on condition.
3. Make home smarter with door condition by automating switching on/off of home appliances, during exit and entry.
4. Resolve society problems at homes while someone forgets to switch off appliances.
5. Making the domestic appliances smarter by incorporating smart door locking system.

7 Methodology

The procedures adopted to meet the objectives are:

1. Atomizing all domestic appliances.
2. Controlling the devices by the condition of door and person.
3. The keypad matrix is fixed at the entrance of the house.
4. Door is opened when password entered by the user is correct.
5. IR sensor is used to detect the persons inside the house.
6. If presence is detected the appliances turns on.
7. Lights turn on when light levels are low; they switch off when light levels are high.
8. Appliances are switched off automatically, when door is closed and presence of person is not detected.

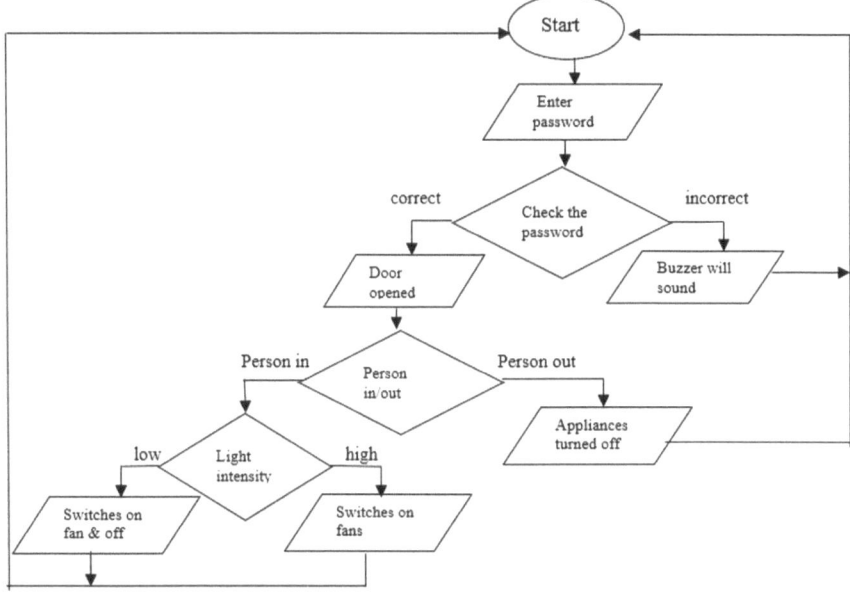

Fig. 4 Flowchart depicting the operation of smart door locking system

8 Flowchart Construction

Flowchart of Fig. 4 explains the procedure of operation; when one enters into the house password is entered using keypad, and if entered password is correct then the door is opened, else the buzzer starts buzzing to alert the user. After the door is opened, the fan is turned on only if the person's motion is detected by the IR sensor. Then LDR sensor will check the light intensity; if it is low the light will be switched on, else light will not be switched on [31]. While leaving the house the user can unlock the door by using mobile app with the help of Bluetooth module. When the user leaves home and there is no presence of person the appliances will be turned off.

9 Implementation and Working of the Smart Home System

Implementation of smart door locking system is shown in Fig. 5. The various devices are connected to check the functionality of the various devices like fan, solenoid lock is connected to the L298N motor driver and Arduino Uno, bulb is connected to 1-channel relay and Arduino Uno, keypad is directly connected to door and Arduino Uno, Bluetooth module is connected to the Arduino Uno to the Rx and Tx keys, and buzzer is connected to the Arduino Uno [32, 33].

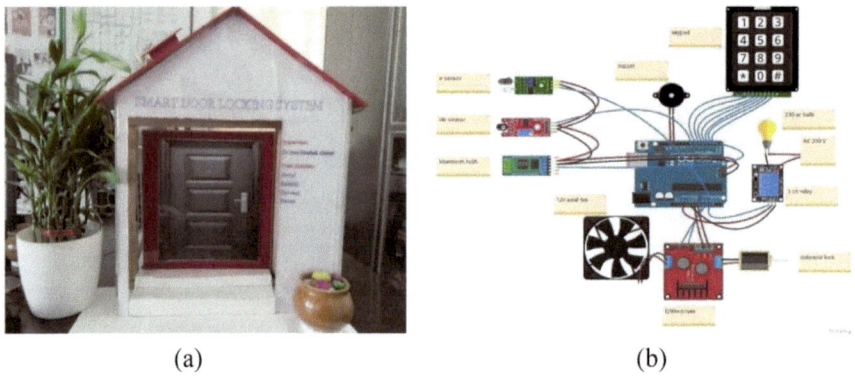

(a)   (b)

Fig. 5 **a** Model of smart door locking system and **b** schematic diagram showing the interconnection of devices of smart door locking system in the proposed work

The operation of the smart door is explained by configuring into four parts: door open—appliances on, door open—appliances off, door close—appliances on and door close—appliances off. The first phase is a situation where residents are present in the house and at same time the devices in working condition. The second phase is a situation where people are present in/out of the house, while the door is opened [34]. The third situation is where door is closed while appliances are on. All the above situations are where the residents are present and devices can be controlled. The present work is to implement the fourth phase, i.e., door closed while the appliances are kept on. Here the circuit is designed to see that the devices automatically switch off while residents leave the house [35, 36]. This is normal situation where the people are present in the house, at same time making use of devices.

The present work is the focus toward the switching off the appliances while the residents leave the house. In this work the design is carried out to see that the devices are switched off automatically when the devices are on and residents have left the house by forgetting. When the door is opened, the IR detects the person then fan turns on, and if light intensity is low, bulb is turned on as shown in Fig. 6a. When the door is opened and IR detects no person then the appliances will turn off as shown in Fig. 6b. When the door is closed and IR detects the person then the fan turns on, and if light intensity is low, bulb turns on as shown in Fig. 6c. When the door is close and IR detects no person, the appliances turns off as shown in Fig. 6d.

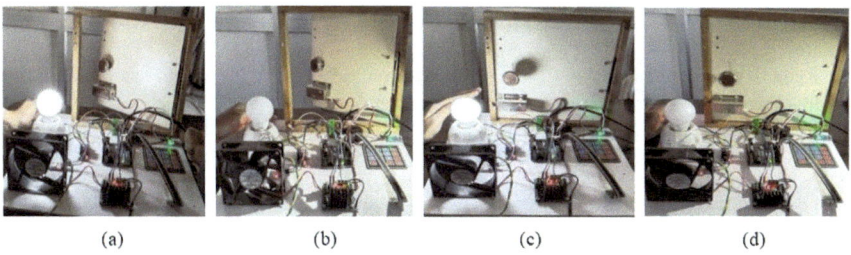

(a)	(b)	(c)	(d)

Fig. 6 **a** Door opened and appliances are turned on, **b** door opened and appliances are turned off, **c** door closed and appliances turned on and **d** door closed appliances turned off

10 Conclusion

This paper presents a basic prototype of a home with door locking and appliances controlling automatically. The concept introduced is to give an idea of a smart home system. The implementation helps the residents to save their tariff when they forget to switch off the appliances. The implementation is carried out with help of Arduino by incorporating various sensors such as IR for person detection to turn on the fan, LDR to check the light intensity to turn on or off the bulb during day/night and Bluetooth which controls the door digitally. This project can be implemented in apartments, office rooms and villas, where people are busy usually forgetting to switch off. The module developed is a low cost and easily affordable. Our idea incorporated as prototype is implemented in this work. In the future it can be extended to any of the places where there is a need to control of appliances, i.e., switching off appliances when not in use, thereby saving electricity, tariff, time and cost.

Acknowledgements The authors acknowledge Mr. Sridhar, Head, Center of Design, and Mr. Mettupelly Kumaraswamy, Center of Design, SR Engineering College, Warangal, for their support in preparing the model of the project. The authors also extend their profound thanks to the principal and management of our institution in taking up this project work in the campus.

References

1. Rudregowda S (2019) A smart door lock system using fingerprint sensor. Int J Mod Trends Sci Technol 05(02):36–38
2. Motwani Y, Seth S, Dixit D, Rajesh P (2021) Multifactor door locking system
3. Ponmalar A, Anand J, Dharshini S, Aishwariya K (2021) Smart phone-controlled fingerprint door look system. In: Advances in parallel computing technologies and applications
4. Al Rakib MA, Rahman MM, Uddin S, Anik MSA (2022) Fingerprint based smart home automation and security system
5. Naga Vishnu Vardhan J, Antra D, Eswari Antitha D, Sumathi B A biometric lock using bluetooth technology working through Android app. Adv Intell Comput Commun 17–26
6. Karim B, Nawzad H (2018) Smart home system based on GSM network. Kurdistan J Appl Res 3:17–21

7. Saxena D, Bisen P, Bhoyerkar S (2019) Intelligent security and automation system. Int J Adv Res
8. Kamelia L, NoorHassan A (2014) door automation system using bluetooth based android for mobile phone. 9
9. Saroha A, Gupta A, Bharagava A, Singh H (2022) Biometric authentication based automated secure, and smart IOT door lock system. In: IEEE India council international subsections conference (INDISCON)
10. Datar VP, Tankasali A, Chavan K (2021) Smart door lock and lighting system using Internet of Things. IARJSET Int Adv Res J Sci Eng Technol 8(8)
11. Khalimov R, Rakhimbayeva Z, Shokayev A (2022) Development of intelligent door locking system based on face recognition technology. The system operates through a combination of Arduino UNO and Android-based smartphone. In: 11th international conference on mechanical and aerospace engineering (ICMAE)
12. Sayem IM, Chowdhury MS (2018) Integrated face recognition security system with IOT using raspberry pi to a camera module for face recognition. In: International conference on machine learning and data engineering (ICMLDE)
13. Park YT, Pyun JY (2009) Smart digital door lock for the home automation. In: TENCON IEEE region 10 conference
14. Mhatre A, Jawle C, More V Security door lock for mains supply, to provide safety to the engineer, the feeder box is equipped with RFID sensor and RFID reader switch plays an important role. Int J Adv Res Sci Commun Technol (IJARSCT)
15. Tuyen NT, Ngoc NQ, Hung NX (2021) An application of node MCU Esp8266 in opening and closing the laboratory door opening. Global J Eng Tech Adv
16. Chandramohan J, Nagarajan R, Satheeshkumar K, Ajithkumar N, Gopinath PA, Ranjithkumar S (2017) Intelligent smart home automation and security system using Arduino and Wi-Fi. Int J Eng Comput Sci
17. Kumar S (2014) Smart home system using android application. Int J Comput Networks 6
18. Kadhim HJ, Ameen MJM (2020) Design and implement a smart system to detect intruders and firing using IOT. Int J Electri Comput Eng (IJECE) 10:5932–5939
19. Prasad DD, Mallika GJ, Farooq SU, Tanmaie A, RadhaKrishna D, Pramod S (2021) Voice controlled home automation. In: Second international conference on electronics and sustainable communication systems (ICESC), pp 673–678
20. Sulaiman RB (2018) Voice controlled home automation. University of Bed ford shire
21. Akula CS, Prathima C, Srinivasulu A (2022) Reliable smart grid framework designs through data processing and analysis process. In: Proceedings of the international conference on computer vision, high performance computing, smart devices and networks: CHSN-2020. Springer Nature Singapore, Singapore, pp 189–194
22. SuryaNarayana G, Kolli K, Ansari MD, Gunjan VK (2021) A traditional analysis for efficient data mining with integrated association mining into regression techniques. In: Kumar A, Mozar S (eds) ICCCE 2020. Lecture notes in electrical engineering, vol 698. Springer, Singapore. https://doi.org/10.1007/978-981-15-7961-5_127
23. Srinivasa Babu K, Jaya Sri K, Jyothi C (2022) Sensor-based radio frequency identification technique for the effective design process of diverse applications using WSN. In: Proceedings of the international conference on computer vision, high performance computing, smart devices and networks: CHSN-2020. Springer Nature Singapore, Singapore, pp 181–187
24. Ahmed SM, Kovela B, Gunjan VK (2020) IoT based automatic plant watering system through soil moisture sensing—a technique to support farmers' cultivation in Rural India. In: Gunjan V, Senatore S, Kumar A, Gao XZ, Merugu S (eds) Advances in cybernetics, cognition, and machine learning for communication technologies. Lecture notes in electrical engineering, vol 643. Springer, Singapore. https://doi.org/10.1007/978-981-15-3125-5_28
25. Bhowmik T, Majumdar S, Choudhury A, Banerjee A, Roy B (2023) Importance of internal and external psychological factors in digital learning. In: Digital learning based education: transcending physical barriers. Springer Nature Singapore, Singapore, pp 119–132

26. Nandy D, Majee D (2023) Socio-economic relations in digital education: a comparative study between Bangladesh and Nepal. In: Digital learning based education: transcending physical barriers. Springer Nature Singapore, Singapore, pp 103–117

27. Rudra Kumar M, Pathak R, Gunjan VK (2022) Machine learning-based project resource allocation fitment analysis system (ML-PRAFS). In: Kumar A, Zurada JM, Gunjan VK, Balasubramanian R (eds) Computational intelligence in machine learning. Lecture notes in electrical engineering, vol 834. Springer, Singapore. https://doi.org/10.1007/978-981-16-8484-5_1

28. Kar S, Kar S (2023) Pedagogical considerations in the new normal: from tradition to technology. In: Digital learning based education: transcending physical barriers. Springer Nature Singapore, Singapore, pp 97–102

29. Prasad PS, Sunitha Devi B, Janga Reddy M, Gunjan VK (2019) A survey of fingerprint recognition systems and their applications. Lecture notes electrical and engineering, vol 500. pp 513–520

30. Srivastava V, Srivastava MK, Singhal RK (2023) Enhancing shoppers' loyalty by prioritizing customer-centricity drivers in the retail industry. In: Deep learning technologies for the sustainable development goals: issues and solutions in the post-COVID era. Springer Nature Singapore, Singapore, pp 227–246

31. Rudra Kumar M, Pathak R, Gunjan VK (2022) Diagnosis and medicine prediction for COVID-19 using machine learning approach. In: Computational intelligence in machine learning: select proceedings of ICCIML 2021. Springer Nature Singapore, Singapore, pp 123–133

32. Verma A (2023) Automation of brain tumor segmentation using deep learning. Deep learning technologies for the sustainable development goals: issues and solutions in the post-COVID era. Springer, Singapore, pp 189–202

33. Kumar S, Ansari MD, Gunjan VK, Solanki VK (2020) On classification of BMD images using machine learning (ANN) algorithm. In: ICDSMLA 2019: proceedings of the 1st international conference on data science, machine learning and applications. Springer Singapore, pp 1590–1599

34. Pathak R (2020) Support vector machines: introduction and the dual formulation. In: Gunjan V, Senatore S, Kumar A, Gao XZ, Merugu S (eds) Advances in cybernetics, cognition, and machine learning for communication technologies. Lecture notes in electrical engineering, vol 643. Springer, Singapore. https://doi.org/10.1007/978-981-15-3125-5_57

35. Rashid E, Ansari MD, Gunjan VK, Ahmed M (2020) Improvement in extended object tracking with the vision-based algorithm. Modern approaches in machine learning and cognitive science: a walkthrough: latest trends in AI, pp 237–245

36. Pathak R, Soni B, Muppalaneni NB (2023) Role of blockchain in health care: a comprehensive study. In: Gunjan VK, Zurada JM (eds) Proceedings of 3rd international conference on recent trends in machine learning, IoT, smart cities and applications. Lecture notes in networks and systems, vol 540. Springer, Singapore. https://doi.org/10.1007/978-981-19-6088-8_13

Dynamic Game Difficulty Adjustment Based on Facial Emotion Recognition

Harish Akula, Dinesh Rayala, and Morarjee Kolla

Abstract Gaming has become a highly popular mode of entertainment, with games often eliciting strong emotional responses from players. As such, there is a need to develop new approaches that can enhance the overall gaming experience. Given that humans frequently undergo changes in emotional states without being fully conscious of them, proper emotion control is imperative. To address this, we propose the development of a game that changes difficulty with player's emotions. This game can record user emotions in real-time and modify the game's difficulty accordingly. This emotion-adaptive game operates by adapting to the emotional states of the player, with the difficulty level changing based on these emotions. For the game, a desktop-based space shooter game has been selected, where the player must shoot down enemy ships. Players who exhibit negative emotions are subjected to a penalty, resulting in an increase in game difficulty. The machine learning algorithm has been chosen for emotion prediction, which has been trained on a dataset of labeled facial expressions and their corresponding emotions. To predict emotions in real-time during gameplay, a machine learning model has been integrated into the game, and the game and emotion predictor are run concurrently using threads. The emotion information obtained from the predictor is then used by the game to dynamically adjust the difficulty level. By employing this approach, emotion-adaptive gaming can not only improve the overall gaming experience but can also enhance emotion control and self-awareness.

Keywords Facial expressions · Gaming · Emotion control · Difficulty · Real-time

H. Akula · D. Rayala · M. Kolla (✉)
Department of Computer Science and Engineering, Chaitanya Bharathi Institute of Technology, Hyderabad, Telangana, India
e-mail: morarjeek@gmail.com

H. Akula
e-mail: ugs18096_cse.harish@cbit.org.in

D. Rayala
e-mail: ugs18093_cse.dinesh@cbit.org.in

223

A. Kumar et al. (eds.), *Proceedings of the 4th International Conference on Data Science, Machine Learning and Applications*, Lecture Notes in Electrical Engineering 1038, https://doi.org/10.1007/978-981-99-2058-7_22

1   Introduction

Emotions are complex mental and physiological states that can be associated with a wide array of thoughts, feelings, and behaviors and can be experienced as positive or negative depending on the individual and the situation. Facial expressions refer to a set of specific movements of the small muscles in the face that can be used to identify or infer a person's particular emotional state, such as happiness, anger, or sadness. Although there is ongoing debate among scientists about whether animals possess emotions, a growing body of research indicates that many species do exhibit behaviors and physiological responses that are indicative of various emotional states [1]. Nonetheless, it is widely recognized that human beings have evolved more complex and sophisticated mechanisms for the expression and control of emotions in comparison to other animals [2]. It is not uncommon for individuals to lose control of their emotions, particularly in situations that trigger intense emotional responses. When this occurs, the individual may display behaviors that are inconsistent with their usual personality and values and may experience negative consequences, such as damaged relationships, effective decision-making, and reduced well-being. Thus, it is imperative for individuals to exercise appropriate control over their emotions in order to mitigate the potential negative impact.

Gaming is a popular form of entertainment and has the potential to evoke a wide range of emotions in players [3, 4]. However, humans can experience a variety of emotions without being fully aware of them, which can impact their ability to control their emotions in a healthy way. This can be particularly problematic for players who struggle with uncontrolled emotions, such as anger or anxiety, and can negatively affect their gaming experience. Therefore, the problem that this paper aims to address is the need for a game that can adapt to a player's emotions and provide a more tailored, enjoyable, and beneficial experience.

As technology has advanced, there has been a growing interest in using machine learning and computer vision to enhance gaming experiences. The availability of data on people's faces and their corresponding emotions has presented an opportunity to use this information to improve gaming experiences. By combining a game with a computer vision module, it is possible to read a player's emotions and adjust the game's difficulty in real-time, providing a more personalized and engaging experience.

The significance of this project is that it has the potential to help individuals develop emotional control and self-awareness, which can lead to healthier emotional responses and stronger relationships. People who struggle with uncontrolled emotions, such as those with anxiety disorders or short tempers, could benefit from playing a game that helps them practice managing their emotions. Additionally, this game provides a fun and engaging way for players to assess their emotional control and awareness. The objective of this paper is to describe the design and development of a game that uses computer vision to adapt to a player's emotions and provide a more personalized gaming experience. By doing so, this paper hopes to contribute to the development of games that promote emotional intelligence and control.

2 Related Work

There has been a growing interest in the use of machine learning and computer vision to recognize emotions in recent years. This has led to the development of many commercial applications that use emotion recognition to improve user experience, such as chatbots, virtual assistants, and personalized advertising. However, the use of emotion recognition in gaming is still a relatively new and underexplored area of research. By creating a game that uses computer vision to recognize a player's emotions, this project aims to contribute to this emerging field.

The authors of the article [5] used wearable biofeedback sensors to measure players' peripheral physiological signals, such as heart rate and skin conductance, and analyzed several physiological indices to determine their correlations with anxiety. They then used these data to infer the player's probable anxiety level and adjust the game difficulty level in real-time based on the emotion state.

This paper [6] describes the implementation of a dynamic game difficulty adjustment (DDA) method in Tetris using Active Shape Model (ASM) and Hidden Markov Model (HMM) to recognize players' emotion states from a camera feed. They utilized a Kalman filter to dynamically detect players' experience and adjust the game speed accordingly. Experimental results showed that the DDA method improved players' game experience.

The authors created a horror game, Caroline, [7] that uses the player's biometric data to adapt the difficulty level based on their stress levels. They explored the impact of this approach on players' cognitive, emotional, performative, and decision-making challenges, as well as their intrinsic motivation, and compared it to the base game without any DDA. Their results showed that players felt more motivated when the gameplay was adjusted according to their heart rate, and the DDA only affected the decision-making challenge [8, 9].

3 Methodology

We proposed a solution that involves the creation of an emotion-adaptive game that utilizes real-time emotion recognition program to capture and store the player's emotional state in a database. This information is then utilized by the game to dynamically adjust the game's difficulty level as necessary. Specifically, we have chosen to develop a space shooter game to showcase this emotion-adaptive functionality, in which the player maneuvers a spacecraft laterally and fires at incoming enemies [10].

Our study explores two approaches for setting the difficulty level in an emotion-adaptive game. The first approach involves giving the player an easier difficulty when experiencing negative emotions such as sadness, anger, or fear, with the expectation that this will improve their mood. However, we found this approach to be ineffective, as it can lead to a stagnation in the player's emotional state, with the player being incentivized to show negative emotions to achieve a higher score [11, 12].

The second approach involves penalizing the player for showing negative emotions by increasing the difficulty level, which challenges the player to control their emotions. We believe this approach to be more effective in helping players to regulate their emotions [13]. Our proposed game will include three different modes, each designed to elicit a specific emotional response from the player. Overall, our goal is to create an engaging and constructive gaming experience that helps players to manage their emotions in a healthy way [14].

We propose three types of modes for the respective emotions.

1. Easy (A)—happy and neutral.
2. Medium (B)—angry and sad.
3. Hard (C)—fear and surprise.

There are two main modules involved in this game, and they are:

1. Emotion prediction module: The emotion predictor module is designed to recognize faces in real-time, crop and resize the image, convert it to grayscale, and predict the associated emotion using computer vision techniques. The prediction model is trained beforehand, with the resulting model being used to predict human emotions and store the data in a database. Real-time images are captured using OpenCV with Haar cascade classifier, and the model is used to accurately predict emotions [15, 16].
2. Game module: In conjunction with the emotion predictor module, the game module operates parallelly with the use of threads. The game module accesses the emotion data stored in the database by the predictor and adjusts the difficulty level accordingly [17]. To ensure an optimal gaming experience, the player begins with a relatively easy difficulty level that gradually increases as the game progresses.

The system can be represented as a block diagram as shown in Fig. 1, where the player inputs keypresses through a keyboard, which are then transmitted to the game. The game processes these inputs and moves the player's character accordingly, with the output being displayed on a monitor visible to the player. Meanwhile, a camera captures the player's facial expressions in real-time and transmits them to an emotion prediction model. The model generates emotion predictions, which are then stored in a database [18]. The game reads the database and adjusts its difficulty level based on the frequency of emotions detected.

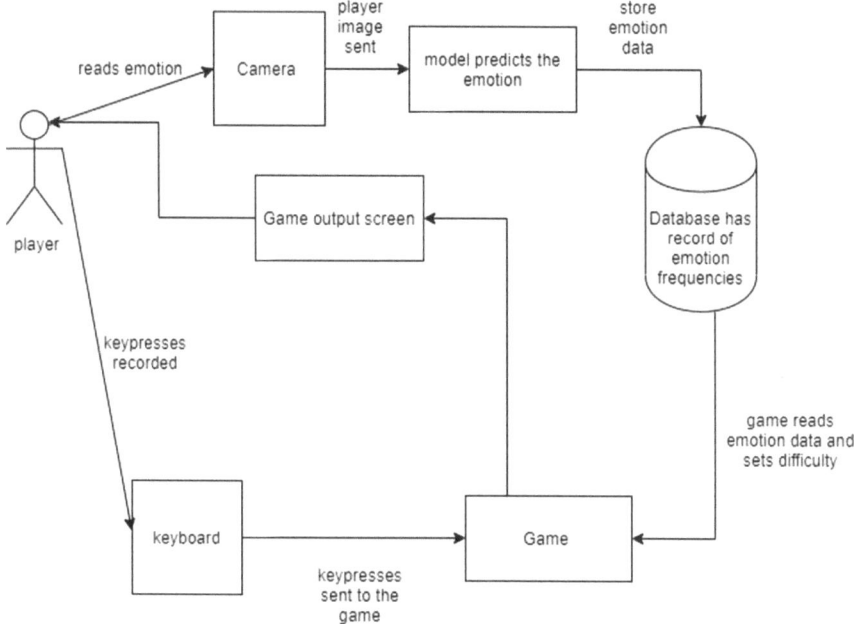

Fig. 1 Block diagram of the proposed game

4 Implementation Details

The process of predicting emotions involves using a Convolutional Neural Network (CNN) model. An appropriate architecture is selected and hyperparameters are set before training the model. Once trained, the model is evaluated on both the training and testing datasets. To improve accuracy, the methods are fine-tuned. After obtaining a high-performing model, it is saved. A system is then implemented to capture facial images of a person and send them to the saved model in real-time to predict emotions.

To achieve this, we first select a dataset for training the model. For this purpose, we chose the Kaggle FER 2013 dataset, which consists of grayscale images of people's faces labeled with six different emotions: angry, happy, neutral, surprise, fear, and disgust. The dataset contains 35,887 images with dimensions of 48 × 48 pixels.

We utilized a CNN model that was implemented using the Keras API in Tensor-Flow. The model consisted of repeated combinations of Convolutional, Batch Normalization, and Max pooling layers, with an activation function of ReLU. Our initial training on this model yielded a prediction accuracy of only around 63%. Upon further analysis, we discovered issues within the dataset that impacted our accuracy [19, 20]. Firstly, the dataset only contained grayscale images with dimensions of 48 × 48 pixels. Training on grayscale images with only a single channel of pixel intensity resulted in less detailed information compared to training on the more commonly available three-channel RGB images. Additionally, while larger image sizes generally provide more precise features, training on larger images becomes more difficult when there is a large amount of data to process. Furthermore, predicting emotions on large images in real-time can become problematic. Therefore, in this particular task, 48 × 48 pixel images were deemed sufficient [21].

We collected additional data and used data augmentation techniques to improve the dataset. Further, we reduced the skewed data by removing "disgust" emotion which is less relevant. Improving upon all these factors gave us about 74% of accuracy. The fact that the accuracy is still not near perfect is because a significant amount of detail is lost when images are 48 × 48 resolution. Moreover, some emotions had different interpretations in dataset and the model predicted reasonably, meaning that the model is good to be used [22].

The space shooter game is implemented in Python with Pygame module. The main game loop starts, and it also starts a thread in parallel that reads emotions. The game initially starts in the A mode. It generates some enemies for A mode, and the emotion predictor reads all the emotion data. When player defeats all the enemies for A mode, new wave of enemies is generated, and the game difficulty mode is now set based on the emotions displayed by the player. The mode can now be A, B, or C based on emotions shown. Every time a wave of enemies is approaching, the predictor records the emotion information, and the game reads the emotion information when the wave of enemies is zero and creates a new wave of enemies based on the recorded emotions [23].

Since the emotions fear and surprise are not that frequent, they are given more weightage than other emotions. Sad and fear emotions are given medium weight. Normal and happy emotions are given least weights as they are the emotions that occur frequently. The emotion data are the number of frames the model has predicted the user showing a particular emotion, while a wave is going on as shown in Fig. 2.

5 Result Analysis

The locally saved CNN model is utilized in conjunction with a camera module, which implements OpenCV, PIL, and Haar cascade classifier to capture facial images. The captured image is cropped and resized and subsequently fed into the emotion predictor for the generation of real-time emotion data. The generated data are then stored in an SQLite database every 30 frames. The game and predictor operate concurrently in separate threads. The game retrieves the emotion data from the database and resets the emotion counts to 0 upon each retrieval, allowing for the loading of new data by the emotion predictor. Our emotion prediction when ran in parallel with game gave us around 55 frames per second on a computer with Intel Core-i5 2400 CPU and 4 GB DDR3 RAM. Some of the images of the game are shown.

Our game successfully manages to utilize real-time emotions of the player and use it to set the game difficulty as shown in Figs. 3 and 4.

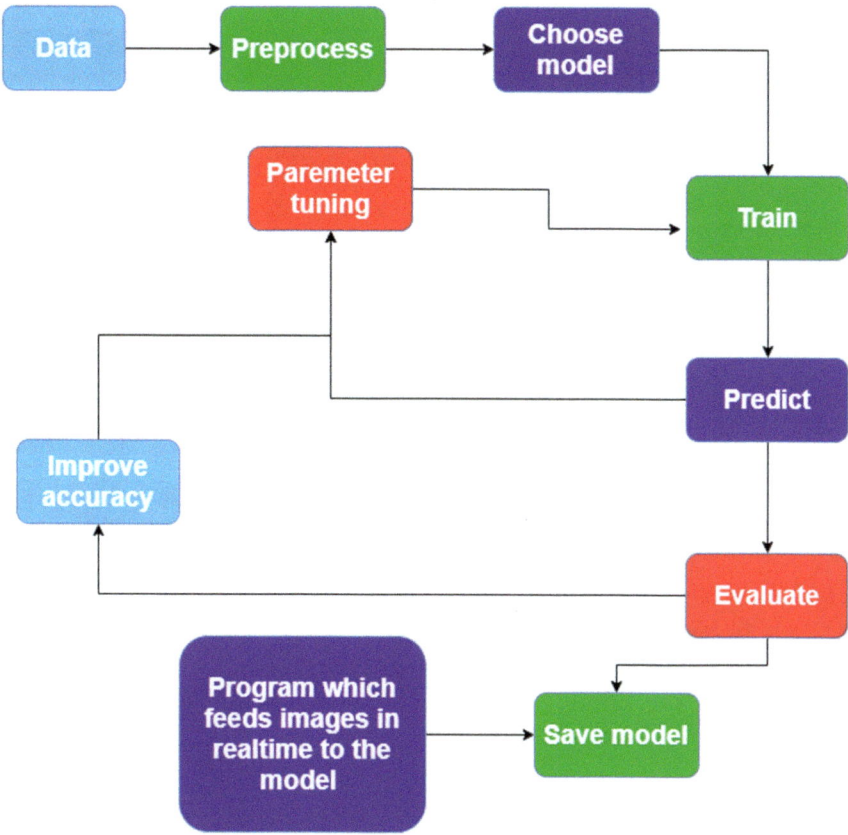

Fig. 2 Machine learning training process flowchart

6 Conclusion

In this paper, we discussed the design and implementation of an Emotion-Adaptive Space Shooter Game, which incorporates real-time emotion prediction to optimize the game's difficulty level. Our results show that our proposed game system can accurately predict the player's emotional state and adjust the game's difficulty level accordingly, resulting in a more engaging and challenging user experience. Moreover, our system's approach of penalizing negative emotions rather than comforting them showed significant improvement in emotion regulation and control. Overall, our Emotion-Adaptive Space Shooter Game provides a promising avenue for future research in the field of game design, particularly in developing more personalized and emotionally intelligent gaming experiences.

Furthermore, the potential benefits of this project extend beyond the gaming industry. Emotional control and self-awareness are important skills in many aspects of life, including relationships, work, and personal well-being. By providing a fun

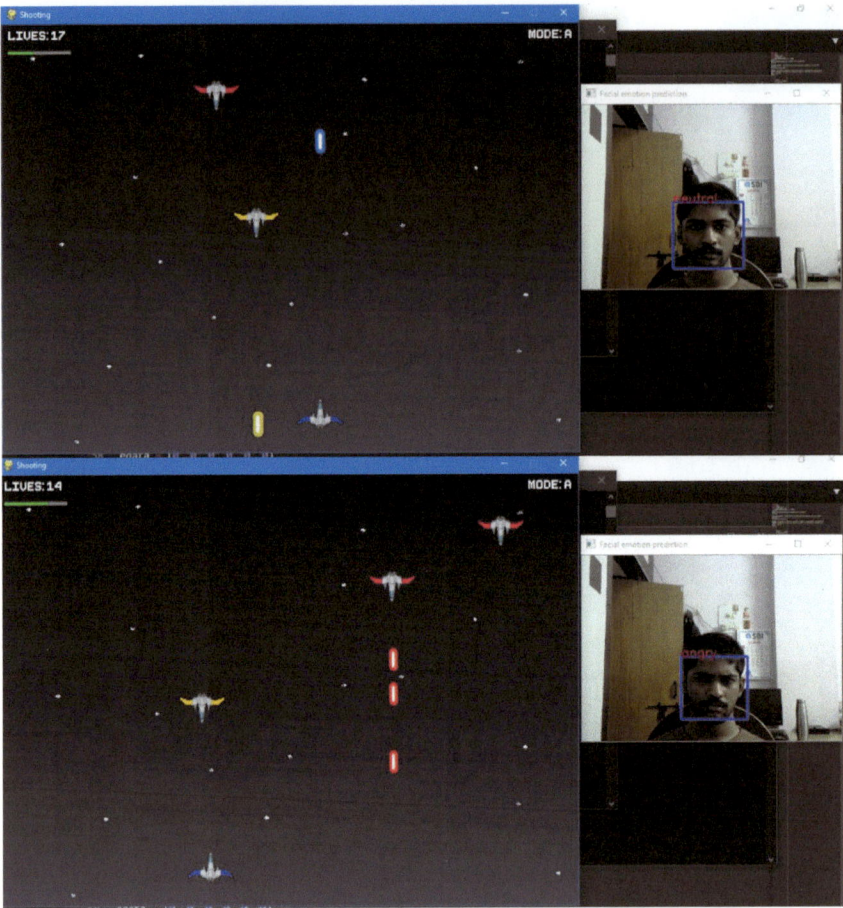

Fig. 3 Gameplay, modes set to mode A in both the images, and the player showing neutral and angry emotions

and engaging way for players to practice managing their emotions, this game has the potential to promote emotional intelligence and self-awareness beyond the world of gaming.

In summary, the motivation behind this project is to leverage advancements in technology to create a game that adapts to a player's emotions and provides a more personalized and engaging experience. This has the potential to promote emotional control and self-awareness in players, while also contributing to the emerging field of emotion recognition in gaming.

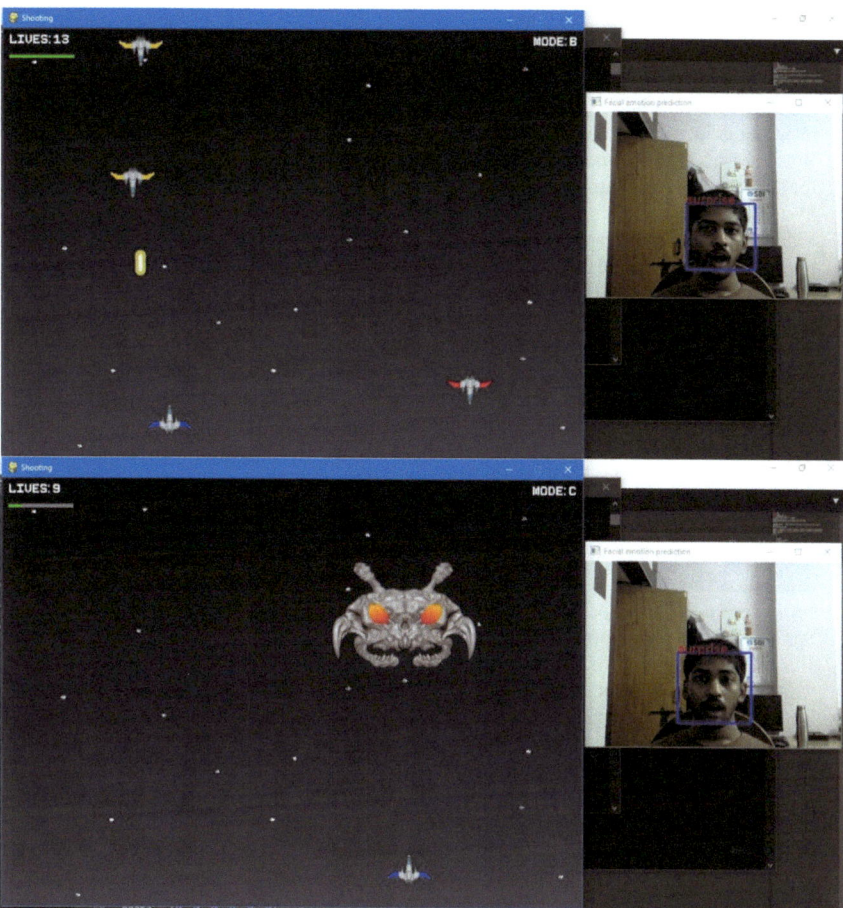

Fig. 4 Gameplay, modes set to mode B in the first and mode C in the second image, and the player showing surprise emotion

References

1. Ferretti V, Papaleo F (2019) Understanding others: emotion recognition in humans and other animals. Genes Brain Behav 18(1):e12544. https://doi.org/10.1111/gbb.12544
2. Kumar A, Vohra R (2023) Impact of deep learning models for technology sustainability in tourism using big data analytics. In: Deep learning technologies for the sustainable development goals: issues and solutions in the post-COVID era. Springer Nature Singapore, Singapore, pp 83–96
3. Jyothirmayi T, Satyananarayana C (2022) Identification of brain tumors using deep learning techniques. In: Machine learning and Internet of Things for societal issues. Springer Nature Singapore, Singapore, pp 127–136
4. Belmon AP, Auxillia J (2022) IoT-based continuous glucose monitoring system for diabetic patients using sensor technology. In: Machine learning and Internet of Things for societal issues. Springer Nature Singapore, Singapore, pp 35–41

5. Singh N, Gunjan VK, Nasralla MM (2022) A parametrized comparative analysis of performance between proposed adaptive and personalized tutoring system "seis tutor" with existing online tutoring system. IEEE Access 10:39376–39386

6. Grizzard M, Francemone CJ (2018) Research on the emotions caused by video games demands integration. In: Video games, 1st edn. Routledge, pp 60–73

7. Bouazzi I, Zaidi M, Usman M, Shamim MZM, Gunjan VK, Singh N (2022) Future trends for healthcare monitoring system in smart cities using LoRaWAN-based WBAN. Mob Inform Syst 2022

8. Video games evoke emotions that even music and cinema fail to reach (n.d.) Aalto.Fi. Retrieved 18 Feb 2023, from https://www.aalto.fi/en/news/video-games-evoke-emotions-that-even-music-and-cinema-fail-to-reach

9. Kumar S, Gunjan VK, Ansari MD, Pathak R (2022) Credit card fraud detection using support vector machine. In: Proceedings of the 2nd international conference on recent trends in machine learning, IoT, smart cities and applications: ICMISC 2021. Springer, Singapore, pp 27–37

10. Xiang N, Yang L, Zhang M (2013) Dynamic difficulty adjustment by facial expression. In Lecture notes in electrical engineering. Springer, London, pp 761–768

11. Gaddam DKR, Ansari MD, Vuppala S, Gunjan VK, Sati MM (2022) Human facial emotion detection using deep learning. In: ICDSMLA 2020: proceedings of the 2nd international conference on data science, machine learning and applications. Springer, Singapore, pp 1417–1427

12. Akula CS, Prathima C, Srinivasulu A (2022) Reliable smart grid framework designs through data processing and analysis process. In: Proceedings of the international conference on computer vision, high performance computing, smart devices and networks: CHSN-2020. Springer Nature Singapore, Singapore, pp 189–194

13. Gunjan VK, Singh N, Shaik F, Roy S (2022) Detection of lung cancer in CT scans using grey wolf optimization algorithm and recurrent neural network. Heal Technol 12(6):1197–1210

14. Bhowmik T, Majumdar S, Choudhury A, Banerjee A, Roy B (2023) Importance of internal and external psychological factors in digital learning. In: Digital learning based education: transcending physical barriers. Springer Nature Singapore, Singapore, pp 119–132

15. Mokhlesabadifarahani B, Gunjan VK (2015) EMG signals characterization in three states of contraction by fuzzy network and feature extraction. Springer

16. Kar S, Kar S (2023) Pedagogical considerations in the new normal: from tradition to technology. In: Digital learning based education: transcending physical barriers. Springer Nature Singapore, Singapore, pp 97–102

17. Ahmed M, Ansari MD, Singh N, Gunjan VK, Santhosh Krishna BV, Khan M (2022) Rating-based recommender system based on textual reviews using IoT smart devices. Mob Inform Syst 2022

18. Singh A (2023) Transportation management using IoT. In: Kadyan V, Singh TP, Ugwu C (eds) Deep learning technologies for the sustainable development goals. Advanced technologies and societal change. Springer, Singapore. https://doi.org/10.1007/978-981-19-5723-9_14

19. Siddiquee KNEA, Islam MS, Singh N, Gunjan VK, Yong WH, Huda MN, Naik DB (2022) Development of algorithms for an iot-based smart agriculture monitoring system. Wirel Commun Mob Comput 2022:1–16

20. Sachdev D, Pokhriyal SK, Nechully S, Sundaram SS (2023) Healthcare 4P: systematic review of applications of decentralized trust using blockchain technology. In: Deep learning technologies for the sustainable development goals: issues and solutions in the post-COVID era, pp 133–156

21. Pathak R, Soni B, Muppalaneni NB (2023) Role of blockchain in health care: a comprehensive study. In: Gunjan VK, Zurada JM (eds) Proceedings of 3rd international conference on recent trends in machine learning, IoT, smart cities and applications. Lecture notes in networks and systems, vol 540. Springer, Singapore. https://doi.org/10.1007/978-981-19-6088-8_13

22. Rama Santosh Naidu P, Lavanaya Devi G, Kondapalli VR, Neelapu R (2022) A novel and self adapting machine learning approach of ECG signal classification in association with cardiac arrhythmia. In: Proceedings of the international conference on computer vision, high performance computing, smart devices and networks: CHSN-2020. Springer Nature Singapore, Singapore, pp 195–207

23. Moschovitis P, Denisova A Keep calm and aim for the head: biofeedback-controlled dynamic difficulty adjustment in a horror game. IEEE Trans Games. https://doi.org/10.1109/TG.2022.3179842

Parallel Implementation of PageRank Based K-Means Clustering on a Multithreaded Architecture

Eedi Hemalatha

Abstract PageRank is the most famous algorithm used for ranking of the search results which has been developed and used by the most famous search engine Google. It is used to order the list of web pages which are specific to the query. This algorithm's simplicity makes it generic and a very powerful tool that can be used in a broad range of various applications. PageRank was most importantly designed to use the structure of a graph to measure the importance of nodes in the graph. It is this intuition that drives it to find its application in another most widely used algorithm, K-Means clustering algorithm. K-Means algorithm uses the centroids to cluster the nodes present in the graphs. PageRank can be used to find these centroids, that is, to find the nodes that have high centrality measure to cluster the nodes. Our work uses the PageRank as the method for finding the node that has the highest centrality for a particular cluster and uses it as the centroid to find the clusters for new iteration in K-Means algorithm. It uses Voronoi cells to determine the clusters iteratively using the centroids formed by the PageRank algorithm. We use the NetworkX PageRank module to determine the K center nodes that take the NetworkX graph as the input. The NetworkX module Voronoi cells is used to find the Voronoi cells, that is, clusters of the graph, which takes the centroids and NetworkX graph as input. The datasets used in this project are social circles from Facebook, Wikipedia who-votes-on-whom network, email communication network from Enron, Gnutella peer-to-peer network from August 4, 2002, and from August 6, 2002. Our methodology implements clustering of graphs using K-Means and PageRank which greatly improved the performance over the standard K-Means clustering for graphs. Our methodology achieved a considerable speed up by implementing the algorithm in a parallel paradigm. With parallel implementation, the speed up achieved on the above datasets ranged from 1.33 to 4.87 ms.

Keywords PageRank algorithm · K-Means clustering algorithm · Voronoi cells · NetworkX · Multithreading · Shared memory architecture

E. Hemalatha (✉)
Department of Computer Science and Engineering, JNTUH University College of Engineering, Science and Technology, Hyderabad, Telangana, India
e-mail: hemamorarjee@jntuh.ac.in

© The Author(s), under exclusive license to Springer Nature Singapore Pte Ltd. 2023
A. Kumar et al. (eds.), *Proceedings of the 4th International Conference on Data Science, Machine Learning and Applications*, Lecture Notes in Electrical Engineering 1038,
https://doi.org/10.1007/978-981-99-2058-7_23

1   Introduction

There has been rapid growth in the data that is getting generated every day, and the information that can be extracted from this data is also growing tremendously, but the importance that this information holds is also very high. This knowledge gained can be used in various domains to analyze what improvements need to be made, what more enhancement can be brought up, and what new dimension can be developed. There are various ways of representing data and extracting information from this data. One such way of representing data is using graphs [1]. To extract useful knowledge from this data, we can use a very famous graph analysis method called graph clustering [2].

In this project, an algorithm has been developed using K-Means clustering algorithm [2] and making few modifications to improve the performance. Voronoi cells have been used to create new clusters based on the centroids [3]. To generate new centroids, PageRank has been used. This algorithm has been used for unweighted graphs. This algorithm has the scope of applying parallelism. Hence, this algorithm is also implemented using a parallel paradigm using the multithreading concept. Using PageRank implementation in a multithreaded environment gives an increase in the performance for the standard K-Means clustering algorithm [2, 4].

Using the clusters generated by this algorithm, analysis can be made on the graph which has been clustered. Since graphs are used to model a wide variety of domains, the information inferred from clusters generated can vary from domain to domain. But the main objective is to generate clusters that can be used to group nodes that have similar properties, and these properties can be used to make useful decisions. Also to generate these clusters, this algorithm is designed to perform faster using modules that take advantage of the underlying structure of the graphs.

1.1   *Graph and Its Applications*

In the real world, there are various problems that are represented in terms of entities and connections between them [4]. For instance, in an airline route map, we might be interested to know what is the fastest route to go from Bangalore to Delhi? or what is the cheapest way to go from Bangalore to Delhi? To answer these questions, we will need information about the connections between entities related to the context. Graphs are the data structures that are used for analyzing and solving such problems [1, 4].

By definition, a graph is a pair (V, E) where V represents a set of entities or nodes called vertices and E represents a set of connections between these nodes, and it is a collection of pairs of nodes, and they are called edges. Graphs come in a variety of forms depending on how they are categorized. Graphs can be categorized in one way based on the direction of their edges. Directed graphs, like routed networks, are those in which the edges have directions. They are known as undirected graphs, such as the flight network, if the edges are undirected. The graphs are separated into weighted and unweighted graphs if the weights are taken into account. Weights are typically used to depict costs and distances in weighted graphs. Graphs are divided into sparse and dense graphs based on the quantity of edges. For a given number of vertices, sparse graphs have a small number of edges. Dense networks have a small number of missing edges for a given set of vertices [1, 4].

Every abstract data type has a representational form to use the data structure in a useful way. Similarly, graphs have three representational forms, which are Adjacency Matrix, Adjacency List and Adjacency Set. Based on the definition of requirement and feasibility, graphs can be represented using any of these three forms [1, 4].

To solve the problems involving graphs, there needs to be a mechanism for traversing the graphs. Graph traversal algorithms are also termed as graph search algorithms. The graph traversal algorithms involve starting at one of the source nodes and then searching the entire graph by going through the edges and marking the vertices. There are two ways of graph traversal. They are: Breadth First Search and Depth First Search [1, 4].

Graphs can be used in a wide range of real-world applications, since they are used to represent real-world objects and their relationships. The domain includes transportation networks such as highway networks and flight networks, computer networks such as local area network, internet and web and electronic circuits wherein the components and their relationships are represented using graphs [1, 4].

1.2 Importance of Clustering Graphs

Clustering [2] is the process of forming groups of items or entities that are closer to each other than from remaining entities based on some similarity measure. Detecting elements with such similar properties is of great importance, where it is very important and crucial to find specific structures or patterns very quickly.

Graphs have been used to model many types of interactions such as social interactions between people, communications between the objects of computer networks and relations between biological species [2, 4]. Although there are types that do not appear to be graphs, these types of data can be converted into graphs using certain operations, and analysis can be performed on these graphs to retrieve useful information on the underlying data. To perform such analysis, understanding the relationships present in the data is important. This can be done by clustering the graphs [3].

Graph clustering are divided into two categories: node clustering and graph clustering [5]. In case of node clustering algorithms, a single graph is clustered, which means the nodes are clustered given a single graph as an input. In case of graph clustering algorithms, more number of graphs available are clustered based on the underlying behavior of the graph structure. Clustering graphs and graph nodes are a challenging task due to the dynamic changing behavior of the graph with varied structural properties [5].

Graph clustering is becoming predominant in various fields of engineering, physics, image processing, social sciences and the medical field. As the graphs are massive in nature due to millions of vertices and trillions of edges, it is essential to apply partitioning and clustering techniques before processing the graph. It also helps to analyze and examine the graph cluster individually [5].

1.3 K-Means Clustering Algorithm

Generally, K-Means is an unsupervised learning algorithm which is used for clustering and solves multiple problems present in Data Science and Machine Learning fields. It is used to group the unlabeled dataset into various clusters. In the name K-Means, K represents the clusters count that are supposed to be created by the algorithm; this value is predefined. It allows to group the data into different clusters and is a suitable way of discovering different categories of groups present in the unlabeled dataset on its own without the need for any of the training [2].

K-Means is a centroid-based algorithm, in which every cluster has a centroid corresponding to it. The objective of the algorithm is to minimize the sum of distances between the data points present in the clusters. The algorithm takes the dataset that is unlabeled as the input and also the value of K which is predefined and divides the dataset into K number of clusters and performs this process repeatedly until the best clusters are not found [2].

1.4 PageRank

PageRank is a method for determining the ranks of the web pages that was developed by the founders of Google Larry Page and Sergey Brin at Stanford University [6].

In the case of graphs, the PageRank function can be defined on the nodes of a graph as a centrality measure. For a directed graph $G(V, E)$, the PageRank function $PR:V \rightarrow R$ is defined for each and every vertex v belonging to V as:

$$PR(v) = \frac{1-\alpha}{n} + \alpha \sum_{u_j} \frac{PR(u_j)}{Out\_deg(u_j)},$$

where n is the number if nodes in the given graph, G is the set of neighbor nodes of node u_j, $Out\_deg(u_j)$ is the set of nodes connected to u_j by out edges leaving u_j, and $0 < \alpha < 1$ is called the damping factor, which is typically set to 0.85.

1.5 Contributions

In this project, the contribution made was to utilize the PageRank algorithm efficiently into the clustering algorithm. In Sect. 4.1, a sequential algorithm is proposed which implements the clustering of nodes of the graphs by utilizing the PageRank feature for finding the mean of the cluster in the generic K-Means clustering algorithm which ensures that it uses the underlying structure of the graph and then finds the node with high centrality measure which more efficiently represents a central node of a cluster. The other contribution made is to implement this sequential algorithm using a parallel paradigm in Sect. 4.2. Multithreading concept was utilized to achieve parallelism. The main contribution made is to show the speed up in Sect. 6 achieved from sequential to parallel algorithms which is necessary for clustering large graphs.

2 Related Work

K-Means clustering algorithm [2, 7] can also be generalized to be used for clustering the graphs. The nodes of the graphs can be considered as the data points used in K-Means clustering algorithm. The graph can be clustered by predefining the number of clusters and iteratively finding the center nodes and refining the clusters. There has been various works done on implementing clustering of nodes of the graphs using K-Means clustering algorithm. In the graph clustering technique K-Means, PageRank has been substituted for mean [6–8]. The intention behind the construction of the PageRank algorithm, which is to assign an importance score to each online page that may be viewed as a node in the expansive web graph, was one of the key intuitions behind its use. The websites were then ranked according to this aspect. This technique makes use of the web's vast underlying graph structure. The PageRank vector can be defined for both directed and undirected graphs, and this was designed to be computed on massive graphs. Hence, PageRank is utilized in this project to gain additional speed. The works include adaptations of the K-Means algorithm to community detection in parallel environments [9], K-Means-based approaches to clustering nodes in annotated graphs [10] and Adapting K-Means for graph clustering [11]. Because of its simplicity and generality, this algorithm has been used in this project to form the basis.

3 Sequential K-Means Clustering Algorithm Using PageRank

In this implementation, initially a network graph is created from the input. The value of k is predetermined. For this graph G, the Adjacency Matrix is determined to use it when it is required to find whether an edge exists between two pairs of nodes. Initial cluster center nodes are determined. In this algorithm, the initial cluster nodes are determined randomly [7, 8].

Then the iteration starts, the termination criteria chosen in this algorithm are when the new cluster center nodes and initial center cluster nodes become similar or when the number of iterations has reached a predefined value. In each iteration, the initial cluster nodes are initialized with previous cluster nodes. Then the Voronoi cells are determined, which takes the graph and the center nodes as the arguments and returns the partition of the graph, that is, it returns the clusters of the graph based on the center nodes. The Voronoi cells partition the graph based on the center nodes that are given to it as input.

Input:

    Graph G (with N nodes)
    k = number of clusters

Algorithm:
center_nodes = initial random set of cluster centers
WHILE termination condition is not met
DO

    initial_center_nodes = center_nodes
    clusters = voronoicells(G, center_nodes) //returns partition of graph based on cluster center nodes
    count = 0
    FOR i in cells

        create subgraph gi of cluster cell i
        pagerank[] = pagerank(gi,0.85) // finding the pagerank value of each and every node present in the subgraph
        max = max(pagerank[]) // finding the node with max pagerank value
        center_nodes [count] = max
        count = count + 1

    IF initial_center_nodes == center_nodes

        break

    return cluster

In every iteration when the center nodes are refined, the clusters formed also get refined. The Voronoi cells method gives the dictionary as output with each center node as key, and their respective cluster or related partition centered around that center node is the value. This output then should be formatted to form the subgraph. Each partition is formed as a subgraph based on the edges present in the main graph. Then for each cell or each cluster formed by Voronoi cells, the subgraph that is determined is given as input to the PageRank method. This PageRank method determines the new center node which has the highest centrality based on the edges that are relationships present in the graph. In this way, the node with the highest centrality is chosen as new center nodes to represent the cluster. This process repeats until the newly formed cluster center nodes and the initial center nodes are converged or the number of iterations reaches the predefined value [8].

This algorithm uses Voronoi cells to efficiently determine the clusters based on the central nodes that have the highest centrality measure, and the PageRank method determines the nodes that have the highest centrality measure. These two methods recursively are used to obtain the refined clusters.

3.1 Parallel K-Means Clustering Algorithm Using PageRank

Input:

Graph G (with N nodes)
k = number of clusters

center_nodes = initial random set of cluster centers
WHILE termination condition is not met
DO

initial_center_nodes = center_nodes
clusters = voronoicells(G, center_nodes) //returns partition of graph based on cluster center nodes
count = 0
FOR i in cells

t = threading.Thread(target = ClusterCenter,args = (i,center_nodes,count))
t.start()

threads.append(t)

count = count + 1
FOR thread in threads:

thread.join()

IF initial_center_nodes = = center_nodes

break

return cluster
ClusterCenter(i,center_nodes,count):

create subgraph gi of cluster cell i
pagerank[] = pagerank(gi,0.85) // finding the pagerank value of each and every node present in the subgraph
max = max(pagerank[]) // finding the node with max pagerank value
center_nodes [count] = max
return

In this implementation, similar to sequential implementation, initially a network graph is created from the input. The value of k is predetermined. For this graph G, the Adjacency Matrix is determined to use it when it is required to find whether an edge exists between two pairs of nodes. Initial cluster center nodes are determined. In this algorithm, the initial cluster nodes are determined randomly.

Similar to sequential implementation the iteration starts [12], the termination criteria chosen in this algorithm are when the new cluster center nodes and initial center cluster nodes become similar or when the number of iterations has reached a predefined value. In each iteration, the initial cluster nodes are initialized with previous cluster nodes. Then the Voronoi cells are determined, which takes the graph and the center nodes as the arguments and returns the partition of the graph, that is, it returns the clusters of the graph based on the center nodes. The Voronoi cells partition the graph based on the center nodes that are given to it as input.

In every iteration when the center nodes are refined, the clusters formed also get refined. The Voronoi cells method gives the dictionary as output with each center node as key, and their respective cluster or related partition centered around that center node is the value. This output then should be formatted to form the subgraph. Each partition is formed as a subgraph based on the edges present in the main graph. Then for each cell or each cluster formed by Voronoi cells, the subgraph that is determined is given as input to the PageRank method. In the parallel implementation, the cells are processed parallely using multithreading where in each cell is handled by a separate thread. Each center node is determined simultaneously thereby reducing the time to wait for each cell to determine the center nodes after others have determined in an iterative manner.

This PageRank method determines the new center node which has the highest centrality based on the edges that are relationships present in the graph. In this way, the node with highest centrality is chosen as new center nodes to represent the cluster. This process repeats until the newly formed cluster center nodes and the initial center nodes are converged or the number of iterations reaches the predefined value.

By utilizing multithreading, the time to determine new center nodes for each cluster can be divided among multiple threads, and the new center nodes can be calculated simultaneously which gives an improvement in the performance. This way parallel implementation can be used to achieve speed. Improved parallel implementation with sophisticated famous algorithms can be used to achieve more speed and performance improvement. This can be done in the future implementations.

Data set	Nodes	Eges
Wikipedia voting on promotion to administratorship	7115	103689
Direcred Gnutella P2P network from August 6 2002	8717	31525
Direcred Gnutella P2P network from August 4 2002	10876	39994
Email communication network from Enron	36692	367662
Social circles from facebook	4038	88221
email-Eu-core	1005	25571

Fig. 1 Dataset list

4 Implementation Details

4.1 Platform

Platform used is Param Shavak, an HPC system which has 2 Intel Xeon (R) Gold 6145 CPU 2.00 GHz having 40 threads each, having a memory of 92.9 GB and has CentOS Linux 7 on it as the operating system. Python is used as a programming language, and a threading library is imported to support multithreading.

4.2 Dataset

For both sequential and the parallel versions, real-world datasets were used from "Stanford Large Network Dataset Collection" [13] including small graphs to large graphs. The graphs used are unweighted graphs. Most of the graphs used are directed graphs, i.e., nodes in the graphs point from one node to another can only send and receive information in one direction. A few of them are undirected graphs, i.e., nodes in the graphs can receive and send information from their connected nodes. The datasets used in this project are:

In Fig. 1, the list of datasets used in this project is presented. There are six datasets used. The names of the datasets and nodes present in each of the dataset along with the number of edges are listed in the above figure.

4.3 Dataset Description

The dataset taken from SNAP [13–16] consists of two columns and rows equal to the number of edges present in the graphs. Each row represents an edge from the first column node of the row to the second column node of the row.

4.4 Libraries Used

NetworkX [14, 15] is a Python-based framework implemented for the design and exploitation of the dynamics, structure and use of complex networks. NetworkX is used in the study of complex networks structures, represented as graphs. We are able to create a variety of random and conventional networks, study their structure, generate network models, create new network methods and even sketch them.

NETWORKX VORONOI CELLS,
voronoi_cells(G, center_nodes, weight = 'weight')[source]

Returns the Voronoi cells centered at center_nodes with respect to the shortest-path distance metric. If C is a set of nodes in the graph and c is an element of C, the Voronoi cell centered at a node c is the set of all nodes v that are closer to c than to any other center node in C with respect to the shortest-path distance metric.

5 Result Analysis

This algorithm was run on six datasets using both sequential and parallel implementations which included 1000, 5000, 8000, 10,000 and 36,000 nodes, and these were the observations made from the output obtained.

Figure 2 shows the results obtained by running the sequential algorithm on the datasets. It shows the time taken to run each of these datasets. The runtime is given in seconds.

Figure 3 shows the results obtained by running the parallel algorithm on the datasets. It shows the time taken to run each of these datasets. The runtime is given in seconds.

Figure 4 shows the results obtained by running the standard K-Means algorithm for clustering the graphs on the datasets. It shows the time taken to run each of these datasets. The runtime is given in seconds.

Nodes	Sequnetial Runtime(secs)
email-Eu-core	14.3056
facebook_combined	142.6804
Wiki-Vote	435.5041
p2p-Gnutella06	425.3163
p2p-Gnutella04	561.6017
Email-Enron	5945.0804

Fig. 2 Sequential algorithm runtimes

Nodes	Parallel Runtime(secs)
email-Eu-core	6.5189
facebook_combined	85.8164
Wiki-Vote	156.2729
p2p-Gnutella06	319.4566
p2p-Gnutella04	242.7493
Email-Enron	1222.0274

Fig. 3 Parallel algorithm runtimes

Nodes	Standard Runtime(secs)
email-Eu-core	90.8345
facebook_combined	700.3256
Wiki-Vote	1389.2399
p2p-Gnutella06	1582.2087
p2p-Gnutella04	1683.4524
Email-Enron	6103.7754

Fig. 4 Standard K-Means algorithm runtimes

Figure 5 shows the bar chart for the sequential algorithm. It is plotted against the number of nodes and the runtime of the algorithm for a particular dataset.

Figure 6 shows the bar chart for the parallel algorithm. It is plotted against the number of nodes and the runtime of the algorithm for a particular dataset.

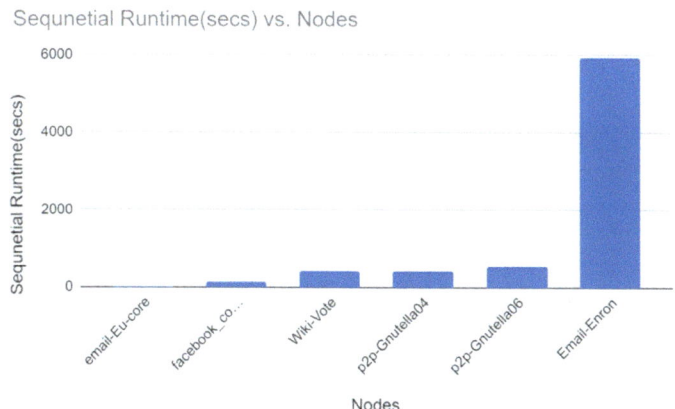

Fig. 5 Bar chart for sequential algorithm

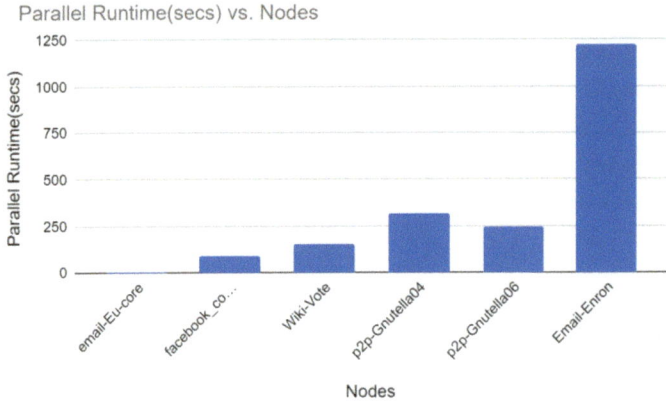

Fig. 6 Bar chart for parallel algorithm

Figure 7 shows the bar chart for the standard K-Means clustering algorithm for graphs. It is plotted against the number of nodes and the runtime of the algorithm for a particular dataset.

Standard K-Means Versus Sequential Versus Parallel Algorithms

From Fig. 8, it can be observed that there is a lot of improvement achieved using the PageRank algorithm as central measure and Voroni cells to iteratively generate the new clusters. This improvement has been escalated by using the parallel algorithm on top of this sequential algorithm.

Figure 8 shows the comparison of runtime between the K-means clustering algorithm using PageRank that runs sequentially, K-means clustering algorithm using PageRank that runs parallely and the standard K-Means clustering algorithm for graphs on the datasets. It shows that the sequential algorithm performs better than

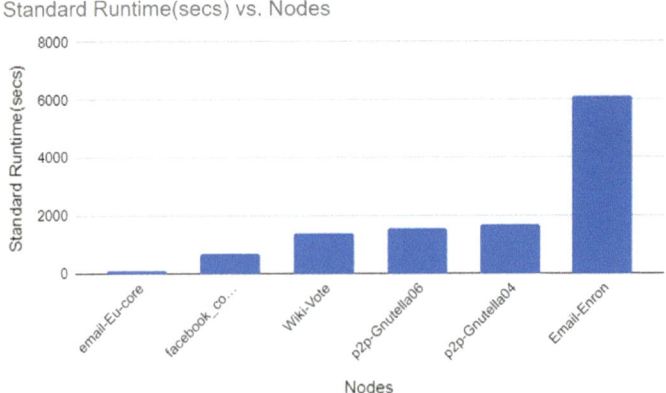

Fig. 7 Bar chart for standard K-Means algorithm

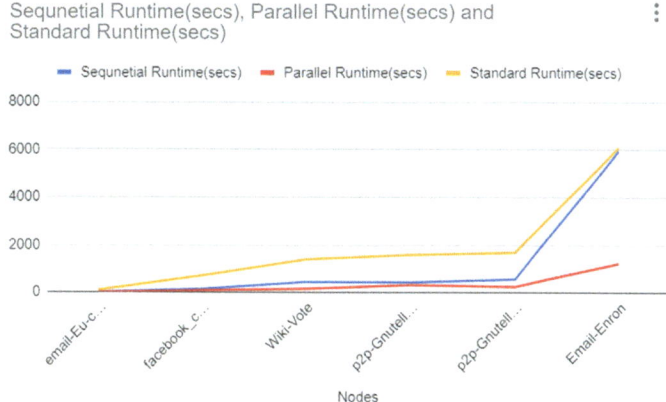

Fig. 8 Standard K-Means versus sequential versus parallel

the standard algorithm, and the parallel algorithm performs better than the sequential K-Means clustering algorithm using PageRank.

Speed-Up

Speed-up achieved is the fraction of speed that we have gained from parallelizing the sequential algorithm. Simply, it is the ratio of sequential runtime to the parallel runtime.

Figure 9 shows speed-up achieved on implementing parallelism on the K-Means clustering using PageRank for graphs. It shows that there has been 1.3–4.8 times improvement made from sequential K-Means clustering algorithm using PageRank to parallel algorithm.

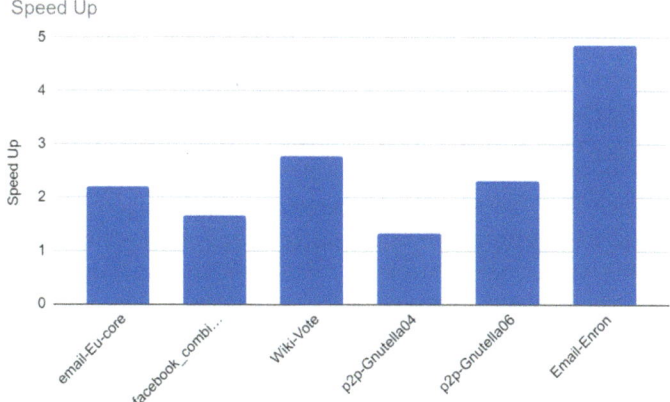

Fig. 9 Parallel versus sequential speed-up

Fig. 10 Sample output

Sample Output

The results shown in Fig. 10 show the cluster center nodes for each iteration, and once the iteration count reaches the predefined value or when the initial cluster nodes are similar to new cluster nodes, the algorithm terminates. The figure shows the final K centers and the final K clusters obtained centered around these K center nodes.

The dataset used in this above is email-Eu-core network. It consists of around 1000 nodes. There is an edge (u, v) in the network if person u sent person v at least one email. The collection does not include any incoming or outgoing emails to or from the rest of the world; instead, the emails exclusively represent communication among institution members (the core). The output shows the clusters of the nodes which frequently send emails to each other. There are 15 clusters formed in this output.

6 Conclusion

With increasing data every day, there is a need to analyze this data and use it efficiently to utilize the information obtained by the analysis to understand the data and use it to its full capacity. This project aimed to provide one such contribution by implementing an algorithm to cluster the large network graphs that are used to represent the real world. The method defined in this project utilizes PageRank and K-Means to cluster the graphs efficiently. The parallel implementation of the algorithm has also been performed to achieve performance improvement. The K-Means clustering algorithm using PageRank implemented sequentially and K-Means clustering using PageRank implemented parallelly are compared with the standard K-Means clustering algorithm for graphs as well. It shows that using PageRank as a measure to find the mean of the cluster improved the performance of the standard algorithm. This has a lot of significance in case of massive graphs. The parallel implementation showed better performance than the sequential algorithm. Speed-up achieved was from 1.33 to 4.8 for various datasets used. Hence, utilizing the feature of the PageRank that uses the underlying structure of graphs ensures that the clusters formed have quality, and applying parallelism gives speed-up. The centrality measure used in this project is PageRank; other centrality measures can also be used and tested. The comparison of all the centrality measures used for clustering the graphs can be made to determine the better measure to be used.

References

1. Kolhe NP, Dhote KK, Parihar A (2023) Mapping the access of internet facility to schedule tribe population—an approach towards digital education. In: Digital learning based education: transcending physical barriers. Springer Nature Singapore, Singapore, pp 1–20
2. Muthmainnah, Ganguli S, Al Yakin A, Ghofur A (2023) An effective investigation on YIPe-learning based for twenty-first century class. In: Choudhury A, Biswas A, Chakraborti S (eds) Digital learning based education. Advanced technologies and societal change. Springer, Singapore. https://doi.org/10.1007/978-981-19-8967-4_2
3. Prasad PS, Sunitha Devi B, Janga Reddy M, Gunjan VK (2019) A survey of fingerprint recognition systems and their applications. Lecture notes in electrical engineering, vol 500, pp 513–520
4. Al Yakin A, Ganguli S, Cardoso L, Asrifan A (2023) Cybersocialization through smart digital classroom management (SDCM) as a pedagogical innovation of "Merdeka Belajar Kampus Merdeka (MBKM)" curriculum. In: Digital learning based education: transcending physical barriers. Springer Nature Singapore, Singapore, pp 39–61
5. Rudra Kumar M, Pathak R, Gunjan VK (2022) Diagnosis and medicine prediction for COVID-19 using machine learning approach. In: Computational intelligence in machine learning: select proceedings of ICCIML 2021. Springer Nature Singapore, Singapore, pp 123–133
6. Das N (2023) Digital education as an integral part of a smart and intelligent city: a short review. Digit Learn Educ Transcending Phys Barriers 81–96
7. Rashid E, Ansari MD, Gunjan VK, Ahmed M (2020) Improvement in extended object tracking with the vision-based algorithm. In: Modern approaches in machine learning and cognitive science: a walkthrough: latest trends in AI, pp 237–245
8. Nandy D, Majee D (2023) Socio-economic relations in digital education: a comparative study between Bangladesh and Nepal. In: Digital learning based education: transcending physical barriers. Springer Nature Singapore, Singapore, pp 103–117
9. Lakshmanna K, Shaik F, Gunjan VK, Singh N, Kumar G, Shafi RM (2022) Perimeter degree technique for the reduction of routing congestion during placement in physical design of VLSI circuits. Complexity 2022:1–11
10. Usman M, Wajid M, Shamim MZ, Ansari MD, Gunjan VK (2021) Threshold detection scheme based on parametric distribution fitting for optical fiber channels. Recent Adv Comput Sci Commun 14(2):409–415
11. Galluccio L, Michel O, Comon P, III Hero AO (2012) Graph based k-means clustering. Signal Processing 92(9):1970–1984
12. Bouazzi I, Zaidi M, Usman M, Shamim MZM, Gunjan VK, Singh N (2022) Future trends for healthcare monitoring system in smart cities using LoRaWAN-based WBAN. Mob Inform Syst 2022
13. Erwig M (2000) The graph voronoi diagram with applications. Networks 36(3):156–163
14. Vijayalakshmi R, Rajiah P, Lakshmi Sangeetha A, Balaji Ganesh A (2022) Sleep quality analysis using motion signals and heart rate. In: Proceedings of the international conference on computer vision, high performance computing, smart devices and networks: CHSN-2020. Springer Nature Singapore, Singapore, pp 41–49
15. Kotha MK, Pavan KK (2022) Deep learning for object detection: a survey. In: Proceedings of the international conference on computer vision, high performance computing, smart devices and networks: CHSN-2020. Springer Nature Singapore, Singapore, pp 61–84

16. Pathak R, Gupta SS (2020) A study on natural computing: a review. In: ICDSMLA 2019: proceedings of the 1st international conference on data science, machine learning and applications. Springer Singapore, Singapore, pp 1975–1983

ZACube-2 Mission Operations Analysis

Gregory J. Naidoo, Robert van Zyl, and Gunjan Gupta

Abstract This research entails a comprehensive analysis of ZACube-2 operational mission data which addresses on-orbit operations. The mission operations evaluation criteria that will be investigated include all available telemetry, tracking, and command datasets at system and subsystem levels, with the orbital dynamics of the system and the ground segment operations. An auxiliary analysis will investigate space weather and its effects on the mission about how the system reacts in the presence of solar activity over particular areas of the Earth, which will provide an assessment of overall system robustness. The research output will give an objective evaluation of the operational performance of the system comprising space and ground segments against the mission design specifications. It will compare the as-designed features of the satellite with actual system performance. Design improvements will be recommended for incorporation in the design cycle of future missions.

Keywords Telemetry · ZACube-2 · Mission operations · System performance · Mission analysis · Orbital dynamics

1 Introduction

Long-range radio (LoRa) is mainly used for machine-to-machine (M2M) and the Internet of Technology (IoT); however, they are gaining attention for satellite communication [1–3]. In [4], a survey of satellites launched, and timelines have been conducted. In [5–7], the launch of satellites with different payloads and controller types is compared. The ZACube-2 mission is the second satellite developed in the Cape Peninsula University of Technology (CPUT) Satellite Programme, hosted by French South African Institute of Technology (F'SATI) with funding from the Department of Science and Innovation. The satellite is a 3-unit (3U) CubeSat and

---

G. J. Naidoo (✉) · R. van Zyl · G. Gupta
French South African Institute of Technology, Department of Electrical, Electronics and Computer Engineering, Cape Peninsula University of Technology, Cape Town, South Africa
e-mail: naidoogr@cput.ac.za

A. Kumar et al. (eds.), *Proceedings of the 4th International Conference on Data Science, Machine Learning and Applications*, Lecture Notes in Electrical Engineering 1038, https://doi.org/10.1007/978-981-99-2058-7_24

was launched in 2018 with the prime objective of providing sovereign maritime-domain awareness services in support of Operation Phakisa, an initiative supporting the National Development Plan (Operation Phakisa). A novel fire detection imager was also demonstrated as a secondary mission objective.

The technology and mission roadmap of the CPUT Satellite Programme aims to continually support national imperatives, such as the National Development Plan. The immediate development is toward a constellation of satellites that deems to complement and further develop the maritime communications services, as demonstrated by ZACube-2. The mission, therefore, is of critical strategic importance to the South African space industry as it serves as the baseline design for future missions and programs.

To this end, evaluating the satellite's operations against its primary mission objectives and to validate and improving the design approach that will be incorporated in future missions is imperative. By doing so, more effective and reliable methods across the full-design lifecycle of the mission can be adopted, contributing to the longevity of the South African space industry.

2 Methodology for Resource Budgets Analysis

The purpose of this analysis is to validate the various budgets that were used to design ZACube-2 operations. These budgets include power, thermal, mass, and link budgets, and the mass budget does not require verification through telemetry analysis.

Validation of the various budgets is essential as it guides the development of the next iteration of CubeSats, relaying information on areas of improvement. A prime example would be comparing different power profile versions, with one being calculated or simulated and the other being calculated directly from the telemetry.

2.1 Power Budget

ZACube-2 has four power profiles describing the calculated power used when that profile is activated, either when the CubeSat is in a minimal solar condition or experiencing the maximum eclipse (Table 1).

To calculate the telemetry-based values for the various profiles, a random day when the profile activation criterion was met was chosen, and the average power was calculated in the period in which the respective profile was active (Table 2).

The subsystems that are vital in the basic operation of ZACube-2 require power to operate

- UHF transceiver.
- Electronic power supply.
- Battery.

Table 1 ZACUBE-2 power profiles description

Power profile	Description
Safe sun/eclipse	Basic safe mode with most of the bus in idle and payload off
Safe TTNC sun/eclipse	Safe mode with TTNC session and most of the bus in idle and payload off
Downlink sun/eclipse	Downlink mode with TTNC session, data downlink, and most of the bus in idle and payload off
Payload sun/eclipse	Payload operations with TTNC session, data downlink, and most of the bus in idle and payload on for one pass

Table 2 ZACUBE-2 power profile condition

Power profile	Non-vital subsystem active
Safe sun	Only the vital subsystem
Safe TTNC sun	S-band transmitter
Downlink sun	S-band transmitter, UHF transceiver
Payload sun	Software-defined radio, K-line imager, S-band transmitter

- On-board computer.
- Attitude determination control system.

In addition to validating the power profiles, analysis of both the battery and the solar array is necessary as these two subsystems are essential in ensuring power to the spacecraft.

For the battery voltage, close inspection will be taken to the depth of discharge, which is in the power budget and, following the manufacturer's recommendations, was described at a maximum of 20%, which for an 8.4 V capacity battery is 6.72 V shown in Fig. 1.

Various "profiles" were described for the solar array; however, these profiles were configurations with each configuration being a scenario on what panels were illuminated and how much power was being produced. The minimum power generated was described as 6.4717 Wh shown in Fig. 2. To calculate this, the same method that was used to calculate the power profiles is used.

2.2 Thermal Budget

The thermal budget is used to simulate the thermal working conditions at which the satellite will operate and verify if there is enough thermal headroom for the various subsystems to operate effectively and efficiently.

To simulate this environment, two methods were employed, namely [8, 9], respectively, with the latter being the most commonly used in industry. While these two

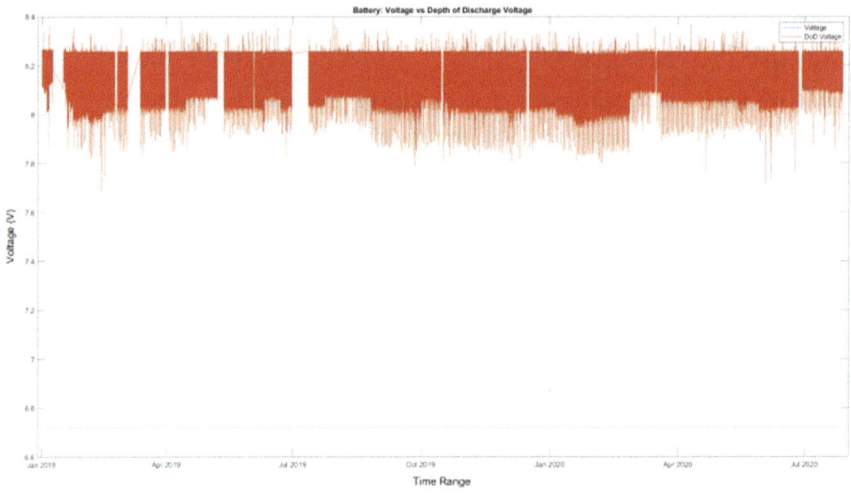

Fig. 1 Battery voltage versus depth of discharge voltage

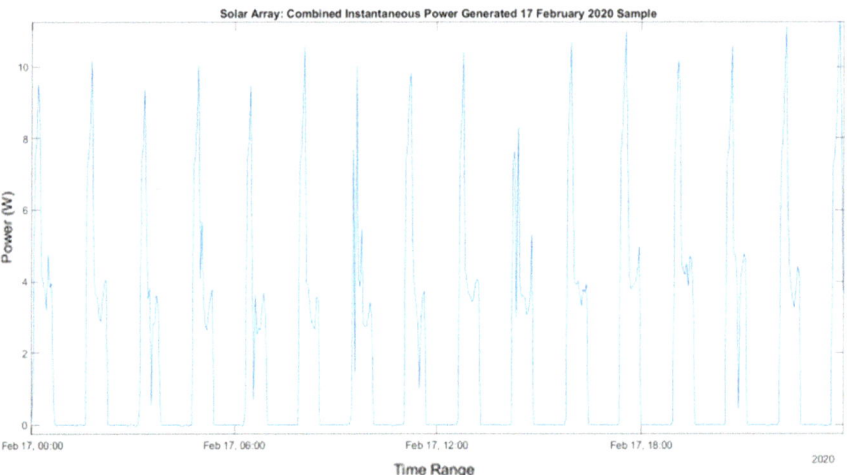

Fig. 2 Solar array combined instantaneous power generated February 17, 2020, sample

methods are similar, the methodology described in [8] is more comprehensive as it is based on actual telemetry instead of being altitude and inclination dependent.

In the aforementioned method, both yield an approximate thermal balance when the satellite is sunlit and in eclipse conditions which makes the satellite experiencing maximum and minimal heat dissipation. Once the maximum and minimum temperatures were calculated, it was compared with individual thermal telemetry channels to check whether the telemetry exceeded the calculated temperatures and the manufacturer's thermal limits [10–14].

2.3   Link Budget

A link budget is a set of calculations to consider whether the satellite makes a good link with the ground station at various elevation angles, ensuring no packet loss. An accurate means of measuring the link integrity is to analyze the RSSI or received signal strength indicator telemetry [15].

The ground station could communicate is already known as data was being sent and received, and the only things need to be learned are the performance of the link. To verify this, the RSSI telemetry, which is given in voltage, was cross-referenced with Table 3 that showed the RSSI value signal strength equivalent [16].

2.3.1   Resource Budgets Results

After completing the power, thermal, and link budget, the telemetry-based results were compared with the designed resource budgets.

2.3.2   Power Budget

In Table 4, various power profiles are compared. It should be noted that there is telemetry data for the eclipse profile due to no operation.

Table 3 UHF RSSI voltage value to signal strength

RF level (dBm)	Voltage (V)		
	Minimum	Typical	Maximum
− 118	N/A	0.3	0.8
− 68	0.7	1.1	1.8
− 23	1.2	1.8	2.5

Table 4 Comparison of power profiles

Profile name	Power consumed (Wh)		
	Power budget	Telemetry	Difference
Safe sun	1.759	1.705	0.054
Safe eclipse	1.068	1.049	0.019
Safe TTNC sun	2.576	2.101	0.354
Safe TTNC eclipse	1.885	1.583	0.274
Downlink sun	3.576	2.848	0.728
Downlink eclipse	1.432	1.408	0.024
Payload sun	4.976	2.297	2.679
Payload eclipse	4.285	N/A	N/A

Table 5 Solar array configuration calculation results

Profile name	Power (Wh)
Calculated value from sample	3.3869
Sun on 3U	6.4717
Difference	− 3.0848

Table 6 Comparison of thermal telemetry limits for various subsystems

Parameter	Temperature (°C)		
	Maximum	Minimum	Range
Battery	31.9865	− 8.4731	40.4596
K-line imager	29.1500	− 15.1250	44.2750
S-band transmitter	37.2314	13.8672	23.3642
UHF transceiver	28.0000	− 11.0000	35.0000
Fortescue	49.2125	− 21.4518	80.6643
Gilmore	43.3290	− 21.0940	64.4230

The battery voltage, which has taken over the span of the available telemetry, was compared to the calculated maximum depth of discharge (DoD) described in the power budget, which is 6.72 V.

The solar panel average power is generated on a random day to see if enough power is available to operate both the spacecraft and charge the battery for operation when in eclipse.

Using the figure above, a value can be calculated of the average generated power when the satellites are in view of the sun. This telemetry value can then be compared to the calculated stipulated in the power budget. This can be seen in Table 5.

2.3.3 Thermal Budget

In Table 6, the maximum and maximum telemetry temperatures for the various subsystems are compared. This is done to show the thermal range that each respective subsystems experiences over temporal span that the telemetry was taken from.

2.3.4 Link Budget

A combination of a Gaussian probability function and a cumulative density function (CDF) was generated to verify the integrity of the link [17].

3   Discussion of Results

With the results being generated from the aforementioned data analysis, comments on the performance can be made on whether the telemetry metrics fit within the system's expectations derived from the resource budgets.

3.1   Power Budget

From Table 4, it can be seen that overall, the power budget values are higher than the ones calculated using the respective telemetry and within a 1 Wh margin. The most significant difference between the telemetry-derived values and the power budget can be seen when the payload sun is active, where only 46.1616% of the power specified in the power budget was used [18].

The reason behind the significantly lower power value was that during the calculation, the different subsystems had different data logging times, leading to the recording of power values in one subsystem, while the other subsystem is not logging data. While linear interpolation can be used to replace missing values, it is not an accurate representation of the data recorded; additionally, this brings up an issue that not all the subsystems data are logged at the same time interval [19, 20].

The battery used for ZACube-2 has a charge limit of 8.4 V; therefore, this equates to a minimum battery charge of 6.72 V or a discharge value of 1.98 V. However, the maximum and minimum values were found to be 8.3917 V and 7.6849 V, respectively, indicating that the maximum discharge voltage experienced was 0.7068 V which is far smaller than the specified discharge value of 1.98 V.

Maximum discharge, when calculated as a percentage of the maximum charge limit of 8.4 V, gives a value below 8.42%, and likewise, the minimum discharge was calculated as 8.51%. Both of these telemetry values are less than that of the maximum designed depth of discharge value of 20%.

The solar array calculation showed the instantaneous peak power generated is approximately between 8 and 8.5 W; however, according to the telemetry calculations, only about 52.3340% of the required power was generated when comparing to the value required in the power budget. The exact process was replicated using a different time period to perceive if the first instance was an anomaly. The results from these two sample calculations are seen in Table 7.

Table 7 Difference in solar array telemetry configuration

Sample number	Date of telemetry sample (UTC)	Power (Wh)
1	February 17, 2020	3.3869
2	March 15, 2019	4.4171
	Difference	1.0302

From Table 7, the second sample calculation is 130.4171% larger than the total power generated from sample 1 which as clarified before is only 52.3340% of the minimum required power with sample 2 value only being 68.2525% of the aforementioned minimum required power stated in the power budget. Two pieces of information can be extracted from the sample calculations and could explain as to why these values differ and/or not being close to the required power value.

The first piece of information is that from the two sample calculations, the power being generated by the solar array does not meet the minimum required power in the power budget. However, the battery's voltage depth of discharge should be greater than the value specified in the power budget, i.e., 20% if the solar array failed to meet this minimum required value. Yet, the calculated depth of discharge was found to be 8.42%, well within the 20% margin. In this case, telemetry data was logged at an interval of 3 min.

The second piece of information is that sample values variate at 1.0302 Wh. This variance can be accredited to the combination of two factors: solar declination and/or whether the sun is in solar minimum or maximum. Due to the Earth's rotational axis being off-center as the Earth rotates around the sun, the declination angle variates over the year. This affects the angle at which the solar rays hit the solar panels and in turn the power produced [21].

The second factor is whether the sun is in solar minimum or maximum also referred to as the solar cycle. The solar cycle is experienced by the sun where approximately every 11 years, the density of sunspots increases or decreases, respectively, where the number of sunspots directly affects the strength of the UV radiation and as such the power generated. In the telemetry range that was used to calculate the sample values, it was found that the sun experienced solar minimum, and subsequently, there was minimal number of sunspots, therefore less power being generated [22].

From Table 5, the power budget values are higher than the ones calculated using the respective telemetry and within a 1 Wh margin. The largest difference between the telemetry-derived values and the power budget can be seen when the payload sun is active, where only 46.1616% of the power specified in the power budget was used. While linear interpolation can be used to replace missing values, it is not an accurate representation of the data recorded; additionally, not linearly interpolating brings up an issue that not all subsystem data is logged at the same time interval [23].

3.2 Thermal Budget

From Table 6, which shows the available thermal telemetry and said range of the various respective subsystems, i.e., battery, K-line imager, S-band transmitter, and the UHF transceiver. What can be seen is that the S-band transmitter experiences the highest temperature of 37.2314 °C with the lowest temperature being experienced by the K-line imager of 15.1250 °C. These values co-inline with the operational behavior of the satellite as the S-band transmitter provides a high-speed downlink, which requires more power to the ground station, therefore incurring a higher thermal value.

While the K-line imager being the hosted payload is not used as frequently in comparison with the communications-based subsystem, therefore having a lower thermal value. While temperature can be somewhat be seen as a measure of how frequent and/or how much power said subsystem consumes, this is only an implied notion as multiple factors such as thermal control and operational scheduling play a vital role in which affects the subsystem temperature.

The results in Table 6, in particular the last two rows, additionally show that the method that employs calculation techniques from [9] has a larger thermal range when compared to the method that employs calculation techniques from [8], suggesting that both are valid approaches. The method described by Fortescue et al. [9] is more conservative than that described by Rashid et al. [8], e.g., the latter method produces results that are more in line with actual performance, and as such, there is not much thermal headroom [10].

Telemetry from the battery, UHF transceiver, and S-band transmitter showed that the respective subsystems thermally performed within the manufacturer's specifications with additional headroom. What this means practically is that the passive thermal control techniques employed on ZACube-2 worked as designed.

Using Tables 8, 9, and 10, the averaged thermal performance of the respective subsystem is well within the limits specified by the manufacturer, especially in the case when looking at the values described in Table 9, with there being more thermal headroom when the respective subsystems are experiencing maximum temperature in comparison with the minimum temperature thermal headroom.

Two approaches were taken to manually calculate the thermal range that would be experienced by ZACube-2, while it is vital to compare these approaches to one another, and a comparison between the two approaches and the thermal telemetry is needed as well.

From Table 11 in relation to the thermal telemetry from ZACube-2, the hand calculation that employs [8] method is more in line with the thermal telemetry

Table 8 Comparison of battery thermal telemetry and manufacturer specifications

Parameter	Temperature (°C)		
	Telemetry	Manufacturer	Difference
Maximum	31.9865	50	18.0135
Minimum	− 8.4731	− 10	1.5269

Table 9 Comparison of UHF transceiver thermal telemetry and manufacturer specifications

Parameter	Temperature (°C)		
	Telemetry	Manufacturer	Difference
Maximum	28	61	33
Minimum	− 11	− 25	14

Table 10 Comparison of S-band transmitter thermal telemetry and manufacturer specifications

Parameter	Temperature (°C)		
	Telemetry	Manufacturer	Difference
Maximum	37.2314	61	23.7686
Minimum	13.8672	− 25	11.1328

Table 11 Comparison of thermal hand calculation techniques in relation to temperature telemetry

	Thermal hand calculation temperatures (°C)			
	Fortescue		Gilmore	
	Minimum	Maximum	Minimum	Maximum
Thermal telemetry (°C)	− 21.4518	49.2125	− 21.0940	43.3290
Minimum	− 15.1250		− 15.1250	
Maximum		37.2314		37.2314
Difference	6.3268	11.9811	5.9690	6.0976

compared to the method described by Fortescue et al. [9] in addition to having a smaller thermal headroom when comparing maximum temperatures in relation to the telemetry temperatures.

3.3 Link Budget

When cross-referencing the values in Table 3 in relation to Figs. 3 and 4, it can be seen that the lowest RSSI value is 1V which means that anything below that value would be considered noise because the transceiver does not have the functionality to process signals weaker than − 116 dBm. Therefore, by looking at the x-axis of the UHF transceiver, RSSI cumulative probability function graph the percentage of the received signal which is noise can be determined.

By looking at the graph at 1 V, this correlates to a cumulative value of 89.1896%, what this means is that less than 89.1896% of signals processed by the transceiver is noise with only 10.8104% of the signals be interpreted or useful. This value can be seen as how well the transceiver works seeing as the ground segment did indeed receive the data from ZACube-2.

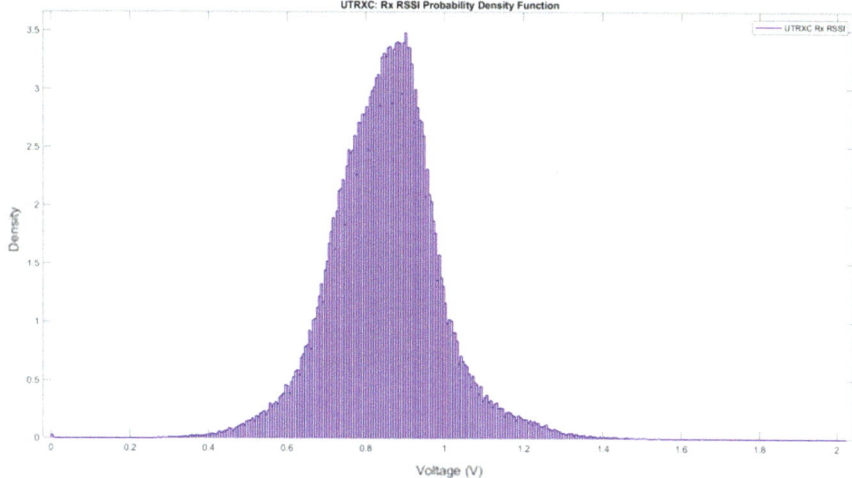

Fig. 3 UHF transceiver RSSI probability density function

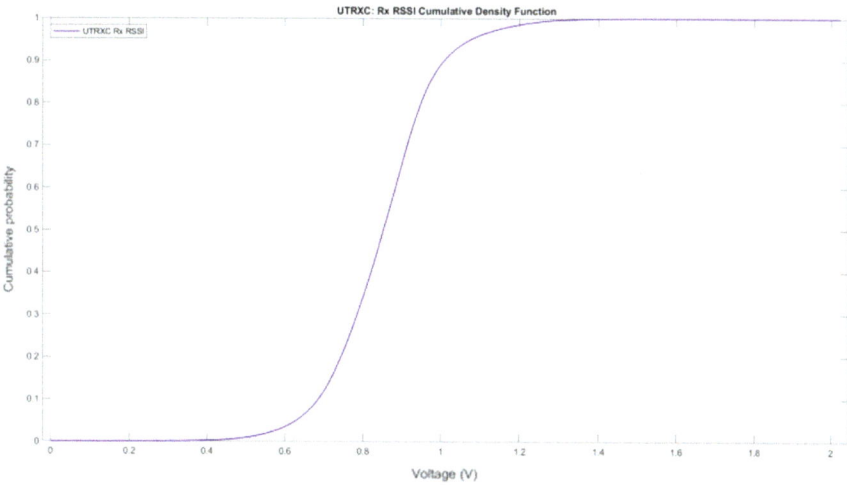

Fig. 4 UHF transceiver RSSI cumulative density function

4 Conclusion

Using the output of this assessment, design improvements can be recommended for incorporation in the design cycle of future missions. Therefore, not only improving the satellite regarding how effective and efficient mission operations are executed but also further improving upon the satellite design expertise of the engineering team.

From the discussions that were had in the previous sections regarding the performance at a systems-level, e.g., the various resource budget as well as at a subsystems level. Additionally, problems that arose will be a discussion from an objective view.

4.1 Power Budget

ZACube-2 had six power profiles and two power-related profiles which were the orbital periods and minimum power generated by the solar array. All six of the power profile values, apart from the payload eclipse as the imager was not operated in eclipse, were lower than the values calculated in the power budget suggesting that the power budget was conservative.

As for the conservative metric, on the lowest end of the spectrum, the safe eclipse telemetry-based value was only 1.7790% smaller, while on the highest end of the spectrum, the payload sun telemetry-based value was 53.8384% smaller than the power budget value. The reason why there was such a drastic difference in power level is that there was no common interval at which the telemetry data was logged causing missing respective values across the respective subsystems. This dissimilarity was accredited to the variating solar declination that is experienced throughout the year as the sample was calculated on March 15, 2019, UTC and February 17, 2020, UTC, respectively.

Furthermore, the difference could be a trait of the solar cycle progression as the range of telemetry value that was used in calculation occurred when at solar minimum, e.g., transition phase between solar cycle 24 and solar cycle 25.

With both telemetry sample calculations conveying that not enough power was being generated to sustain the needs of the satellite, there should be large amounts of discharge from the battery to provide that remaining power needed. However, battery discharge was minimal at 8.42% conveying that either method in which the power values were calculated is incorrect or the problem lies within the telemetry values.

4.2 Thermal Budget

From the telemetry, the largest thermal range experienced by a single subsystem was the K-line imager of 44.2750 °C with a maximum of 29.15 °C and a minimum of −15.1250 °C.

The S-band transmitter had the maximum overall temperature from any subsystem of 37.2314 °C with the K-line imager having the lowest temperature logged from any subsystem. All the subsystems performed well within the thermal limits specified by the manufacturer implying that passive thermal management techniques employed on ZACube-2 worked as desired leaving thermal headroom for either switching subsystems on longer, if there is power available, or if the stacking order of the subsystems needs to be changed.

Telemetry-based thermal values were compared to the two methods, the first method described by Maharaj et al. [2] and the second method described by Gupta and Van Zyl [1].

The first method had the largest thermal range of 80.6643 °C with the second method having a thermal range of 64.4230 °C this directly translated to the [1] variance in maximum and minimum values being within 6.0976 °C and 5.9690 °C, respectively, whereas [2] maximum and minimum values were 11.9811 °C and 6.3268 °C respectfully.

From this, it can be seen that to give a rough idea of the thermal range as well as the expected maximum and minimum temperatures, the satellite should experience before a computer simulation, and the method described by Gupta and Van Zyl [1] should be used.

4.3 Link Budget

Knowing that the UHF transceiver does make successful contact with the ground station as there is communication, it was a matter of verifying how well the UHF transceiver link works.

Using cumulative probability function, the Rx RSSI of ZACube-2 UHF transceiver was plotted, in conjunction using the signal strength to voltage table provided in the UHF transceiver hardware report, and it was found that 89.1896% of the RSSI signal was noise.

5 Recommendations for Future Work

While objectively it can be said that ZACube-2 works as the system did produce enough power to still be operational, communicated with the ground station, and never experienced any difficulties that are associated with poor thermal management, there were points of interest that could be implemented which would make it easier for future telemetry analysis on the next iteration of satellites.

- Recalculation of solar array results as the information relayed by the telemetry does not accurately depict the actual performance.
- Examination of a broader spectrum of space weather events that possibly caused the same events of no telemetry or outlier data to be present in multiple telemetry channels.
- Possibility of synchronizing the interval at which telemetry across multiple channels is logged.
- Pre-set exposure threshold to account for antenna reflections or consider either relocating the ADCS to mitigate issues related to reflection from the sun.

- Complete a computer simulation of ZACube-2 thermal performance to confirm that all thermal parameters are met and to confirm that the method described by Gupta and Van Zyl [1] is a good basis prior simulation.
- Have individuals who have more experience in the respective field to view the data and draw conclusions separately.

References

1. Gupta G, Van Zyl R (2022) NOMA-based LPWA networks. In: Expert clouds and applications. Lecture notes in networks and systems, vol 209. pp 523–530. https://doi.org/10.1007/978-981-16-2126-0_42
2. Maharaj R, Balyan V, Khan MTE (2022) Optimising data visualisation in the process control and IIoT environments. Int J Smart Sens Intell Syst 15(1):1–14
3. Maharaj R, Balyan V, Khan MTE (2022) Design of IIoT device to parse data directly to scada systems using LoRa physical layer. Int J Smart Sens Intell Syst 15(1):1–13
4. Gupta G, Van Zyl R (2022) Survey based on configuration of CubeSats used for communication technology. Adv Intell Syst Comput 1235:59–71. https://doi.org/10.1007/978-981-16-4641-6_6
5. Prasad PS, Sunitha Devi B, Janga Reddy M, Gunjan VK (2019) A survey of fingerprint recognition systems and their applications. Lecture notes electrical engineering, vol 500. pp 513–520
6. Jan YW, Chiou JC (2005) Attitude control systemfor ROCSAT-3 microsatellite: a conceptual design. Acta Astronaut 56(4):439–452
7. Rudra Kumar M, Pathak R, Gunjan VK (2022) Diagnosis and medicine prediction for COVID-19 using machine learning approach. In: Computational intelligence in machine learning: select proceedings of ICCIML 2021. Springer Nature Singapore, Singapore, pp 123–133
8. Rashid E, Ansari MD, Gunjan VK, Ahmed M (2020) Improvement in extended object tracking with the vision-based algorithm. In: Modern approaches in machine learning and cognitive science: a walkthrough: latest trends in AI. pp 237–245
9. Fortescue P, Swinerd G, Stark J (2011) Spacecraft systems engineering, 4th edn. Wiley, Chichester
10. Wertz J, Larson W (1999) Space mission analysis and design, 3rd edn. Microcosm, Torrance, California
11. Lakshmanna K, Shaik F, Gunjan VK, Singh N, Kumar G, Shafi RM (2022) Perimeter degree technique for the reduction of routing congestion during placement in physical design of VLSI circuits. Complexity 2022:1–11
12. Kaminskiy M, Kashem N (2015) CubeSat data analysis revision—revision 2015. National Aeronautics and Space Administration, Maryland
13. Gaddam DKR, Ansari MD, Vuppala S, Gunjan VK, Sati MM (2022) A performance comparison of optimization algorithms on a generated dataset. In: ICDSMLA 2020: proceedings of the 2nd international conference on data science, machine learning and applications. Springer Singapore, pp 1407–1415
14. Kaminskiy M (2015) Cubesat data analysis revision. National Aeronautics and Space Administration, Greenbelt, Maryland
15. Kolhe NP, Dhote KK, Parihar A (2023) Mapping the access of internet facility to schedule tribe population—an approach towards digital education. In: Digital learning based education: transcending physical barriers. Springer Nature Singapore, Singapore, pp 1–20
16. Muthmainnah, Ganguli S, Al Yakin A, Ghofur A (2023) An effective investigation on YIPe-learning based for twenty-first century class. In: Choudhury A, Biswas A, Chakraborti S

(eds) Digital learning based education. Advanced technologies and societal change. Springer, Singapore. https://doi.org/10.1007/978-981-19-8967-4_2
17. Ganguli S, Al Yakin A (2023) An effective investigation on YIPe-learning based for twenty-first century class. In: Digital learning based education: transcending physical barriers. Springer Nature Singapore, Singapore, pp 21–38
18. Pathak R, Gupta SS (2020) A study on natural computing: a review. In: ICDSMLA 2019: proceedings of the 1st international conference on data science, machine learning and applications. Springer Singapore, Singapore, pp 1975–1983
19. Chatterjee R, Bandyopadhyay A, Chakraborty S, Dutta S (2023) Digital education: the basics with slant to digital pedagogy-an overview. In: Digital learning based education: transcending physical barriers. pp 63–80
20. Das N (2023) Digital education as an integral part of a smart and intelligent city: a short review. In: Digital learning based education: transcending physical barriers. pp 81–96
21. Pathak R, Soni B, Muppalaneni NB (2023) Role of blockchain in health care: a comprehensive study. In: Gunjan VK, Zurada JM (eds) Proceedings of 3rd international conference on recent trends in machine learning, IoT, smart cities and applications. Lecture notes in networks and systems, vol 540. Springer, Singapore. https://doi.org/10.1007/978-981-19-6088-8_13
22. Nandy D, Majee D (2023) Socio-economic relations in digital education: a comparative study between Bangladesh and Nepal. In: Digital learning based education: transcending physical barriers. Springer Nature Singapore, Singapore, pp 103–117
23. Bhowmik T, Majumdar S, Choudhury A, Banerjee A, Roy B (2023) Importance of internal and external psychological factors in digital learning. In: Digital learning based education: transcending physical barriers. Springer Nature Singapore, Singapore, pp 119–132

A Novel Approach for Speech Emotion Recognition with Facial Expression Analysis

Uma N. Dulhare, Shaik Rasool, Gautam Kumar, Gazna Khan, and Vinit Kumar Gunjan

Abstract An effective communication involves interaction by comprehending emotions through voice and face recognition. Communication and confidence are base for any business and organization. Accurate prediction may yield in success leading to profits and growth. Thus, it is very crucial to recognize emotions in vital communication. It was always very challenging to recognize such crucial emotions with high accuracy as people are diverse and exhibit heterogeneous expressions according to situation. This area of emotion recognition has gained interest from several researchers to explore artificial intelligence and machine learning techniques to build an apt solution with high accuracy. A work has been proposed in this paper based on these research lines to utilize the audio signals from human voice and validate the emotions through facial images. It can be used to have an efficient and enhanced communication between computers and humans with high accuracy to ease several daily tasks where it was extremely difficult for systems to understand the emotions of humans and act accordingly especially in robotics. To acquire extremely significant features from speech signal, multiscale transformation techniques are utilized which decomposes the signals into multi-frequency bands. Mutually, the amplitude and phase are deemed as essential features and are processed. The proposed approach adopts support vector machine algorithm as a supervised learning algorithm and principal component analysis used for facial images. An accuracy of 86.66% is achieved. The system is trained and tested with speech audios of 15 actors taken from the RAVDESS database. Recall rate of 73.33% is detected.

U. N. Dulhare (✉) · G. Khan
Muffakham Jah College of Engineering and Technology, Hyderabad, Telangana, India
e-mail: prof.umadulhare@gmail.com

G. Kumar
Department of Computer Science and Engineering, Manipal University, Jaipur, India

S. Rasool
Methodist College of Engineering and Technology, Hyderabad, Telangana, India

V. K. Gunjan
CMR Institute of Technology, Hyderabad, India

© The Author(s), under exclusive license to Springer Nature Singapore Pte Ltd. 2023
A. Kumar et al. (eds.), *Proceedings of the 4th International Conference on Data Science, Machine Learning and Applications*, Lecture Notes in Electrical Engineering 1038,
https://doi.org/10.1007/978-981-99-2058-7_25

Keywords Machine learning · Facial expression · Emotion recognition · Speech analysis · Signal processing · Multiscale transform

1 Introduction

Human beings possess impulse of emotion recognition. The backbone of human communication is emotional intelligence which is encapsulation of recognition, interpretation, and expression of emotions. One of the biggest problems in recognizing emotions of different individuals is that they may express the same emotion differently. The objective is complex even in a case where we have people speaking in same language as they differ in how they produce speech signal. Every individual speaks with a different or variable accent and their pitch. Quality of speech also varies person to person. It becomes tedious task for the computer to process speech signal and differentiate language from the features that are used to identify a person based on his expression. Thus, it becomes very difficult to interpret what was said unless the system is trained on a dataset of specific person or the dataset may be like most. Hence, there is a necessity to design effective speech-based emotion recognition system through which the HCI performance will increase. This system can be used in educational field, health, psychology, and cognitive sciences as the system will work as a therapist system advising the user with better suggestions to reduce stress and improve their performance in the respective fields. This can help reduce the stress rates in individuals and help them lead a happy life ahead. Emotions are generated based on a sudden change in mental or physical state of a person that may be disturbing and lead to a specific reaction to that situation. It may also be recognized as a complex progression of several characteristics of a human like his behavior and feelings to react differently in diverse situations.

- Conventional speech emotion recognition methods have been used in the past which did not effectively recognize the emotions from speech.
- Linear vocal and constant speech signal are essential for any speech processing approaches.
- Speech signals present a challenging task as they are viable. Every individual has their own style of speech, with different accent, pace, and changing emotional state adapting to the situation around them. There could also be variations in speech due to the surrounding constraints like noise and hardware for recording.

To achieve more accurate classification in the case of emotion recognition through speech signals, a new speech-based emotion recognition system considering the spectral feature of speech signals and machine learning algorithms is considered.

- To extract the most important features from the speech signal, this work accomplished a multiscale transform which decomposes the signals into multi-frequency bands. Here, both the amplitude and phase are considered as the required features and are processed.

- To achieve more accuracy, both for known and unknown emotional data, the proposed approach adopts the support vector machine algorithm [1, 2].

2 Related Work

Speech emotion recognition (SER) is evolving as research area that could support and ease the process in human computer interactions. The periodic nature of the speech makes it a challenging task for emotion recognition. Humans are gifted with capability to easily understand emotion from face and gestures, but computing machines require advanced capabilities to perform the same. Superior frameworks are needed for accurate emotion recognition. Wavelet packet features proposed by Firoz Shah A. and Babu Anto P. used a dataset built on nine emotions and extreme learning machine for pattern classification. The data surges in unidirectional from source to outcome traveling hidden layers. The source and the outcome tier points are stable. The predictive power varies with increase in the no of hidden nodes. The system achieved 76.66 accuracy [3]. Another model was built incorporating energy features [4]. It constituted five modules and involves deep analysis of speaker's emotion.

Another model speaker emotion recognition system was able to identify emotions utilizing a general characteristic set-in application [5]. Continuous classifiers HMM and LIBSVM were used. Better recognition rates can be achieved using LIBSVM in comparison with continuous HMM when there are few emotions. Better recognition accuracy was achieved by employing discrete wavelet transform for extracting the features. The emotional characteristic features are provided by line spectral frequency. Emotion detection is achieved by characterizing features like epochs and residual parameters using sound reduction mechanism. Mel-frequency cepstral coefficients (MFCCs) can be utilized when high recognition rate is not desired. It does not require additional sound features. It is novel to encapsulate features of articulation and prosodic to recognize emotions. By extracting the features like Teager energy cepstral coefficients, instantaneous frequency weighted energy cepstral coefficients and using LIBSVM classifier, a better accuracy of 54.71% is achieved. LIBSVM can be applied on stored data.

Another approach [6] was able to recognize emotions involving surprise, sadness, angry, kindness, and rude. The study employed Naive Bayes classifier and zero crossing rate features. It showed an accuracy of 78% for the emotion recognition. Statistical values are obtained from the base features that are extracted from input speech. Training and testing are the two phases of entire process. Testing phase produced an output dataset containing 15 features. For emotion recognition, classification techniques were used after processing the sample speech in testing data. 16-bit pulse code modulation was used to code and sample the input signals at 16 kHz. Each frame was subjected to hamming window of 25 ms size and shift of 10 ms. To extract the energy value in each frame, short-term energy function was applied resulting in statistics of energy. Linear predictive coding was used to extract format frequencies.

A dataset of three hundred speech from thirty persons was used. The results showed an 80% accuracy. Naïve Bayes algorithm is best suited when the dataset is small.

Energy entropy was calculated with introduction of a feature parameter on wavelet packet frequency values [7]. Classical Gaussian mixture model was employed for classification and modeling tasks. Two datasets, Berlin Emotional Speech and Speech Under Simulated Emotion, were utilized in the system. Higher accuracy of 74–76% was achieved in Speech Under Simulated Emotion in comparison with 51–54% of Berlin Emotional Speech dataset. Wavelet packet energy may lead to invalid results as it does not signify feature effectively.

Another system used support vector machine for classification where mel-frequency cepstral coefficients are employed to obtain spectral features and discrete wavelet transform for decomposing the speech signals [8]. Heterogeneous frequencies of sinusoids are obtained by segregating the signals using Fourier transform. Analysis of dynamic signals like sharp transition of speech, drift, and trends may not be possible using Fourier transform. Wavelets can be utilized to represent signals in time frequency model. Discrete time domain signals are filtered using successive low and high pass computation to obtain discrete wavelet transform. Orthogonality can be achieved by decimating filters output containing number of coefficients in every iteration. At every step of disintegration, estimates are filtered out. A two-channel filter is used in disintegration of estimate coefficients through iterative implementation of discrete wavelet transform. An equal frequency bandwidth is obtained through iterative decomposition of all coefficients through implementation of wavelet packet transform. Encapsulation of discrete wavelet transform and mel-frequency cepstral coefficients resulted in recognition accuracy of 83%. Discrete wavelet transform fails when a shift occurs in the signal as wavelet transform is not shift invariant. Information of time doesn't last and is the major drawback of frequency domain.

3 Proposed Work

Facial expressions of humans are utilized for identification of emotions as part of emotion recognition process. Utilizing the speech for detection of emotions is the emerging area in research that may provide plethora of applications. Emotion recognition systems offer users, improved services by adapting to emotions whether it is human to computer or human to human. The task of detecting the emotions using speech is very narrow. There are two kind of emotion and may be classified as primary and secondary. Surprise, anger, happy, disgust, and fear are part of primary emotions. Secondary emotion is encapsulation of one or more primary emotions. Emotions are discrete without continuity and can be characterized on basis of dimension. These are the perspectives on which the classification of emotions is carried out.

3.1 Speech Emotion Recognition

In this technology advancing world, understanding and developing solution to ease the human computer interaction has become the major and rapidly growing field. Speech forms the basis for having communication. Speech is encapsulation of sounds arranged in order. Complex computations are performed by the human brain to analyze acoustic sound. It generates ideas like instructions, commands, etc., by processing and transforming sounds. There are 3 stages involved in processing of sound:

1. Preprocessing
2. Feature extraction
3. Pattern recognition.

Figure 1 shows these stages in the process. Vowels are most vital part in the speech that signifies the information. They are the sound elements of the speech and are to be separated from the unvoiced elements before proceeding further in signal processing and is mandatory. Pattern recognition is used to extract patterns involving mel-frequency cepstral coefficients, energy, and pitch and are mapped utilizing the ANN, SVM, and HMM models as part of speech emotion recognition.

The entire system constitutes five modules as shown in Fig. 1, as speech input, feature extraction, feature selection, classification, and output module. Speech signal generation is the starting point of the system followed by extraction of essential features which may be vital and contain the information of emotions of the speaker, lastly identifying the patterns by employing apt pattern recognition models that outputs the state of emotion.

3.1.1 Phase-Based Features

Equation 1 shows the computation of Fourier transform of discrete time signal in polar form.

$$X(\omega) = |X(\omega)|e^{j\varphi(\omega)} \tag{1}$$

Here, $|X(\omega)|$ denotes magnitude, and $\varphi(\omega)$ denotes the phase spectrum [9]. Magnitude spectrum is only utilized for speech processing and avoids the use of phase spectrum. Wrapping of phase spectrum and windows position dependency makes useful feature extraction a challenging task [10, 11]. Studies have proven that higher

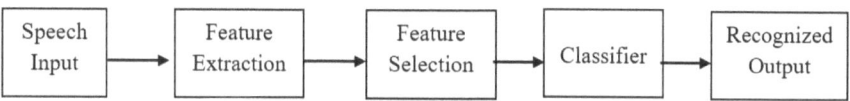

Fig. 1 Process of emotion recognition using speech

performance is delivered by a system that employs phase information. Speech and speaker recognition and environment acoustics detection widely employ the all-pole group delay feature and modified group delay feature [12–14].

3.1.2 Modified Group Delay Feature (MGD)

The negative derivative of the Fourier phase spectrum is the group delay function and denoted as follows [13]

$$rg(\omega) = \frac{-d(\varphi(\omega))}{d\omega}$$

$$d\omega = \frac{K(\omega)FR(\omega) + KI(\omega)FI(\omega)}{|K(\omega)|^2} \tag{2}$$

Here, the angular frequency ω is limited in $[0,2\pi]$, $|X(\omega)|$ is the magnitude of the Fourier transform of $x(n)$, $Y(\omega)$ is the Fourier transform of the signal $y = nx(n)$, and the subscripts r and i indicate real and imaginary parts, respectively. The features obtained by group delay function are discriminative and additive for recognition. Often an inaccurate representation of speech signal is caused by this function. The denominator of the group delay function goes toward 0, when the 0s of the system transfer function are very close to the unit circle in the plane. Hence, group delay function at frequency bins near these 0s inevitably results in false spikes and becomes ill-behaved although it is able to produce a meaningful representation of a signal to a level. To overcome the spiky nature of the group delay feature, a modification of the group delay function is proposed in [15], which is computed as

$$rm(\omega) = \frac{rp(\omega)}{|rp(\omega)|}|(\omega)|^\alpha \tag{3}$$

where

$$(\omega) = \frac{K(\omega)FR(\omega) + KI(\omega)FI(\omega)}{|(\omega)|^{2\gamma}} \tag{4}$$

and $|S(\omega)|$ is a cepstral leveled form of $|X(\omega)|$. The two changing boundaries γ and α rein in the gamut dynamics of the MGD spectrum. Note that $P(\omega) = Xr(\omega) Yr(\omega) + Xi(\omega) Yi(\omega)$ is called the product spectra and incorporates information from both the magnitude and phase spectrum [16]. MFCC computation is identical to that of MGD feature extraction. First, hamming window frames and pre-emphasizes the speech signal. Later, computation of MGD features is carried out. Lastly, decorrelation is done by exposing MGD features to DCT. To overcome the effect of average value, the first coefficient of DCT is discarded.

3.1.3 Group Delay Function of All-Pole Models (APGD)

All-pole models group delay function can be utilized for the recognition of speaker [1]. Accurate interpretation of phase information is possible using APGD, thereby avoiding the need to use MGD parameter adjustment features. Ant extraction has implemented this feature successfully [17]. It was also employed successfully in speaker, music, and environ mental acoustic recognition [18]. APGD features are embedded in the group delay function of parametric all-pole model of speech signal as compared to MGD features computing group delay function. Short-term power spectrum is estimated by linear prediction analysis utilizing the all-pole model.

3.1.4 Mel-frequency Cepstral Coefficients (MFCCs)

Coefficients of MFCC are obtained from audio clip cepstral representation. The difference between the cepstrum and the mel-frequency cepstrum is that the frequency bands are equally spaced on the mel scale in the MFC, which approximates the human auditory system's response more closely than the linearly spaced frequency bands used in the normal cepstrum.

MFCCs are commonly derived by using Fourier transform of signal. The powers of the spectrum are then mapped onto the mel scale, using triangular overlapping windows. Logs of powers are taken at each of mel frequencies. Then, the discrete cosine transform of the list of mel log powers is utilized, as if it were a signal. The MFCCs are the amplitudes of the resulting spectrum.

3.1.5 Support Vector Machines (SVMs)

Classification and regression are performed by a supervised machine learning algorithm known as SVM by analyzing the data. It finds the pattern using algorithms in each large dataset. It generates two parallel lines for acquiring the parallel pattern partitions. All the attributes are utilized in high-dimensional space for every category of data. Linear and flat partitions are obtained by partitioning the space in a single pass. It provides two categories that are widely unidentical. It does this job by employing hyperplane. To partition the given data into classes, SVM employs the hyperplane holding major space in high dimension. The farseeing distance among adjacent data point in these classes is the margin that represents two classes between them. The generalization error of classifier depends on the scope of the margin. Once training is done, then new data is mapped to same space to identify the category and partition data into those sets that they belong. SVM has the advantage of more flexibility as compared to other classifiers.

3.2 Face Emotion Recognition

Identification of faces captured by comparing them with those that are stored in database of all individuals and associating a name or id to each identified face is the process of facial recognition [19]. It has been a challenging task from many years in system visualization to automatically identify and recognize face. A facial image is considered a 3D body that must be identified when illuminated by a light source and should be recognized among the several faces having different features and characteristics. Thus, when it is transformed as 2D image, the characteristics and features differ by huge margin. The recognition process must be robust and accurate.

3.2.1 Principal Component Analysis (PCA)

In any areas of the industry, security and authentication are two crucial elements that decides the adaptation of system. Several approaches are available to achieve this task. Face recognition is one of the available techniques that provides biometric authentication. This is most effective ways of authentication as it has large scope of tracking the change of pattern in face of person. The change in expression of face of a person may not affect the system. It highly robust and can be confidently used to identify the criminals from a large group of individuals.

PCA is linear transformation technique and part of statistics that is utilized to simplify a given dataset. For a given dataset, it selects a new coordinate in way that highest variance of dataset rests on first axis and second highest variance on second axis and so on. PCA reduces dimensionality without effecting dataset characteristics that are integral to variance by considering lower principal components and ignoring higher order. Facial recognition task involves segmentation of input image into different classes. Patterns in input image are to be identified separating the noise caused by environment such as light, shadows surrounding the target, and the pose a person may have. Patterns identified are used in facial recognition process by calculation of distance among the facial objects such as eyes, nose, and mouth. These characteristic features of face are known as eigenfaces. PCA provides transformation of input image into eigenface. It is possible to reconstruct a face when all the eigenfaces of the original face image are available. The reconstructed face image is estimate of original image. The loss that can occur due to omitting eigenfaces may be minimized.

The system architecture shows the basic working of the speech emotion recognition model. The program starts by taking the speech audio as the input. It is then divided into frames and processes for windowing. Fast Fourier transformation techniques are applied with mel spectrum techniques, and magnitude vectors are obtained. Simultaneously, group delay features are applied to the speech audio to obtain phase vectors. Together these magnitude and phase vectors undergo SVM classification that recognizes the emotion in the speech audio. This emotion recognized is again

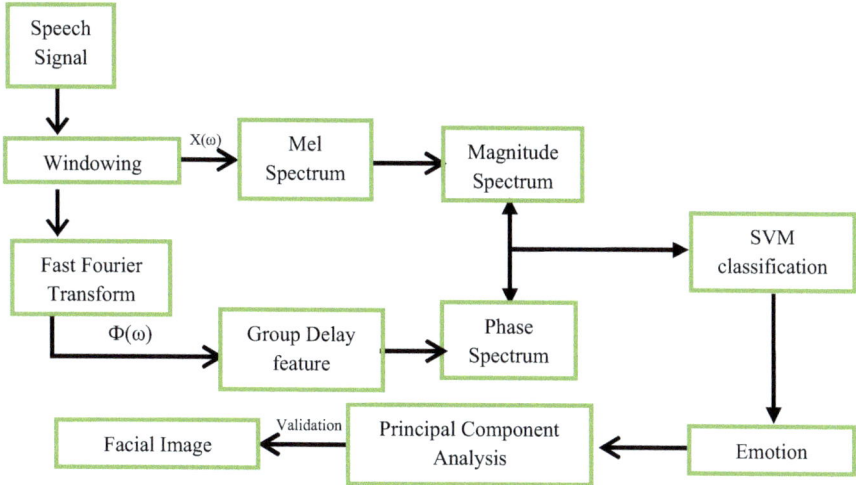

Fig. 2 System architecture $X(\omega)$ = magnitude information $\Phi(\omega)$ = phase information

taken as an input to fetch the corresponding facial image of the user with the emotion. PCA is utilized for this extraction process which can be seen in Fig. 2.

The input to the program is speech signal. It undergoes pre-emphasis that divides the signal into frames. On these frames, hamming window of size 10 ms delay and 25 ms size is applied. Fast Fourier transforms are applied, and mel scale filtering is done. Multiscale transforms are applied to obtain the magnitude feature vector (Fig. 3).

The input to the program is the speech signal. It undergoes pre-emphasis that divides the signal into frames. On these frames, hamming window of size 10 ms delay and 25 ms size is applied. Fast Fourier transforms are applied, and group delay features are applied, namely all-pole group and magnitude delay. Multiscale transforms are then applied to obtain the phase feature vector (Fig. 4).

The input signal is pre-emphasized that is used to balance the frequency spectrum. 10 and 25 ms overlap are obtained by segmenting the audio signal into frames. After the signal is divided into frames, it undergoes windowing, mainly, hamming window. Fast Fourier transform techniques are applied to obtain the magnitude and

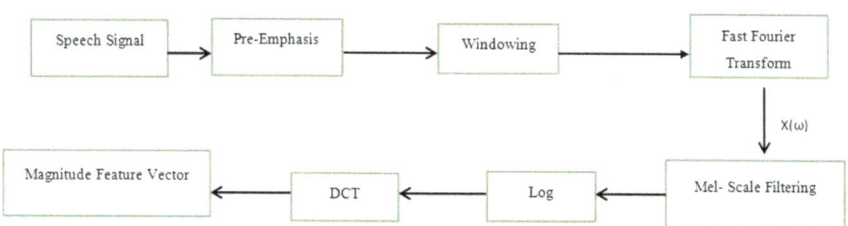

Fig. 3 Magnitude spectrum

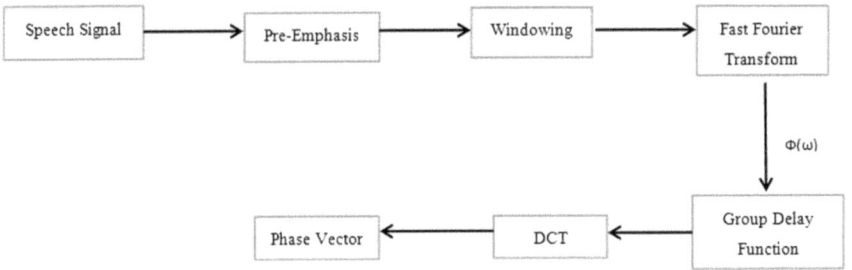

Fig. 4 Phase spectrum

phase spectrum from the audio signals. To obtain the magnitude vectors, the signal undergoes mel scale filtering and normalization, whereas to obtain the phase-based vectors, group delay features are applied, namely MGD and APGD. After obtaining the magnitude and phase vectors, multiscale transform technique, DCT is applied and processed for SVM classification to recognize the emotion from the audio signal. The emotion extracted also fetches the corresponding facial image from the database that undergoes PCA to give the desired output. Using speech emotions are recognized and validated through facial images. Features are extracted using SVM, and MFCC classification algorithm is used. Validation of the emotion recognized through speech is done from facial images that are extracted from the facial database, and validation is performed. A total of 4 emotions—anger, happy, sad, laugh, and fear are determined.

Algorithm: Speech emotion recognition with facial image validation
Input: Audio signal
Output: Recognized emotion with valid facial image from the database Steps Involved:

Step 1. Pre-emphasis is performed on input speech signal.
Step 2. Speech signal is divided into frames using hamming window.
Step 3. Fast Fourier transformation techniques are applied on each frame.
Step 4. Group delay functions are applied to obtain the magnitude and phase vectors.
Step 5. SVM is used as a classifier to identify the emotion.
Step 6. The audio signal is taken as an input, and PCA is applied to obtain the corresponding face from the database.

From step 1, pre-emphasis is performed on the speech signals. Subsequently, frame windowing is accomplished using hamming window with a frame length of 25 ms and frame shift of 10 ms. FFT is performed on the magnitude information to determine the frequency spectrum called short-time Fourier transform. Mel scale filtering is done, and filter banks are created based on the frame size and window length. Discrete cosine transforms are applied, and magnitude feature vectors are obtained.

In step 3, discrete time signal Fourier transform is calculated. Group delay function is computed in step 4. Once this is done, discrete cosine transforms are applied, and phase feature spectrum is obtained. In step 5, once the feature vectors are obtained, SVM classification is done, and the respective emotion is obtained. This emotion obtained from the SVM classifier is then fed to another system where PCA technique for facial image is used, and the resulting facial image is extracted from the facial database. The resulted image proves as the validation for emotion obtained through speech.

4 Experimental Setup

A system with 8 GB RAM and 1 TB HDD with a Windows Operating System (Windows 10 Enterprise Edition) is used. All the speech inputs are recorded with a microphone of 16 bits and 48,000 Hz (studio quality), and all the images are captured with a camera of 720-pixel definition (pd) H.264, AAC 48 kHz from each actor's video song between 8 and 10 s playing at normal speed with varying intensities. The speech emotion recognition system is executed on MATLAB 2015a version. The dataset contains a set of 15 actors, 8 male, and 7 female actors, whose facial images and audios have been taken from the RAVDESS dataset. Each actor represents 4 basic emotions (anger, sad, fear, and happy), and audios are differentiated into two intensities (normal and strong) for each emotion.

Each emotion has varying intensity ranging from 40 to 100 dB. Anger emotion with intensity two is detected when the emotion in beyond 62 dB threshold. Similarly, happy emotion with intensity two is detected when the intensity is beyond 60 dB. The audios are recorded with two statements—kids are talking by the door∥ and—dogs are sitting by the door∥ and have been tested with both and cross intensities for better understanding of emotions [20].

In Table 1, confusion matrix for the emotions (anger, happy, fear, sad) is calculated for statement 1 intensity 1. Out of 15 actors, 11 actors exhibit the emotion anger for an angry audio input whereas 4 actors exhibit angry emotion for a happy audio input. The correct emotions detected by the system for the given emotion input out of 15 actors are 11 for anger, 9 for happy, 8 for fear, and 6 for sad. The individual confusion matrices are generated from the multi-index confusion matrix.

Table 1 Confusion matrix for emotions anger, fear, happy, sad—statement 1 (intensity 1)

Emotions	Angry	Happy	Fear	Sad
Angry	11	2	1	1
Happy	4	9	2	0
Fear	5	0	8	2
Sad	3	0	6	6

Table 2 Confusion matrix for emotions anger, fear, happy, sad—statement 2 (intensity 1)

	Angry	Happy	Fear	Sad
Angry	11	3	1	0
Happy	5	8	1	1
Fear	6	1	1	7
Sad	5	0	9	1

Table 3 Comparison of metrics in %

System	Accuracy	Precision	Detection rate	$F1$-score	False precision rate
Existing	83.6475	68.365	68.75	67.735	11.387
Proposed	86.667	76.72	73.332	73.49	8.887

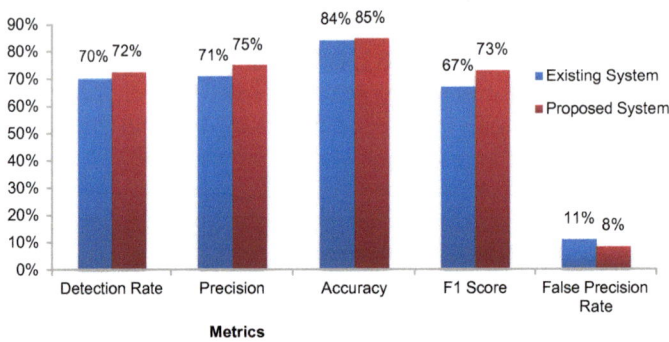

Fig. 5 Overall evaluation of metrics of proposed and existing system

In Table 2, confusion matrix for the emotions (anger, happy, fear, sad) is calculated for statement 2 intensity 1. Out of 15 actors, 11 actors exhibit the emotion anger for an angry audio input, whereas 5 actors exhibit angry emotion for a happy audio input. The correct emotions detected by the system for the given emotion input out of 15 actors are 11 for anger, 8 for happy, 7 for fear, and 9 for sad. The individual confusion matrices are generated from the multi-index confusion matrix.

From Table 3, it can be concluded that the proposed system gave an accuracy 3.02% more than the existing system.

In Fig. 5, the detection rate of the proposed system is 73.33% that varies slightly from the existing system from being 68.75%. Also, precision and the accuracy are calculated to be slightly more than the previous system resulting in 76.72% and 86.66%, respectively (Figs. 6 and 7).

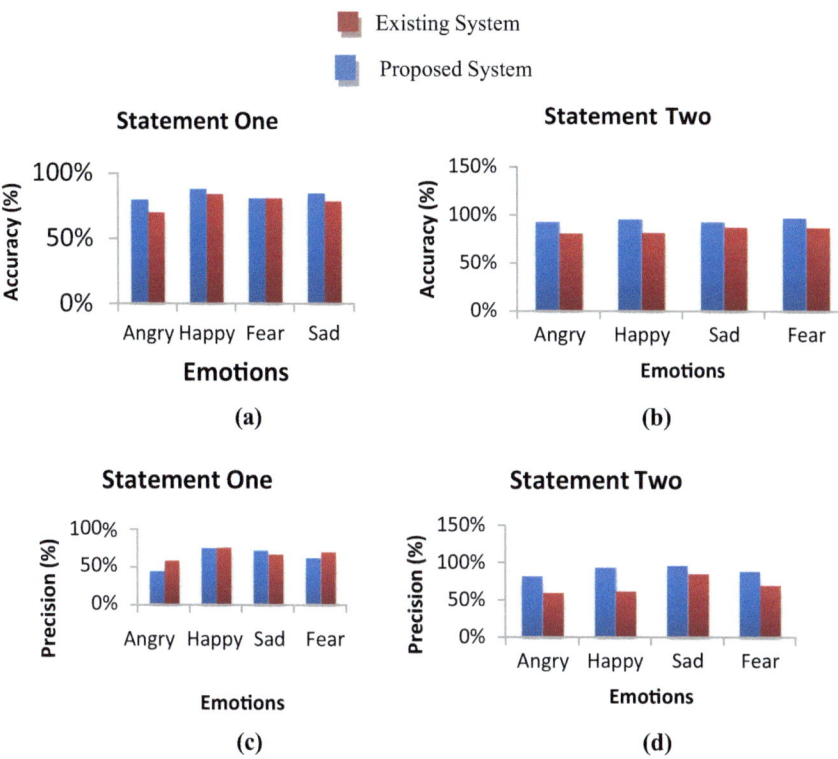

Fig. 6 Accuracy and precision of intensities one and two

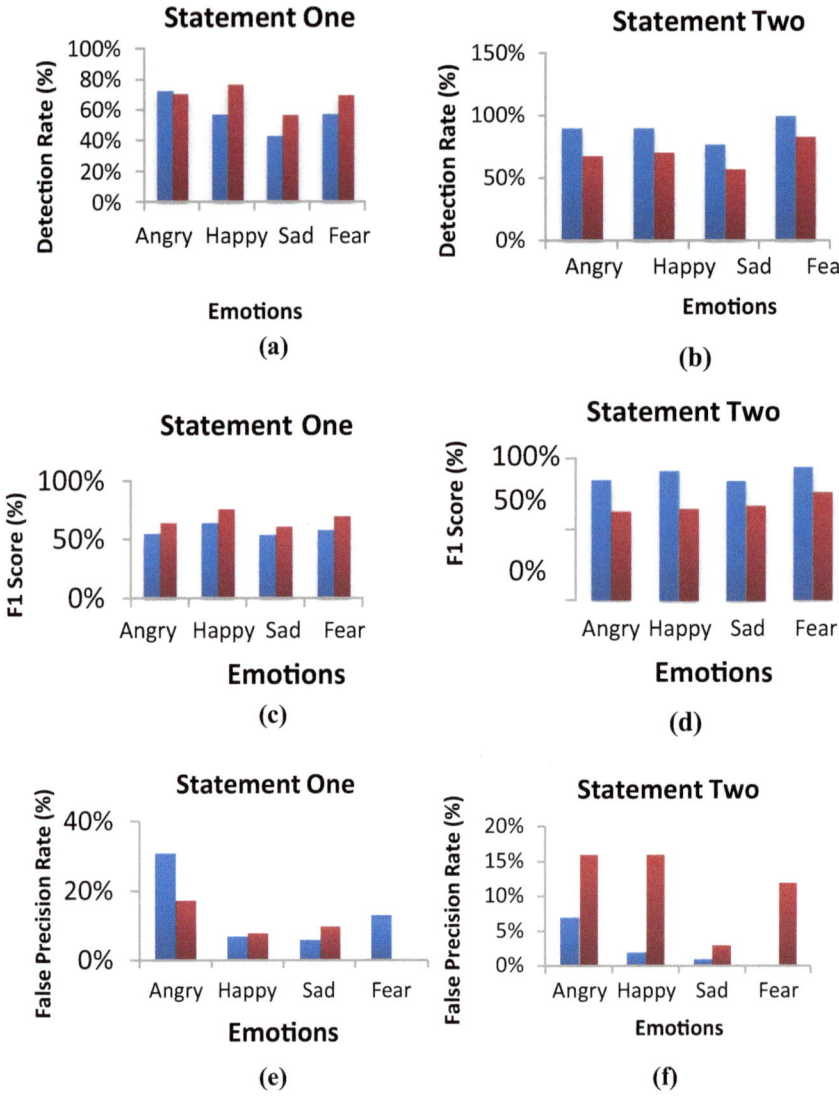

Fig. 7 Detection rate, F1-score, and false precision rate of intensities one and two

5 Conclusion

This work has effectively detected emotions of the actors whose audio speech signals were given as an input and facial image were used to validate the user with the detected emotion. Audio signals of two different intensities ranging between 40 and 100 dB were used to train and test the system. The system was tested with varying combinations of the input signals, i.e.; the system was trained with an audio signal of one intensity and tested with audio signal of another intensity and so on. Also, the system was trained and tested with the audio signals of same intensities. This system can be further enhanced by using real-time data and extending the project to applications like authentication of a user while in a security check, enabling a robot that can recognize the emotions of its user to understand the user better.

References

1. Kumar S, Gunjan VK, Ansari MD, Pathak R (2022) Credit card fraud detection using support vector machine. In: Proceedings of the 2nd international conference on recent trends in machine learning, IoT, smart cities and applications: ICMISC 2021. Springer Singapore, pp 27–37
2. Dulhare UN, Ali MH (2021) Underwater human detection using faster R-CNN with data augmentation. Mater Today: Proc
3. Firoz Shah A, Babu Anto P (2017) Wavelet packets for speech emotion recognition. In: Third international conference on advances in electrical, electronics, information, communication and bio-informatics (AEEICB), Chennai. pp 479–481. https://doi.org/10.1109/AEEICB.2017.7972358
4. Prasad PS, Kumar Gunjan V (2020)Feature descriptors for face recognition. In: 2020 IEEE 17th India council international conference (INDICON), New Delhi, India. pp 1–4. https://doi.org/10.1109/INDICON49873.2020.9342424
5. Das N (2023) Digital education as an integral part of a smart and intelligent city: a short review. In: Digital learning based education: transcending physical barriers. pp 81–96
6. Prasad PS, Varma CE, Gunjan VK (2022) Face recognition and detection algorithm—a review. In: Futuristic trends in networks and computing technologies: select proceedings of fourth international conference on FTNCT 2021. Springer Nature Singapore, Singapore, pp 729–734
7. Muthmainnah, Ganguli S, Al Yakin A, Ghofur A (2023) An effective investigation on YIPe-learning based for twenty-first century class. In: Choudhury A, Biswas A, Chakraborti S (eds) Digital learning based education. Advanced technologies and societal change. Springer, Singapore. https://doi.org/10.1007/978-981-19-8967-4_2
8. Gaddam DKR, Ansari MD, Vuppala S, Gunjan VK, Sati MM (2022) Human facial emotion detection using deep learning. In: ICDSMLA 2020: proceedings of the 2nd international conference on data science, machine learning and applications. Springer Singapore, pp 1417–1427
9. Deng J, Xu X, Zhang Z, Frühholz S, Schuller B (2016) Exploitation of phase-based features for whispered speech emotion recognition. IEEE Access 4:4299–4309. https://doi.org/10.1109/ACCESS.2016.2591442
10. Kar S, Kar S (2023) Pedagogical considerations in the new normal: from tradition to technology. In: Digital learning based education: transcending physical barriers. Springer Nature Singapore, Singapore, pp 97–102
11. Mokhlesabadifarahani B, Gunjan VK (2015) EMG signals characterization in three states of contraction by fuzzy network and feature extraction. Springer, Heidelberg, Germany
12. Nandy D, Majee D (2023) Socio-economic relations in digital education: a comparative study between Bangladesh and Nepal. In: Digital learning based education: transcending physical barriers. Springer Nature Singapore, Singapore, pp 103–117
13. Prasanna K, Seetha M (2015) A doubleton pattern mining approach for discovering colossal patterns from biological dataset. Int J Comput Appl 119(21):41–47
14. Wu Z, Chng E, Li H (2012) Detecting converted speech and natural speech for anti-spoofing attack in speaker recognition. In: 13th annual conference of the international speech communication association 2012, Interspeech 2012, vol 2

15. Hegde RM, Murthy HA, Gadde VRR (2007) Significance of the modified group delay feature in speech recognition. IEEE Trans Audio Speech Lang Process 15(1):190–202. https://doi.org/10.1109/TASL.2006.876858
16. Pathak R, Gupta SS (2020) A study on natural computing: a review. In: ICDSMLA 2019: proceedings of the 1st international conference on data science, machine learning and applications. Springer Singapore, Singapore, pp 1975–1983
17. Yegnanarayana B (1978) Formant extraction from linear-prediction phase spectra. J Acoust Soc Am 63(5):1638–1640. https://doi.org/10.1121/1.381864
18. Diment A, Rajan P, Heittola T, Virtanen T (2013) Modified group delay feature for musical instrument recognition, Marseille, France
19. Dulhare UN, Rasool S (2022) Smart airport system to counter COVID-19 and future sustainability. In: Satyanarayana C, Gao XZ, Ting CY, Muppalaneni NB (eds) Machine learning and Internet of Things for societal issues. Advanced technologies and societal change. Springer, Singapore. https://doi.org/10.1007/978-981-16-5090-1_5
20. Bhowmik T, Majumdar S, Choudhury A, Banerjee A, Roy B (2023) Importance of internal and external psychological factors in digital learning. In: Digital learning based education: transcending physical barriers. Springer Nature Singapore, Singapore, pp 119–132

Effect of Environment on Students Performance Through Orange Tool of Data Mining

Rajesh Tiwari⑩, **Gautam Kumar**⑩, **and Vinit Kumar Gunjan**⑩

Abstract In this paper, we are trying to prove that environment perform main role in student life. To increase overall performance of student, peaceful environment is must, and with the help of orange tool, analyze student performance data. Environment means not only the natural environment but also it means freshness, clean and pollution-free environment, tension-free mind, anxiety-free atmosphere, peaceful atmosphere, educated family, understanding between all family members, financial security, learning and teaching facility, motivational parent, joint and single family, final aim of student life, etc., are part of environment. When student has an aim in case student includes more effort in every field to achieve their goal, and this type of environment can provide by parent only to student. Therefore, first all parents have to understand the importance in the student performance in the environment, and this research paper will be helpful to explain this. So that every parent gives their children a good environment and the students can increase their performance as much as possible. Rural and urban both areas affect the student's performance, just because both environments are different to each other. These ideas are the part orange alert that can be handled using the mining. Data mining is the best way to prove this. Orange tool is more efficient tool of data mining that is why we choose it for this research.

Keywords Data mining · Student performance · Orange tool

R. Tiwari (✉)
CMR Engineering College, Hyderabad, Telangana 501401, India
e-mail: drrajeshtiwari20@gmail.com

G. Kumar
Department of Computer Science and Engineering, Manipal University Jaipur, Jaipur 303007, India
e-mail: gautam21ujrb@gmail.com

V. K. Gunjan
CMR Institute of Technology, Hyderabad, Telangana 501401, India

© The Author(s), under exclusive license to Springer Nature Singapore Pte Ltd. 2023
A. Kumar et al. (eds.), *Proceedings of the 4th International Conference on Data Science, Machine Learning and Applications*, Lecture Notes in Electrical Engineering 1038, https://doi.org/10.1007/978-981-99-2058-7_26

283

1   Introduction

This paper will use Orange tool classification tools and their algorithms like constants, tree, logistic regression, and neural network which help us to prove it. This paper will about student's performance that is depending on environment. The environment is also depending on atmosphere that is generated by human, family, friends, and relatives. This research encourages to all for creating a good environment. When student performance increases it is helpful in the development of the country or particular area because students are considered to be the future of the country, destiny of country. So increasing student performance is not only beneficial for students individually but also it is beneficial for particular area like district, state, and country.

1.1   Dataset

A data has been taken from UCI machine learning repository. This dataset has 34 attributes, and all attributes has different capacity. In this all 34 attributes, we select anyone as a target which is on demand of Orange tool. In this, dataset attributes are in binary form, like 0 and 1. These datasets already used and upload by P. Cortez and A. Silva. The secondary data of school student performance can be best handled using data mining.

2   Feature Selection and Data Cleaning

The accumulated information has 666 cases. Few of the features are unclear. The candidate's position is eliminated in the dataset that is within the information cleansing segment of the information preprocessing. The performance is the main parameter, and others are explanatory parameters. The performance characteristic is the numerical characteristic just in case for k-means clustering, while in other cases, it is called as categorical attribute. The information desk has been used for the information evaluation as given in Table 1. The attributes have been ranked the usage of Info benefit Attribute Eval ranker seek approach of Weka. Info Gain Attribute Eval approach is used for characteristic selection. Entropy calculates the impurity level. It can be presumed that impurity in the dataset is very less.

Clustering, association, and classification rule of data mining in educational datasets is near to zero. The precise characteristic is a characteristic that reduces the maximum entropy and is extremely ranked [1]. The caste turned into the best rank characteristic observed by medium, father occupation, mother occupation, Class XII_Percentage, and Class_X_Percentage of marks. The different excessive ranked attributes are obvious; however, the caste characteristic rank is pretty unexpected

Table 1 Description of attributes with their values

Attribute	Value	Description
Gender	'male', 'female'	Candidate's sex
Caste	'SC', 'ST', 'OBC', 'Gen' SC—schedule caste, ST—schedule tribes, OBC—other backward caste, Gen—open category	Candidate's caste
Coaching	'CWA','COA', 'NO.' CWA—coaching within Assam, COA—coaching outside Assam, NO.—no coaching	The candidate whether attended coaching within the Assam or outside the Assam or not taking any coaching classes
twelve_ education	'AHSEC', 'CBSE', 'OTHERS'	Class XII level board name of the candidate
Class_ten_ education	'SEBA', 'OTHERS', 'CBSE'	Class ten level, candidate's board name
Medium	'ENGLISH', 'OTHERS', 'ASSAMESE'	Class XII level instructions medium
Class_XII_ Percentage	'EXCELLENT', 'VERY GOOD', 'GOOD', 'AVERAGE' If marks % is more than 80, so it is EXCELLENT. If % of marks is in between 80 and 70, then it is VERY GOOD. If the % of marks is in between 70 and 60, then it is GOOD. In all other cases, it is termed as AVERAGE	Candidate's marks % at Class XII level
Class_X_ Percentage	'EXCELLENT', 'VERY GOOD', 'GOOD', 'AVERAGE' If marks % is more than 80, so it is EXCELLENT. If % of marks is in between 80 and 70, then it is VERY GOOD. If the % of marks is in between 70 and 60, then it is GOOD. In all other cases, it is termed as AVERAGE	Candidate's marks % at Class X level
Mother_ occupation	'Bank_Official', 'Doctor', 'College_Teacher', 'School_Teacher', 'Business_Woman', 'Engineer', 'House_Wife', 'Cultivator', 'Others'	Candidate's mother occupation
Father_ occupation	'Bank_Official', 'Doctor', 'College_Teacher', 'School_Teacher', 'Business_Woman', 'Engineer', 'House_Wife', 'Cultivator', 'Others'	Candidate's father occupation
Performance	'EXCELLENT', 'VERY GOOD', 'GOOD', 'AVERAGE' If scored % is in top 100, so it is EXCELLENT. If scored % is in the next 200, so VERY GOOD. If scored % is in next to next 200, so GOOD. In all other cases, it is AVERAGE. Here, the scored % means the % of marks in common entrance examination	Common entrance examination (CEE) performance

with the ranked value 0.51393. The other better ranked characteristics are obvious, but the caste characteristic rank is very amazing with the ranked value of 0.51393.

In this paper, we have used Portuguese dataset and convert it into excel file then input this excel file data into Orange tool set target as well as use tenfold cross validation and apply models like constants, tree, logistic regression, and neural network after applying different model [2]. We will test and score data and check accuracy as well as focusing on the upcoming results of different environments [3–5]. After computing data, we will compare result and by comparing with the incoming result, we find out that when the environment changes, there is also a lot of change in the performance of the student.

3 Methodology

The overall analysis of dataset in the tools can be cleared by Fig. 1.

Finally concludes that environment has the most impact on the performance of children as we can see in the analysis tables and analysis charts [6]. Rural and urban area, parent status, educated and motivational family [7], and use of ICT tools all are part of environment, and in all the above charts, we can see that

- According to Fig. 2, we saw that overall performance of urban students is higher that rural because in rural area students cannot find those facilities which are available in urban areas [8].

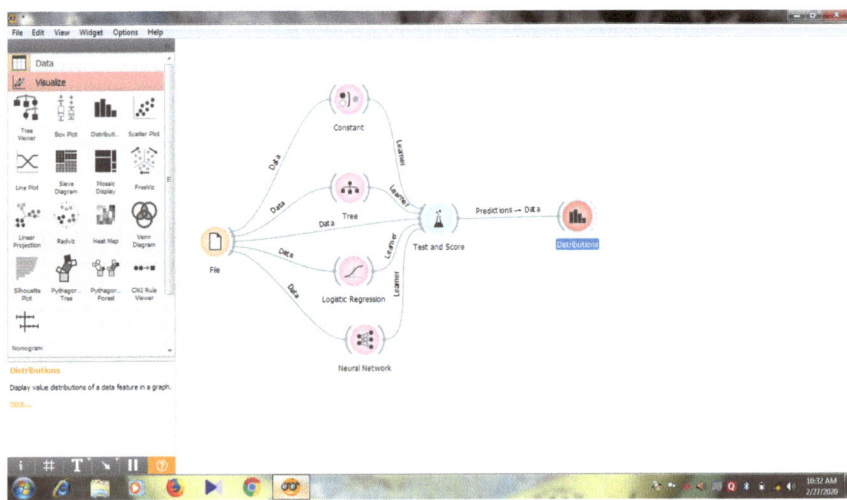

Fig. 1 Screenshot of Orange data mining tool's original model

Fig. 2 Compare rural and urban area

- According to Fig. 3, we saw those children whose parents have good financial condition as well as their mental status is also positive, they provide motivational support to children, this support helps in increasing overall performance of students, such parents are worried about their status, and to maintain that status, they motivate their children all the time [9]. In this chart we can see that those parents whose do not think much about their status and do not pay much attention to their children compared to them, parents who pay attention to their children for maintain status that students have higher performance [10].

- According to Fig. 4, the children are overall performance high whose mother and father are educated and the performance of children of educated parents, just because educated parents understand the importance of children overall performance, at the same time educated parents also help to teach and learn their children a lot [11]. Educated parents understand the importance of peace in the home, and they try to give their children an anxiety-free environment. Educated parents also inspire their children to decide a final aim [12]. Due to which children spend all their efforts to increase their performance to achieve their final aim, and they are also successful in this task due to peaceful environment [13].

- According to Fig. 5, the figure of students using ICT tools to enhance their performance and learn new things are much higher than that of students not using ICT [14]. This chart also shows the impact of ICT on student's performance [15]. Using the tools of ICT, we get the idea of learning and teaching new things. In the chart, we saw how children increase their performance using ICT facility [16] and children who do not know how to use ICT tools or who have not been able to use ICT tools or those who do not know the importance of ICT tools, those

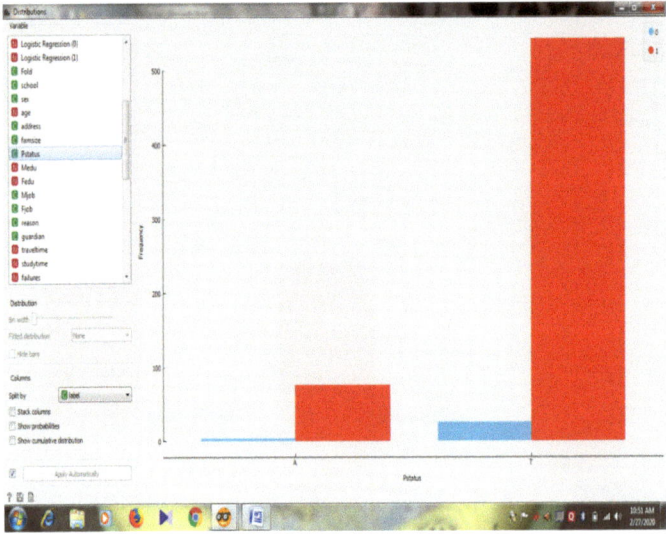

Fig. 3 Compare according Pstatus

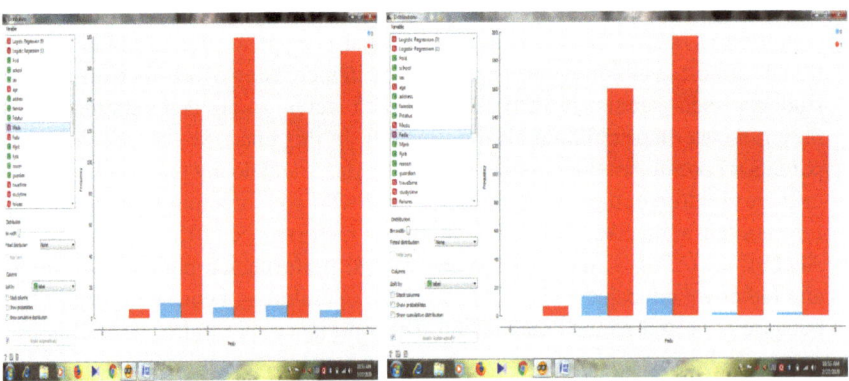

Fig. 4 Compare according to mother and father education

students were somewhere behind to learning and understand new things from students using ICT [17].

Today's ICT tools have become very essential through which students collect study material and new ideas for different creativity, therefore, ICT is considered an important part of the environment ultimately [18], and the conclusion is that the environment has an impact on student performance [19]. The effect of a good

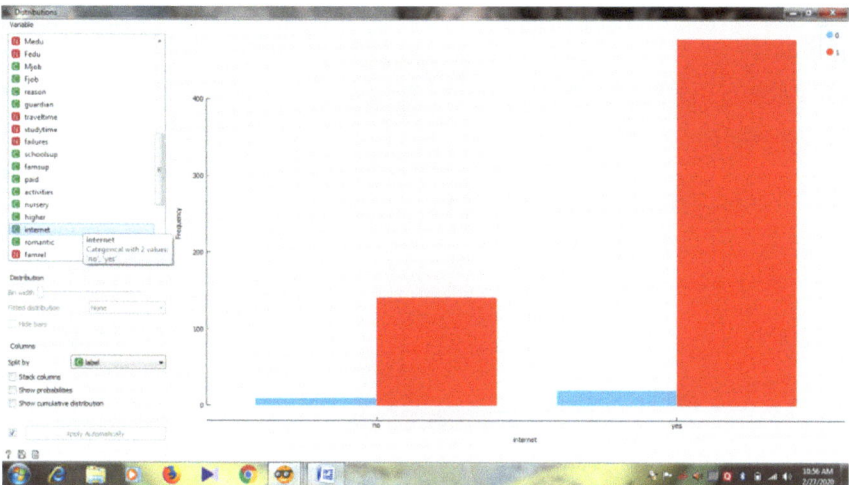

Fig. 5 Compare according to using ICT

environment gives positive results, and the result of poor environment continues to worsen the performance of students [20].

- According to Fig. 6, we saw that those students who are belonging to joint family have more opportunity to learn many things with all family members. But the learning ability of those students who belong to single family is very limited. This is the main reason due to which the performance of the students who belong to the joint family is better than those who belong to the single family.

4 Result

When we adopt all the processes and methodology to analyze data, the result is given in Table 2.

In Table 2, we can see that logistic regression is best model compared to other models because this model gives more accuracy than constants, tree, and neural network model. After checking the accuracy, we will now see the chart, by which we observe the result of different environment. With the help of scatter plotter, we can clearly visualize the effect of mother education (medu) and father education (fedu) on the students' performance in Fig. 7.

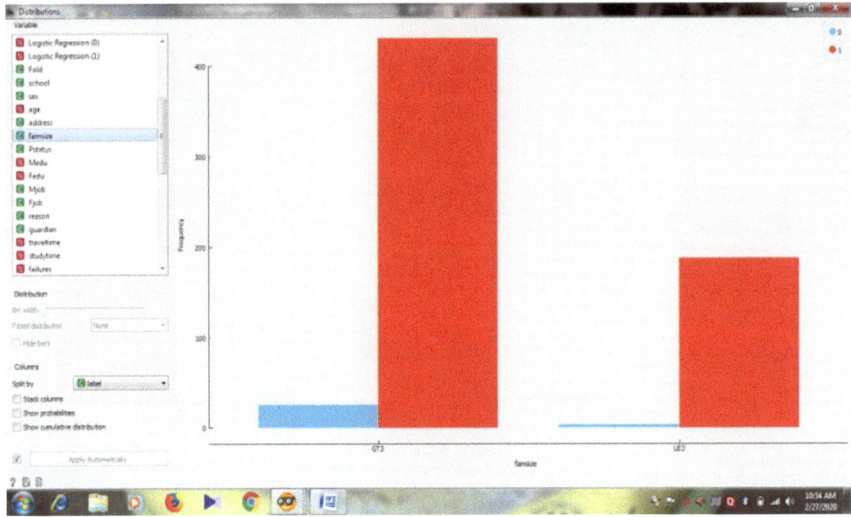

Fig. 6 Compare according to family members

Table 2 Models with their accuracy

S.No.	Used models	Accuracy (%)
1	Constants	91.0
2	Tree	91.0
3	Logistic regression	**98.6**
4	Neural network	97.0

Bold—On the performance scale, the accuracy of logic regression is an optimal

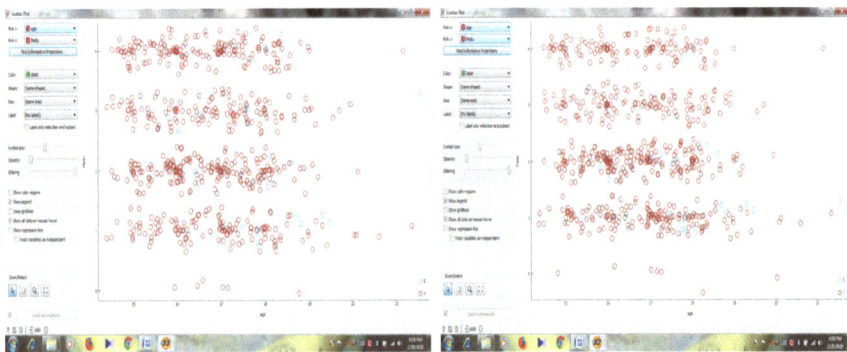

Fig. 7 Scatter plotter figure of medu and fedu

5 Conclusion

Finally concludes that environment has the most impact on the performance of children as we can see in the analysis tables and figures. Rural and urban area, parent status, educated and motivational family, and use of ICT tools all are part of environment. We can see that the rural and urban areas have an impact on student performance. Parent's status affects the student performance. The ICT tools are making a significant importance among an educated mother and father to facilitate the best teaching and learning in overall consideration. After this analysis, we have got positive result only, and this positive result will be responsible for giving positive thinking to the parents and to increase the performance of the student. With the help of scatter plotter, we can clearly visualize the effect of mother education (medu) and father education (fedu) on the students' performance. Family social status plays vital role in the student performance.

References

1. Abrami PC, D'Apollonia S, Rosenfield S (2007) The dimensionality of student ratings of instruction: what we know and what we do not. In: The scholarship of teaching and learning in higher education: an evidence-based perspective. Springer, pp 385–456
2. Adjei SA, Botelho AF, Heffernan NT (2016) Predicting student performance on post-requisite skills using prerequisite skill data: an alternative method for refining prerequisite skill structures. In: Proceedings of sixth international conference on learning analytics & knowledge, ACM, pp 469–473
3. Aghabozorgi S, Mahroeian H, Dutt A, Wah TY, Herawan T (2014) An approachable analytical study on big educational data mining. Springer, International conference on computational science and its applications, pp 721–737
4. Singh N, Gunjan VK, Mishra AK, Mishra RK, Nawaz N (2022) SeisTutor: a custom-tailored intelligent tutoring system and sustainable education. Sustainability 14(7):4167
5. Awantika PM, Tiwari R (2020) A novel based AI approach for real time driver drowsiness identification system using Viola Jones algorithm in MATLAB platform. Solid State Technol 63(05):3293–3303. ISSN: 0038-111X
6. Angeli C, Howard S, Ma J, Yang J, Kirschner PA (2017) Data mining in educational technology classroom research: can it make a contribution? Comput Educ 113:226–242
7. Kumar S, Gunjan VK, Ansari MD, Pathak R (2022) Credit card fraud detection using support vector machine. In: Proceedings of the 2nd international conference on recent trends in machine learning, IoT, smart cities and applications: ICMISC 2021. Springer, Singapore, pp 27–37
8. Sunitha P, Ahmad N, Barbhuiya RK, Gunjan VK, Ansari MD (2022) Impact of Covid-19 on education. In: Kumar A, Mozar S (eds) ICCCE 2021. Lecture notes in electrical engineering, vol 828. Springer, Singapore. https://doi.org/10.1007/978-981-16-7985-8_124
9. Rashid E, Ansari MD, Gunjan VK (2022) Innovation and entrepreneurship in the technical education. In: Kumar A, Mozar S (eds) ICCCE 2021. Lecture notes in electrical engineering, vol 828. Springer, Singapore. https://doi.org/10.1007/978-981-16-7985-8_125
10. Singh N, Gunjan VK, Kadiyala R, Xin Q, Gadekallu TR (2022) Performance evaluation of SeisTutor using cognitive intelligence-based "Kirkpatrick Model". Computat Intell Neurosci 2022, Article ID 5092962:14 p. https://doi.org/10.1155/2022/5092962

11. Gaddam DKR, Ansari MD, Vuppala S, Gunjan VK, Sati MM (2022) Human facial emotion detection using deep learning. In: ICDSMLA 2020: proceedings of the 2nd international conference on data science, machine learning and applications. Springer Singapore, pp 1417–1427

12. Naseer M, Zhang W, Zhu W (2020) Early prediction of a team performance in the initial assessment phases of a software project for sustainable software engineering education. Sustainability 12:4663

13. Pathak R, Prasad PS, Gunjan VK, Solanki VK (2020) Normalization techniques in multi modal biometric. In: ICCCE 2019: proceedings of the 2nd international conference on communications and cyber physical engineering. Springer Singapore, pp 425–431

14. Pozo-Sánchez S, López-Belmonte J, Rodríguez-García AM, López-Núñez JA (2020) Teachers' digital competence in using and analytically managing information in flipped learning. Cult Educ 32:213–241

15. Pathak R, Soni B, Muppalaneni NB (2023) Role of Blockchain in health care: a comprehensive study. In: Gunjan VK, Zurada JM (eds) Proceedings of 3rd international conference on recent trends in machine learning, IoT, smart cities and applications. Lecture notes in networks and systems, vol 540. Springer, Singapore. https://doi.org/10.1007/978-981-19-6088-8_13

16. Mokhlesabadifarahani B, Gunjan VK (2015) Introduction to EMG technique and feature extraction. EMG signals characterization in three states of contraction by fuzzy network and feature extraction. Springer, Singapore, pp 1–9

17. Bondarenko OV, Pakhomova OV, Lewoniewski W (2020) The didactic potential of virtual information educational environment as a tool of geography students training CEUR workshop proceedings 2547:13–22

18. Chorna OV, Hamaniuk VA, Uchitel AD (2019) Use of YouTube on lessons of practical course of German language as the first and second language at the pedagogical university CEUR workshop proceedings 2433:294–307

19. Du L, Feng Y, Tang LY, Kang W, Lu W (2020) Networks in disaster emergency management: a systematic review. Nat Hazards 103:1–27

20. Shan S, Zhao F, Wei Y, Liu M (2019) Disaster management 2.0: a real-time disaster damage assessment model based on mobile social media data-a case study of Weibo (Chinese twitter). Saf Sci 115:393–413

Phishing Email Mitigation Technique Using Back-Propagation Neural Network for Cyber Space

Swapnil P. Goje, Gufran Ahmad Ansari, Mohd Dilshad Ansari, and Sumegh Tharewal

Abstract Nowadays email is gaining significant popularity over the past years. Peoples use email as a one of the fastest formal mediums through which they are sending messages and exchanging some files. Due to tremendous use of email between peoples, the new serious threat comes into picture which will be malicious email. The malicious email can be phishing or spam email. Phishing email means financial loss or identity theft to the person who receive phishing email and through which he/she surrenders or gives his confidential information to the phisher, and spam means unsolicited bulk commercial email or advertising emails. The detection of phishing email is one of the crucial parts, and most of detection schemes are not intelligent as has to be. In this paper, we have proposed a new model for detecting phishing email through the use of artificial neural network (ANN). We have used multilayer feed-forward neural networks for phishing email detection, and both legitimate and phish emails are used for training and testing purpose. We have used optimum set of features which are always present in the phishing email, and using these features, we have got 99.79% accuracy in terms of email classification.

Keywords Social engineering · Spam · Email filtering · Phishing · Neural network · Cyber space

1 Introduction

The social engineering term in cyber security refers to the manipulation of the person to inducing them to do particular actions, and by that action, a person can give certain kind of important information under some warning or due to some fear which was imposed by the attacker. For getting the information from the person as the

S. P. Goje · G. A. Ansari · S. Tharewal
Dr. Vishwanath Karad, MIT World Peace University, Pune, India

M. D. Ansari (✉)
Guru Nanak University, Hyderabad, India
e-mail: m.dilshadcse@gmail.com

© The Author(s), under exclusive license to Springer Nature Singapore Pte Ltd. 2023
A. Kumar et al. (eds.), *Proceedings of the 4th International Conference on Data Science, Machine Learning and Applications*, Lecture Notes in Electrical Engineering 1038,
https://doi.org/10.1007/978-981-99-2058-7_27

attacker need not require a proficient about technical knowledge, but it needs person's attention in terms of interest, consideration, greed, and gullibility.

Phishing is a kind of social engineering attack which can be done using message, phone call, email, etc. As email is one of the fastest and primary media used for communicating between each other and by which we are also sending some data in terms of attachment, majorly email is used for conducting phishing. The phishing email looks same as a legitimate one, and its main objective is to steal the sensitive or confidential information about the user (or receiver of phishing email) which can be useful to the attacker.

First upon the phisher compromise a host and install a phishing Website which looks as close as similar to genuine or legitimate Website. The life of phish Website may be ranged from an hour to a day. The phisher includes the URL of phishing Website into email with using HTML, images, Javascripts, etc., so making it as close to legitimate email. Now, this phishing email is sent to mass mailer, and upon receiving the phishing emasil, user submits his sensitive of confidential information. The sensitive information may include username, password, bank account details, credit card no., etc.

Phishing can also be accompanied with the use of some malicious code or viruses, Trojans can be included as an attachment in email, and those attacks are now considered in targeted malicious email category. These Trojans or viruses are sending the sensitive data of the receiver to attacker anonymously.

The total unique number of phishing reports submitted to APWG [1] during Q3 of 2022 was 1,270,883 which was one of the worst observations ever for all observations till date. The 23.2% phishing attacks are against financial sectors. Fraud scams related to advance fee are increased 1000% in this quarter.

In this paper, we have used neural network for email classification into phishing and ham, i.e., legitimate email. For this purpose, we have used back-propagation neural network [2] for detection of phishing email. In the first phase, email preprocessing of email is done for making it suitable for email feature extraction. The features extracted are used for training purpose, and the features are kept optimum so that we get accurate results. The main objective of our approach is to classify phishing emails using a set of unique properties that remain present across a large number of phishing emails. The features are based on content, link, element, and structure based by which more accurate results we get. In training part, we have given some training samples of phishing and legitimate email along with its desired target value and calculated error value between desired target and actual target and propagate this error value to the hidden layers of network till the error value gets minimized to desired value. After training the network with all samples, we have tested it with some unknown samples of email for classification and validation purpose.

The rest of this paper is structured as follows: We start by presenting related work in Sect. 2, where we have given brief summary of existing approaches and their accuracy. The proposed methodology along with the algorithms and its application for phishing email detection, features used in our approach and algorithm are presented in Sect. 3. The results and discussion are presented in Sects. 4 and 5, respectively.

2 Related Work

The main research in the email classification is devoted to spam email, which is unsolicited bulk email, and their objective is not to steal confidential information of users rather it is used for only advertisement purpose. In this section, we are discussing existing research which is already happened for phishing email classification.

A machine learning approach [2] for phishing Website detection using neural network was proposed. Total 18 features are used for training and testing purpose which includes IP address, URL having @ symbol, prefix and suffix of URL, misuse of HTTPs protocol, etc. The main idea is to find by cross validation how many hidden layer and hidden neurons give mean square error minimum. Total 12 runs are carried out, and best performance is achieved when hidden layer neurons are set to 2 with one hidden layer at 0.7 learning rate. The mean square error got was 0.002234 which was good than another runs.

Reference [3] uses total 25 features which include style marker and structural attributes of phishing email. The style marker features include total number of words, total number of function words, vocabulary richness, etc., along with structure of email body and structure of greetings provided in the email body. The unwanted features can affect the accuracy for that simulated annealing algorithm is used for proper feature subset selection. Total 200 emails are used for training in which 100 are phishing and 100 are ham emails. For classification purpose SVM light this is implementation of vapanik's support vector machines. The experiment is carried out in V runs, and in each run, different features are used for finding accurate results. Removal of structural attributes reduces the accuracy by 20%. Overall accuracy by using this approach is around 95%.

PILFER, a machine learning approach, is used in [4] for email classification purpose. Total ten features of phishing email are used which includes IP-based URLs, number of links, number of domains, number of domains, Javascript, etc. For classification purpose, random forests classifier is used which creates decision tree which is made by choosing attribute randomly split at each level and pruning the decision tree. Overall accuracy achieved by this approach is 99.5%

Clustering of phishing emails proposed by [5] includes orthographic features such that HTML features, text content, document size, and other elements. The proposed system gathers all possible features with adaptive k-means clustering algorithm. The algorithm produces objective function value over a range of values. Total 2048 emails from Australian bank are tested. The accuracy of this approach is based on efficiency of the feature selection and indication of correct clusters.

Reference [6] proposed an automated real-time URL phishingness rating system to protect users against phishing attack. This approach is based on observing phishing URLs with relationship between domains with rest of URL. Seven classifiers tested on function-based (SVM), rule-based (JRip, PART) and tree-based (C4.5, random forest, LMT, and random tree). The classification was made without parameters tuning through a ten-fold cross validation as a first step to select the most promising approach.

The semantic ontology concept was proposed by [7] with adaptive Naive Bayes algorithm. It suggests that every word in email is an attribute and the value of word means its frequency of that word. The stages in this proposed method are eliminating HTML tags, steaming as a part of email preprocessing, and then compiling ontology concept on training dataset after that applying TFV [8] method along with Naïve Bayes classifier for classification purpose. The accuracy approached is 94.87%, but this technique is only depending on one aspect of features as compared with other techniques which are based on multiple aspects of features, and also, the level of accuracy is low.

Amin et al. [9] invented newest class of email category was invented and which was named as targeted malicious email. This kind of attack cannot be detected by conventional detection schemes which was detecting only spam or phishing emails. These email attacks targeting single users or small groups. It also appears legitimate so the received might misguided. They have used total 83 features, but only ten features are used by the random forest classifier. The features are categorized into recipient-oriented and persistent threat. Ten-fold cross validation for validation purpose is performed using NTME1–TME1 dataset, and then, independent TS1 dataset is used for evaluation purpose with 91% accuracy.

A sender-centric approach [10] is proposed for detecting phishing emails. Only banking emails are considered for training and testing purpose. The proposed system contains two steps which include classification of banking and non-banking messages by using SVM classifier, and then in second step, certain rules are applied. Total three rules are defined which include emails from public service providers, sender geographical locations, and authorized sender. Accuracy of SVM classifier was 98.94% and for rule-based classification was 98.7%. The main weakness of this system is that it only detects banking phishing messages.

Reference [11] used hybrid features for phishing email detection. The features used in this approach are based on content-based, behavior-based, and URL-based. Total 500 phishing emails are used in this proposed system, and 97.25% accuracy got in terms of correct prediction.

Reference [12] used five categories of features for phishing email classification, and they are related to subject of email, body of email, attachment-related attributes, forwarding, and reputation. In email subject, they have used seven features, body-related 12 features, attachments-related two features, forwarding-related three features, and reputation-related three features. All features are normalized and enhanced using KM-SMOTE and trained and tested on random forest, decision tree, logistic regression, and support vector machine. The dataset contains 13,916 non-spear phishing emails and 417 spear phishing emails. Author achieved F1 score 97.16%, precision of 98.85%, and recall of 95.56%.

3 Proposed Methodology

To obtain knowledge by learning from the data automatically, through a process of interpretation, or learning from example is one of the rare fields in email classification. The previous researches are mainly focusing on spam email detection, black listing, and white listing of phishing Websites. The use of neural network for email classification makes our approach accurate than other methods.

3.1 Proposed Model

In our approach, we have used labeled phishing and ham emails dataset for training purpose. As the dataset is in raw format and back-propagation neural network accepts an input value ranging from 0 to 1, we need to perform some preprocessing on training dataset and also while testing the dataset for accuracy purpose. For preprocessing, we have written some short scripts using Java for extracting the features as required for the phishing email detection (Fig. 1).

The data preprocessing includes following stages:

(a) *Tokenization*: Firstly, the dataset is not directly applied to back-propagation neural network, so we have written small script using Java for tokenizing the email, i.e., we have separated words from email by finding whitespaces (tabs, spaces, and new lines) as the delimiter.

(b) *Feature Extraction*: The desired features which we have proposed are extracted from email by finding their occurrences in the email.

(c) *Feature Vector Formation*: When desired features are extracted from each email, a matrix is used to store the features which indicates that each row represents one email and each column of that matrix represents feature or attribute value for that email. This feature vector is used for training the back-propagation neural network (BPNN) model.

(d) *Feature Vector Normalization*: As the feature extracted is having value either as binary or may be a number. Binary value 1 indicates presence of that feature, and 0 indicates absence of that feature in that email. For non-binary feature, the value has to be normalized before applying feature vector to BPNN model, and for that, we have normalized the value on the basis of a threshold value for that particular feature.

After preprocessing gets completed for all emails, the feature vector is applied for training back-propagation network, and once back-propagation network is trained, then it is tested on an unlabeled dataset of either phishing or ham emails. While training the labeled dataset, 1 represents phishing email, and 0 represents ham or legitimate email.

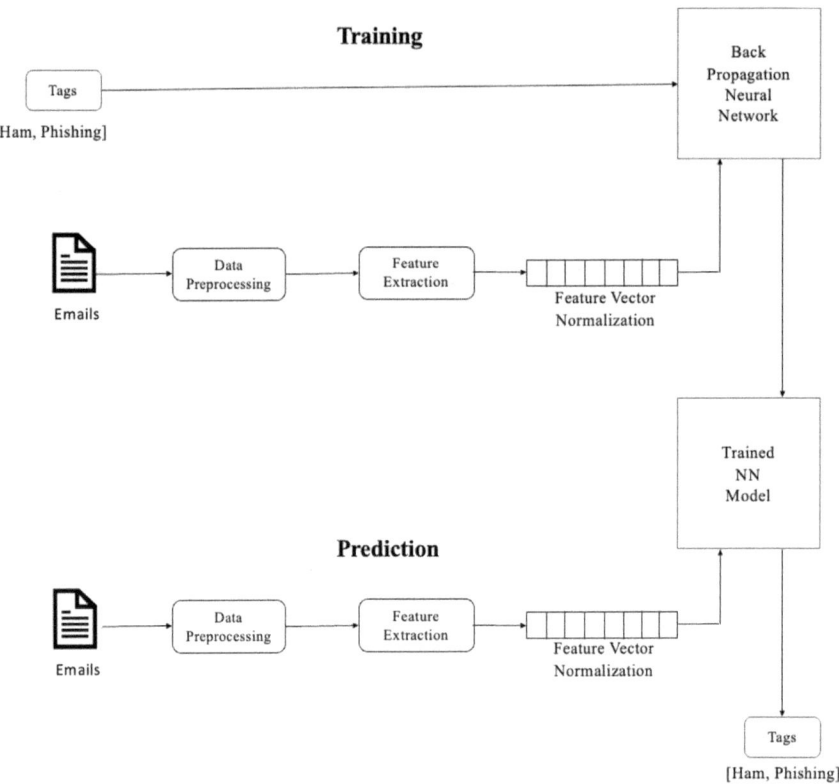

Fig. 1 Proposed model

3.2 Features Used in Classification

There are multiple features which differentiate phishing email from ham email. We have extracted an optimized set of features which denotes the characteristics of the phishing email. Some of the features are already discussed in the previous proposed researches [5, 13, 14].

(a) **HTML Email**: It is a binary feature; this feature is found in the content type parameter which is present in the email header. Majorly the phisher crafted phishing emails using HTML so making the same replica of legitimate email.

(b) **Links Following an Image**: URL-based images included in an email, in order to formulate emails similar to legitimate email and linking this URL to phishing Website.

(c) **Simple Links**: Phisher includes more links into email, so it diverts the mind of receiver to click on to that links.

(d) **Form Tag**: It is the technique by which phisher gathers information from the receiver.

(e) **Use of Javascript**: It is used to help the phisher for creating some pop-up windows or changing the status bar behavior using Javascript.

(f) **Use of IP-Based URLs**: Normally phisher do not have associated a domain name with their phishing Website, so they used IP-based URLs for phishing purpose and includes it into an email.

(g) **Click or Click Here Keyword on Link**: Phisher employs this technique to spoof the receiver by using click or click here word along with the link, so by clicking on that link, they give their own username, password, bank a/c details, etc.

(h) **Deceptive Links**: Some phishers fool the receiver by using visible URLs but pointing to different URLs, so receiver might think the link is authentic one.

(i) **Valid Port**: If the port number included in the URL does not belong to well-known port, then we consider URL belongs to phishing Website.

(j) **Personal and Account Details**: Some keywords which likely to appear in phishing email like username, password, account no., verify, secure, and confirm are used by phisher for making urgency to the receiver. We have extracted those keywords from the email body.

3.3 Artificial Neural Network

A neural network is powerful and important data mining tool which is able to capture and characterize input–output relationships [2, 15]. The neural network is majorly used as a tool for classification and clustering. If neural network is trained with enough set of data, then it can ever perform classification and might find new patterns in data.

Using back-propagation neural network for email content classification becomes our technique distinct and effective. The use of back propagation neural network will help in performing in a better way on adaptive learning, intelligence and self-organization [16].

The main challenge in our approach is to decide the number of hidden layers and number of hidden neurons in each level, and by observing the features, we have used one hidden layer along with five hidden neurons for making calculations easily and correctly. The learning rate also plays an important role in training the network, the higher learning rate results into fast training, but results in wrong predictions, and low learning rate results in slow leaning, so we have used a standard learning rate 0.35. The sigmoid activation function is used for neurons activation [17–24] (Fig. 2).

The algorithm can be decomposed into four stages.

1. Feed-forward computation.
2. Back-propagation to the output layer.
3. Back-propagation to the hidden layer.
4. Weight updates.

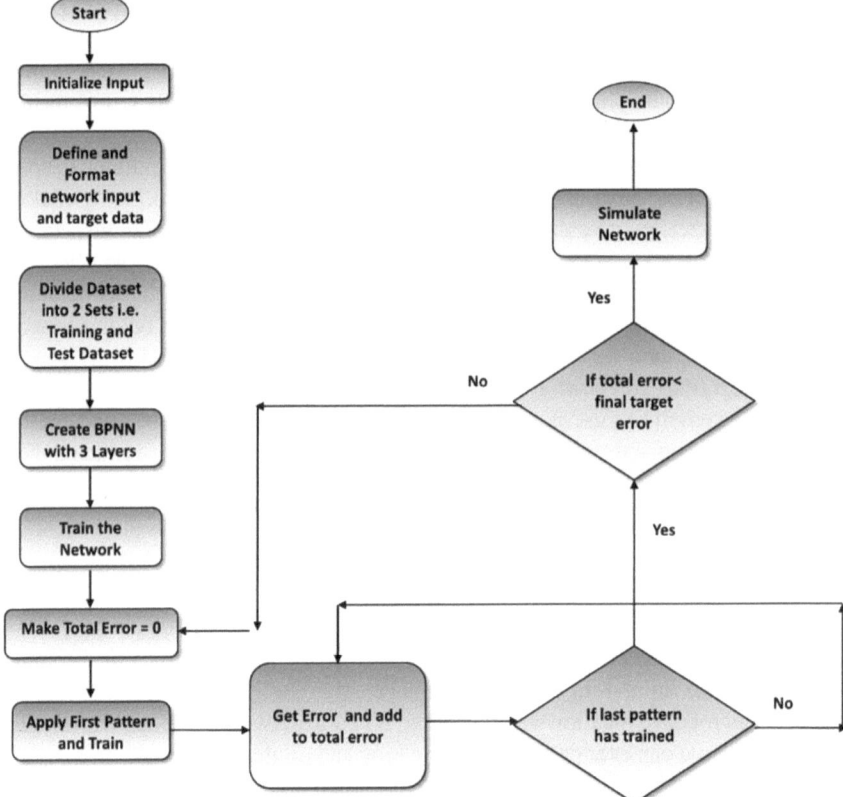

Fig. 2 Proposed flowchart

The back-propagation algorithm is described as following.

Algorithm:

Input:

- D, a dataset which consists of training tuples and their associated target value;
- Learning rate, l;
- Multilayer feed-forward network.

Output:
Trained Neural Network.

3.4 Datasets

Publicly available datasets are used to test our implementation: The phishing datasets are from [25] and ham datasets from [26]. Total 1326 phishing email and 376 ham emails are used for training and testing purpose. We have taken a small quantity of ham emails because the main idea is phishing detection not ham detection. The desired features from each email are extracted by executing some small scripts written in Java.

4 Results and Analysis

In this section, we have shown the implementation of the proposed system. We have used Java framework to implement the phishing email detection system. Figure 3 shows we have imported the train dataset for training purpose. This train dataset is used to train the BPNN. Figure 4 shows the training of the loaded dataset.

Figure 4 shows that train dataset contains both ham and phishing emails. After training, the trained BPNN used to test the test dataset.

Figure 5 shows once the training gets over, it has been tested on single email and to check whether it is ham or phishing (malicious) email.

Figure 6 shows that we have tested the trained BPNN with 1143 phishing and 324 ham emails. Figure 6 shows the actual results we have got after training the BPNN.

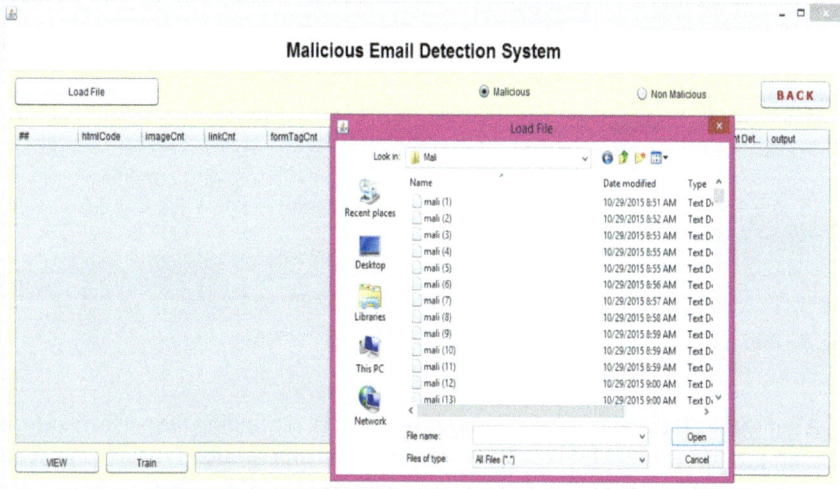

Fig. 3 Importing the train dataset

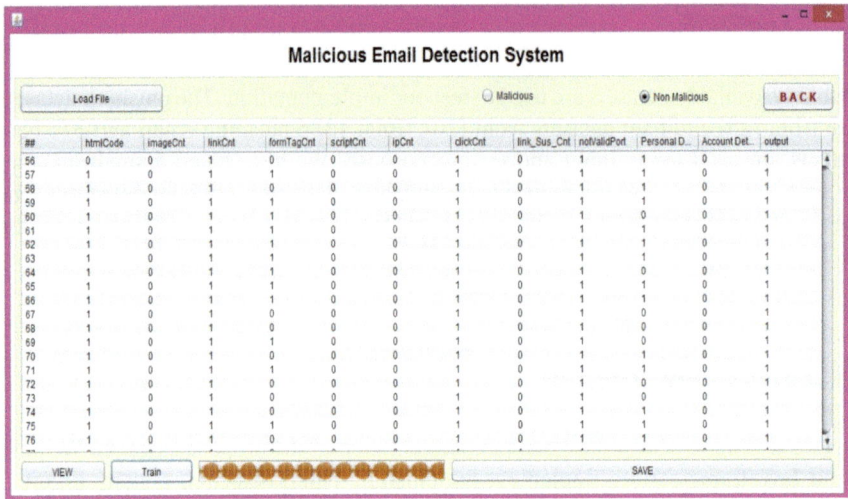

Fig. 4 Training of ham and phishing dataset

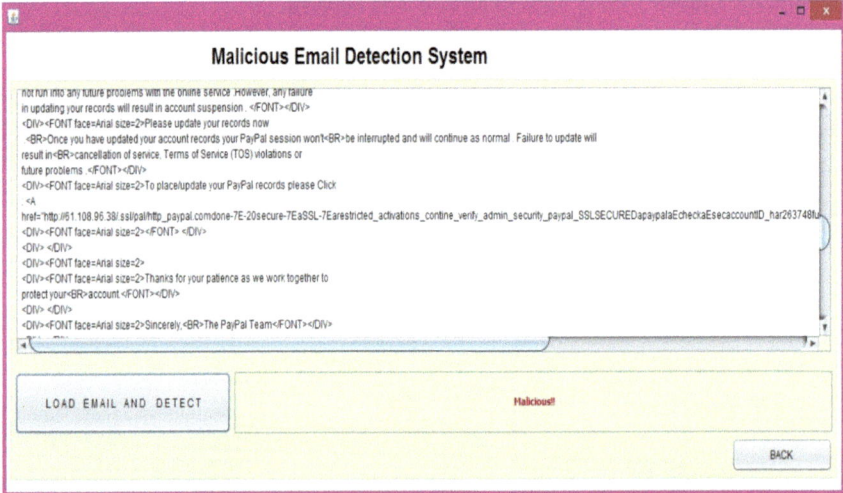

Fig. 5 Testing on single email

After testing the trained back-propagation network for finding the performance, as per Table 1. TN denotes correctly classified ham emails, while FP denotes misclassified ham emails. As FP in our approach is 0 and TN is 324 that means all ham emails are correctly classified. The TP denotes correctly classified phishing emails, and FN denotes misclassified phishing emails and considered as ham emails. Only three emails from phishing are misclassified as normal emails (Fig. 7).

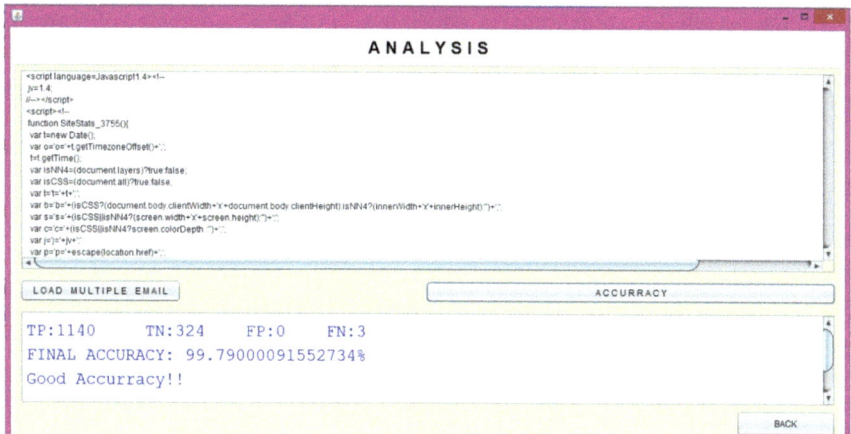

Fig. 6 Test results

Table 1 Result summary

Neural network results					
Total no. of phishing emails	Total no. of ham emails	True positive (TP)	True negative (TN)	False positive (FP)	False negative (FN)
1143	324	1140	324	0	3

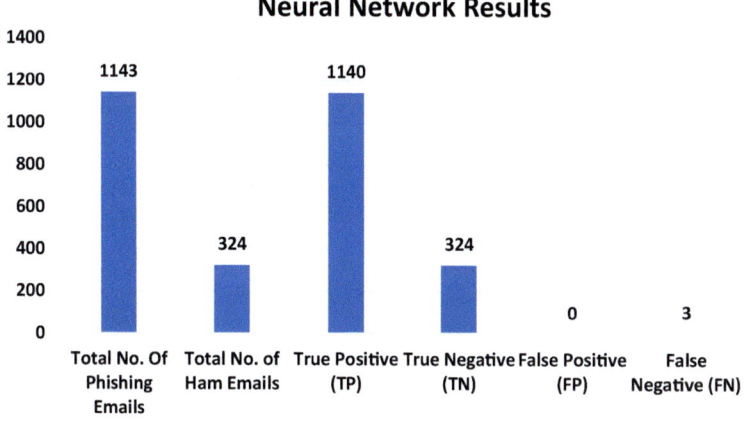

Fig. 7 Neural network results

So, the overall detection accuracy which measures total percentage of correctly classified emails including phishing and ham emails is 99.79%. The figures from Table 1 indicate that false positive rate (FPR) is almost 0% while false negative rate is 0.34%.

5 Conclusion

The accuracy achieved by the proposed system was significantly higher than that of the existing methods. We have not considered the attachment or some malicious code-related features, but it can be implementable in future scope. It may be possible to extract all features from phishing email, but by using some feature selection algorithm along with neural network will be beneficial to find zero-day attack. We have used only a small subset of phishing email for experimental purpose; it may be applicable to train and test a greater number of emails.

References

1. Anti-Phishing Working Group (APWG) (2023) Phishing activity trends report—third quarter 2022. https://docs.apwg.org/reports/apwg_trends_report_q3_2022.pdf. Accessed on 2 Feb 2023
2. Mohammad RM, Thabtah F, McCluskey L, Predicting phishing websites using neural network trained with back propagation. In: 2013 world congress in computer science, computer science engineering, and applied computing, Las Vegas, USA, pp 682–686
3. Fette NS, Tomasic A (2007) Learning to detect phishing emails. In: Proceedings of 16th international world wide web conference (WWW 2007), ACM Press, Banff, Alberta, Canada, pp 649–656
4. InterNIC. Whois search, InterNIC—public information regarding internel domain name registration services. Accessed 16 June 2012. Available http://www.internic.net/whois.html
5. Ma L et al (2009) Establishing phishing provenance using orthographic features. In: IEEE conference, pp 1–10
6. Marchal S, Francois J, State R, Engel T (2014) Phish storm: detecting phishing with streaming analytics. In: IEEE transactions on network and service management
7. Bazarganigilani M (2011) Phishing e-mail detection using ontology concept and Naive Bayes algorithm. Int J Res Rev Comput Sci 2(2)
8. Guzella TS, Caminhas WM (2009) A review of machine learning approaches to spam filtering. Expert Syst Appl 36:10206–10222
9. Amin RM, Ryan JJCH, René van Dorp J (2012) Detecting targeted malicious email. Secur Priv 10(3):64–71
10. Sanchez F (2012) Florida State Univ., Tallahassee, FL, USA; Zhenhai Duan, A sender-centric approach to detecting phishing emails. In: 2012 international conference on cyber security, pp 32–39
11. Form LM, Chiew KL, Sze SN, Tiong WK (2015) Phishing email detection technique by using hybrid features. In: 9th international conference on IT in Asia (CITA), pp 1–5

12. Ding X, Liu B, Jiang Z, Wang Q, Xin L (2021) Spear phishing emails detection based on machine learning. In: 2021 IEEE 24th international conference on computer supported cooperative work in design (CSCWD), Dalian, China, pp 354–359. https://doi.org/10.1109/CSCWD4 9262.2021.9437758
13. Aburrous M (2010) Predicting phishing websites using classification mining techniques with experimental case studies. In: Seventh international conference on information technology, IEEE conference, Las Vegas, Nevada, USA, pp 176–181
14. Bergholz G, Frank SS, Improved phishing detection using model-based features. In: Proceedings of the conference on email and anti-spam (CEAS)
15. Windroe BM, Lehr A (1990) 30 years of adaptive neural networks, vol 78, no 6. IEEE Press, pp 1415–1442
16. Ayodele T, Zhou S, Khusainov R (2010) Email classification using back propagation technique. Int J Intell Comput Res (IJICR) 1(1)
17. Gaddam DKR, Ansari MD, Vuppala S, Gunjan VK, Sati MM (2022) Human facial emotion detection using deep learning. In: ICDSMLA 2020: proceedings of the 2nd international conference on data science, machine learning and applications. Springer Singapore, pp 1417–1427
18. Gaddam DKR, Ansari MD, Vuppala S, Gunjan VK, Sati MM (2022) A performance comparison of optimization algorithms on a generated dataset. In: ICDSMLA 2020: proceedings of the 2nd international conference on data science, machine learning and applications. Springer, Singapore, pp 1407–1415
19. Ahmed M, Ansari MD, Singh N, Gunjan VK, BV SK, Khan M (2022) Rating-based recommender system based on textual reviews using Iot smart devices. Mob Inf Syst
20. Ansari MD, Gunjan VK, Rashid E (2021) On security and data integrity framework for cloud computing using tamper-proofing. In: ICCCE 2020: proceedings of the 3rd international conference on communications and cyber physical engineering. Springer, Singapore, pp 1419–1427
21. Narayana GS, Ansari MD, Gunjan VK (2022) Instantaneous approach for evaluating the initial centers in the agricultural databases using K-means clustering algorithm. J Mob Multimedia 43–60
22. Kumar S, Ansari MD, Gunjan VK, Solanki VK (2020) On classification of BMD images using machine learning (ANN) algorithm. In: ICDSMLA 2019: proceedings of the 1st international conference on data science, machine learning and applications. Springer, Singapore, pp 1590–1599
23. Gunjan VK, Pathak R, Singh O (2019) Understanding image classification using TensorFlow deep learning-convolution neural network. Int J Hyperconnectivity Internet of Things (IJHIoT) 3(2):19–37
24. Gunjan VK, Singh N, Shaik F, Roy S (2022) Detection of lung cancer in CT scans using grey wolf optimization algorithm and recurrent neural network. Heal Technol 12(6):1197–1210
25. Phishing corpus homepage (2006) [Online]. Available: http://monkey.org/%7Ejose/wiki/doku.php?id=PhishingCorpus
26. Spamassassin public corpus (2006) [Online]. Available: http://spamassassin.apache.org/public corpus

Crime Visualization and Forecasting Using Machine Learning

Shivani Sharma, Bipin Kumar Rai, Gautam Kumar, Aakash Prajapati, and Vaibhav Kumar

Abstract Crime is a major issue in many countries around the world. For police forces to deal with inevitable increases in crime rates as a result of urbanization, crime prediction and prevention measures are critical. The crime analysis can assist with overcoming any issues from what information we have to what we need to comprehend crime and assist us in developing computer models for a crime. We want to provide a simple platform to predict crimes as per the requirements of the police officials, like time and type of crimes. These crimes are likely to happen at a place using the dataset collected from verified and official sources such as National Crime Records Bureau (NCRB) and data.gov.in with the help of K-means and logistic regression. We aim to reduce the workload of the investigation by helping them to connect places and people with crimes and find the solutions to curb the crimes. Advanced technologies and novel techniques to enhance crime analytics are needed to safeguard communities and keep society safe from criminals. We propose a platform for creating a visualization by analyzing, detecting, and predicting various crime patterns on three bases, i.e., region, month, and age range (young adults, old).

Keywords Crime analysis · Regression · Prediction · Machine learning · NCRB · Forecasting

S. Sharma · B. K. Rai (✉) · A. Prajapati · V. Kumar
Department of IT, ABES Institute of Technology, Ghaziabad, Uttar Pradesh, India
e-mail: bipinkrai@gmail.com

S. Sharma
e-mail: shivani.shivanisharma@gmail.com

G. Kumar
CMR Engineering College, Hyderabad 501401, India

© The Author(s), under exclusive license to Springer Nature Singapore Pte Ltd. 2023   307
A. Kumar et al. (eds.), *Proceedings of the 4th International Conference on Data Science, Machine Learning and Applications*, Lecture Notes in Electrical Engineering 1038, https://doi.org/10.1007/978-981-99-2058-7_28

1   Introduction

Crimes are a typical social issue influencing the general public's satisfaction and economic development. Crime is a major aspect in determining whether or not people should relocate to some other city and what areas they should avoid when traveling [1]. As the number of infractions increased, law enforcement agencies wanted better spatial data frameworks and new information mining methods to improve crime investigation and protect their organizations. Although breaches could happen all over, offenders must deal with crime openings; they look in the most commonplace territories for them. The increasing population and gaps in society have led to many crimes in metropolitan areas. It is challenging to handle crime without an assessment or a visualized report on crimes, even authorities [2]. The research aims to provide a platform to analyze the crimes in India.

The criminals are active and operate in their comfort zones; this helps predict crimes. Once thriving, they try to recreate the crime in a comparable setting [3]. Crime rates were determined by various elements, including criminal intelligence, location security, etc. The work has followed the steps used in data analysis, in which the crucial phases are data collection, data classification, pattern identification, prediction, and visualization. The proposed system employs a variety of visualization tools to depict crime trends and multiple methods for predicting crime using machine learning algorithms.

Above all the work, the crucial one is the data, both the collection and its source detection. If we were talking about a country like the USA, its crime governing body is the prime source as it provides its data openly on a cloud platform: "Crime Data Explorer" (CDE) [4]; it is free and public. The data is dynamic in many cases for some places. This is not the case in India. We have a body like NCRB to maintain the records, but that is note as dynamic or updated as the FBI CDE but can be used. Similarly, since no crime data is open source in India, techniques like OSINT cannot be involved here as this part of the research itself would become focused on the area or region like the case of the Delhi Police using OSINT and EIIS in 2017 in surveillance projects and huge and time taking.

Even if we try out these techniques to extract data for our research, there is no trustworthy verified source of criminal records that could be used to make our platform dynamic. So, in the end, we decided to use the current data source and mold the data as per our use.

1.1   Work Objective

In our paper, we aim for two things to be accomplished as (i) analyzing previous crime rates and kinds of crime then predicting the possibility and the type of crime most likely to happen and (ii) a platform to find in all probability crime areas and their successive event time.

2 Related Work

The previous works related to this field guide and the research problem. We can show the contributions and limitations we have overcome while doing our research by mentioning them. We have tried to showcase all the research work before placing our piece against the problem statement:

Lisowska-Kierepka et al. (2021): The paper has restructured the spatial algorithm use and plotted the spatial graph on the ten-year data of Wroclaw (Poland) in 2006–2015. The approach is to create an indicator of crime risk in the area in a year. Based on Moran's I and general G, autocorrelation analysis validated this indicator (Getis-Ord General G Statistic) [5].

Zhu and Wang (2021): This paper is quite close to our approach. Its primary outcome includes the model's capacity to successfully replicate hotspots of robbery in the research area, as well as the consistent effects of target definition and strategy on model performance. The study also identifies missing model components that could effectively constrain the displacement of crime opportunities under hotspot policing, which could be the key to resolving the problem—the discrepancy between our simulation results and those of other empirical studies and crime simulations [6].

Shah N., Bhagat N., Shah M., et al. (2021): This paper describes the results of some instances where such approaches inspired them to conduct more research in this area. The authorities before and after statistical observations utilizing such procedures are the fundamental reason for the change in crime detection and prevention. The purpose of this study is to how law enforcement or other authorities can use a combination of machine learning and computer vision to detect criminals, prevent, and solve crimes in a much more accurate and timely manner [1].

Shanjana A. S., Dr. Porkodi et al. (2021): This system can predict regions with a high probability of crime occurrences and visualize crime-prone areas. Using the concept of data mining, we may extract previously unknown, important information from unstructured data. Using the existing datasets predicts the extraction of new information. Crime is a pernicious social problem that affects people all across the world. The quality of life, economic prosperity, and reputation of a country are all impacted by crime [1].

Mahmud, Sakib and Nuha et al. (2021): This study examines Bangladesh's crime rate using several clustering data mining algorithms using the K-nearest neighbor (KNN) algorithm to train the dataset. They used both primary and secondary sources of information. Estimate the prediction rate of various crimes for various places by examining the data, and then use the method to determine the path's prediction rate [7].

Peppesa, T. et al. (2020): The focus of this study is on two elements of a software system for crime prediction and prevention. This project helps law enforcement agencies accept and use emerging technologies like data mining and big data tools, semantic analysis, and visual intelligence in their daily operations. As part of a discussed framework, this research presents a new Semantic Engine with a unique fusion tool and an ontology visualization tool [8].

Yin, Jiarui, Afa, Iduabo et al. (2020): This study presents strategies for fore-casting crimes that are categorized into distinct severity levels. In recent years, it has incorporated visualization and analysis of crime data statistics in Boston. After that, they carried out a comparative study between two supervised learning algorithms, decision tree and random forest, based on the accuracy and processing time of the models to make predictions using geographical and temporal information provided by splitting the data into training and test sets [9].

O. Llaha et al. (2020): This research compares and contrasts data mining approaches and their results in analyzing historical crime data. It compared theo-retically and practically the most relevant data mining approaches for analyzing the acquired data from sources specializing in crime prevention. Gender, age, work position, and crime rate are some of the characteristics of this dataset. Methods are applied to these data to determine their effectiveness in analyzing and preventing crime [3].

C. Vijayalakshmi et al. (2019): The primary goal of this research is to anticipate the type of crime that will occur based on the place where it has previously occurred. Machine learning constructs a model by combining cleaned and altered training datasets. The qualities are implemented with the help of data visualization [10].

Jin Soung Yoo et al. (2019): Using data warehousing and data mining techniques, this study demonstrates how crime log data may be transformed into extremely useful information. Data warehousing ideas are used to organize crime incident data and other pertinent data. Using spatial association rule mining, find interesting local relationship patterns of crime incidents with other spatial features. It also shows how to materialize spatial relationship information between criminal activities and task-relevant other features into data marts and to discover exciting crime patterns from the data mart using a spatial association rule mining technique [2].

Xinagyu Zhao et al. (2018): This article summarizes crime analysis on urban data and provides an outline of major criminological theories. It also examines cutting-edge methods for a variety of computational crime problems. It highlights some intriguing study avenues for taking urban crime studies to new heights [11].

3 Preliminary

3.1 Logistic Regression

We will be using two supervised machine learning algorithms for our research, the KNN algorithm and the logistic algorithm, where one algorithm helps others complete the work. The aim is to get our data classified on the demand of users and then showcase a pattern in the current scenario in crime.

In its most basic form, logistic regression is a statistical model that represents a binary dependent variable using a logistic function; however, there are many more complicated forms. In regression analysis, logistic regression (or logit regression) is a technique for estimating the parameters of a logistic model (binary means two outputs for a dependent value like life or die, 0 or 1). Most of the time, logistic regression is used for binary classification jobs. However, it can also be utilized for multiclass classification [12]. The logistic function, also called the sigmoid function, is denoted as $F(x)$ or $\sigma(z)$ as in the model:

$$F(x) = \frac{1}{1 + e^{-x}} = \frac{e}{e^x + 1}$$

3.2 K-Nearest Neighbor (KNN)

KNN is a supervised machine algorithm used in classification [13]. We can put a KNN model into action by following the steps outlined below:

1. Load the data
2. Initialize the value of k
3. Iterate from 1 to the total number of training data points to determine the projected class.

(i) Determine the distance between each row of training and test data. Because it is the most commonly used method, we will use Euclidean distance as our distance metric here. However, Chebyshev, cosine, and other metrics can also be used.

$$\text{dist}((x, y), (a, b)) = \sqrt{(x - a)^2 + (y - b)^2}$$

(ii) Sort the calculated distances in increasing order based on distance values.
(iii) Get the first k rows from the sorted array.
(iv) Get the most common kind of these rows.
(v) Return the predicted class.

4 Proposed Solution

4.1 Architecture Diagram

A large collection of raw data is presented here which will be processed during the four stages depicted in the Fig. 1 data collection, data cleansing, required input, and prediction using algorithms. If the user does not specify what is needed, then it will be by default provided with the following fields, i.e., age group, gender, and region during the stage of data cleaning. KNN does prediction by taking datasets and user

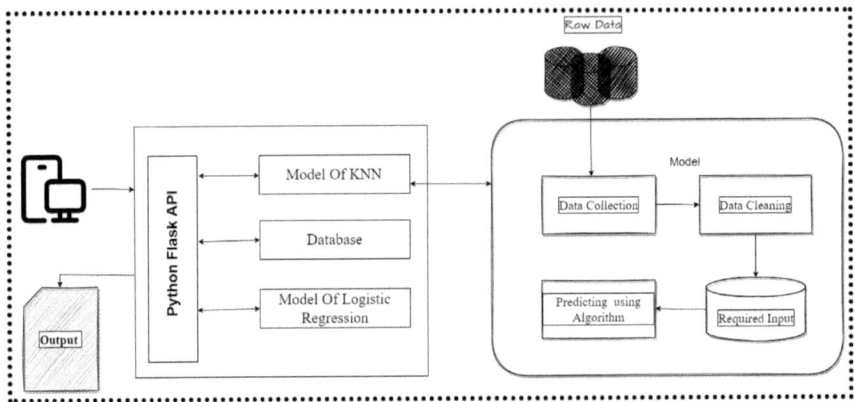

Fig. 1 Architecture diagram of crime prediction and visualization

input, and the same thing is done by using logistics as output. Mathematically, we calculate the mean of KNN and logistics.

4.2 Use Case Diagram

A. **User**: The user is the person from the organization (vigilant criminal authority) who uses the model for visualizing the future prediction of crime that may happen. So that they can take the necessary action to reduce the risk of crime.

    I. **Authentication**: The user provides the details for the login into the application. This section also deals with the registration of new users.

    II. **Category Selection**: The user selects the category in the classified data. We classify all the data into three categories based on age, region, and gender. For the specification in the category, the user selects according to the requirement.

    III. **Visualization**: Its task is to provide the output after all the processes based on the provided input or the model.

B. **System Administrator**: A system administrator is the person or group of members who deal with the backside work, like dealing with the data provided to the machine learning model for the future prediction of crimes. The system administrator also validates the correctness of the past data used in the model.

    I. Authentication:

    The system administrator verifies the user's input details in the login process. Then, the details are inserted into the database for the new users.

Fig. 2 Use case diagram of crime prediction and visualization

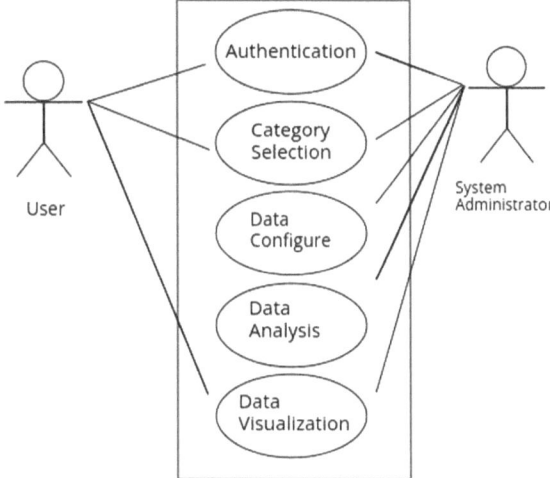

II. Data Configure:
    In this system, administrators configure the past data used to predict the crime. The data is collected from various verified sources. The data can be available in different forms, but the data is needed in the same form in processing, so this is resolved in data analysis [14].

III. Data Analysis:
    In the data analysis phase, our model works and discards the unnecessary data and selects the correct and sound data for prediction. The data analysis phase also helps group the same type of data used in the prediction process for the category selection [15].

IV. Database Configure:
    In the database configure, the processed data is stored in the database after the configure and analysis phase. Then, the visualization of the prediction is done based on this data. As shown in Fig. 2.

4.3 Data Collection

We have collected data for 10 years as provided by the crime authority itself, i.e., National Crime Records Bureau (NCRB). This data would be linked using the API provided by data.gov.in, a government site for datasets through which we are accessing the data. NCRB is an Indian authorities' employer answerable for accumulating and analyzing crime statistics as described using the Indian Penal Code and Special and Local Laws. Our data comes from "Crime in India", NCRB's oldest and most renowned magazine. After each calendar year, the State Crime Records Bureau (SCRB) collects data from the District Crime Records Bureau (DCRB) and provides it to NCRB under reference. Megacities (cities with a population of ten lakh or more according to the most recent census) have their own set of data. The district collects and releases data on several IPC heads independently [16].

4.4 Data Cleaning and Optimization

The data is significant, and with a non-matching format, there is a requirement to optimize all datasets to be used as one and then clean the dataset to remove the unwanted faults the data. However, this is a trivial step as the work depends on the used data.

4.5 Methodology

Dataset:
Our research work required a database that is classified on the categories taken by us: age, gender, and region/state.

We purposely did classification, categorization, and cleaning beforehand so that it could be accessed as per the demands of the user.

4.5.1. The FlaskApi framework connects the database to the dashboard.
4.5.2. This lets us mark the configurations for providing an accurate prediction of crime.

Workflow:
As shown in above Fig. 3 The flow is as follows:
Data Collection → Data Optimization → Model → Visualization → Top Crimes Data Output.

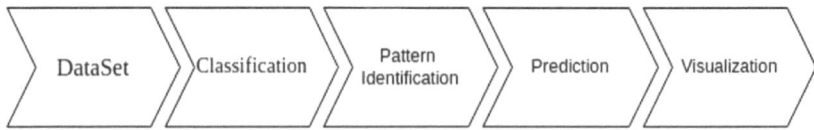

Fig. 3 Visualization workflow of CVF

Data Optimization: The age group has three subcategories: young, adult, and old. In the same way, it is for the gender which has male and female. There are also four regions of the state: north, east, west, and south. After all, these subcategories are the option if the user demands a general answer in some categories like familiar for all gender, the whole of India, or any age.

Model: The Crime Visualization and Forecast (CVF) Model combines two machine learning algorithms, KNN and logistic regression. This concept is used with the concept of using the mean values of the model with high precision for large a value data and small value data. The dataset is applied individually, and the *Total Theorem* predicts individual crime to get average accuracy.

Algorithm (User_decided_data):

1. *K = number of categories*
2. *For each tuple of data:*

   (a) *Calculate the distance between the query data and chosen data*
   (b) *Add distance of chosen data and index of chosen data to an ordered collection using the elbow method*

3. *Sort the ordered collection into smallest to largest*
4. *Pick the first K entries from the sorted collection*
5. *Get the labels of the selected entries*
6. *If regression, return the mean of K labels*
7. *If classification, return mode of K labels*
8. *For every index of an ordered collection, calculate the logistic value of each data tuple.*
9. *Average these with indices values of ordered collection and return the value*

4.6   Data Prediction and Visualization

Data visualization is the graphical representation of information and data. Using visual components like charts, graphs, and maps, data visualization tools make it simple to explore and comprehend trends, outliers, and patterns in data [17]. The predictions are put in a file that the user can take, while the dashboard would create a visual of the output to help the analysis of crimes by the user for further investigation. When forecasting the likelihood of a given result, prediction refers to the output of an algorithm after it has been trained on a previous dataset and applied to new data. For each record in the new data, the algorithm will generate probable values for an unknown variable, allowing the model builder to determine what that value will most likely be [18].

Visualization: Using the Matplotlib and Seaborn Library of Python, create graphs.

5   Result

The results are shown in Figs. 4, 5, 6, 7 and 8 shows the web platform UI for Crime
Predication and Visualization

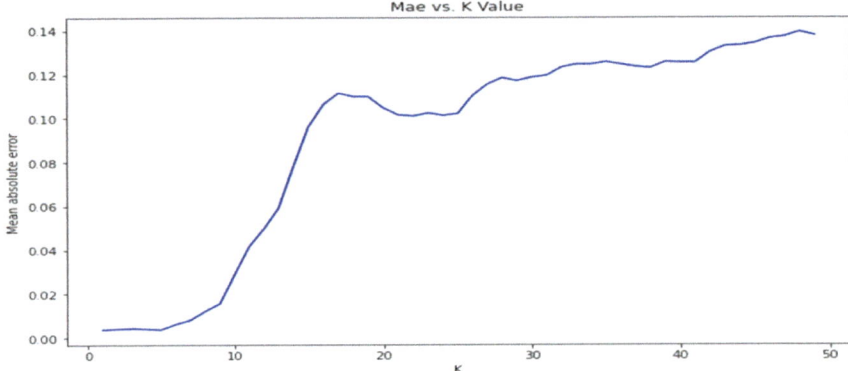

Fig. 4   Mean absolute error versus value of K determined

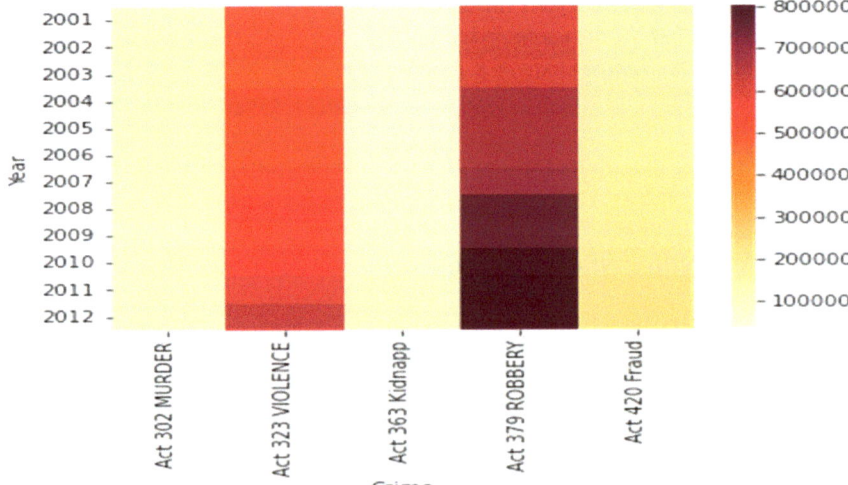

Fig. 5   Heat map to show comparative data of crime per year

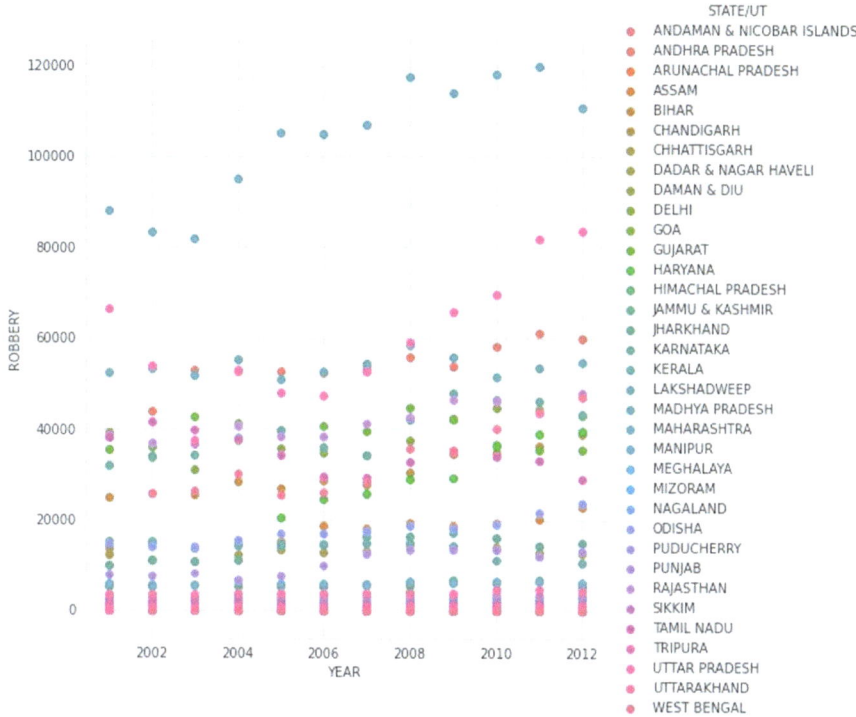

Fig. 6 Robbery in 10 years in respective regions of India

6 Comparison

1. The source of data is vague and unverified in most of the papers. We are using data provided by the government itself.
2. The data provided is mostly city-centric in other papers, while we use nationwide data.
3. We use KNN to classify as it has greater efficiency in most papers.
4. The papers have provided results for overall crime, but we are predicting the possibility of crime related to the category of people and type of crime.

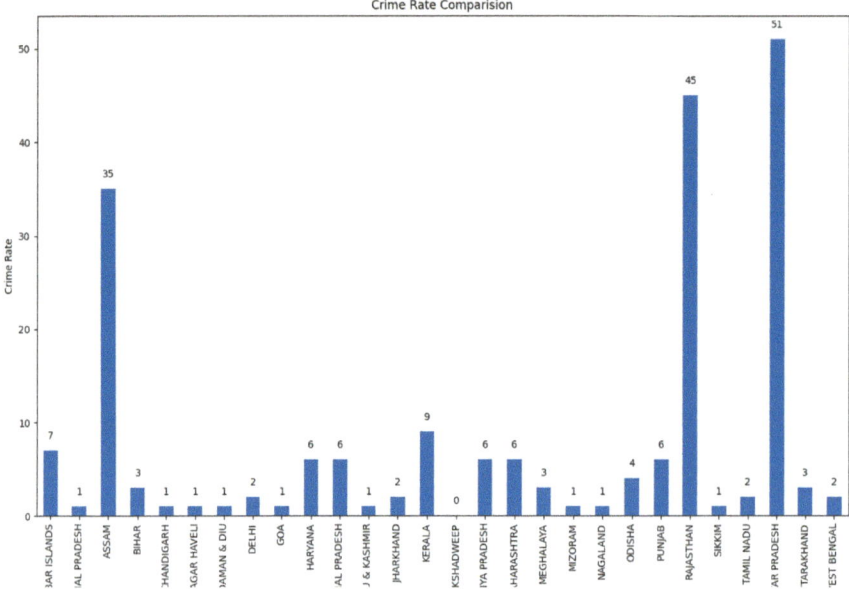

Fig. 7 Crime rates as per options selected by the user for respective regions of India

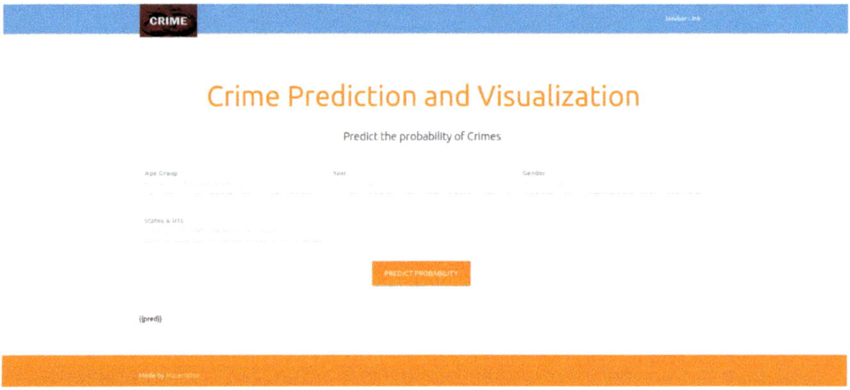

Fig. 8 Web platform UI for crime prediction and visualization

7   Conclusion

The research paper for crime visualization and forecasting seems easy compared to graphs, but the comprehensibility is complex and challenging to implement. As mentioned in the introduction, the Indian vigilant authorities like the police and CBI have started to use new technologies like AI, OSINT, and computer vision, but the actual implementation is complex in reality. In our research paper, we have tried to approach the tech where we have created a dashboard that can utilize the authorities' dataset to visualize and implement ML to predict and curb the crimes in the country. This platform is not so effective for large-scale use now, as the database lacks whole data provided to authorities, and it lacks dynamic aspect due to data absence which would vanish when implemented by the government or in government-aided means. The fear of data security can be kept in check by including the Django framework and other technologies, making it a complex and authenticated used platform.

References

1. Shah N, Bhagat N, Shah M (2021) Crime forecasting: a machine learning and computer vision approach to crime prediction prevention. Vis Comput Ind Biomed Art
2. Yoo JS (2019) Crime data warehousing and crime pattern discovery. In: International conference on data science, E-learning and information systems 2019 (DATA19), December 2–5, 2019, Dubai, United Arab Emirates. ACM, New York, NY, USA, six pages
3. Llaha O (2020)Crime analysis and prediction using machine learning. In: 2020 43rd international convention on information, communication and electronic technology (MIPRO), pp 496–501.https://doi.org/10.23919/MIPRO48935.2020.9245120
4. https://crime-data-explorer.fr.cloud.go
5. Lisowska-Kierepka A (2021) How to analyze the spatial distribution of crime? Crime risk indicator to design an original method of spatial crime analysis. Cities 103403. ISSN 0264-2751
6. Zhu H, Wang F (2021) An agent-based model for simulating urban crime with improved daily routines. Comput Environ Urban Syst 89:101680. ISSN 0198-9715
7. Mahmud S, Nuha M, Sattar A (2021) Crime rate prediction using machine learning and data mining. https://doi.org/10.1007/978-981-15-7394-1_5
8. Peppesa N, Alexakisa T, Adamopouloua E, Remoundoua K, Demestichasa K (2020) A semantic engine and an ontology visualization tool for advanced crime analysis. Procedia Comput Sci 176:1829–1838
9. Yin J, Afa IM, Afa I (2020) Machine learning algorithms for visualization and prediction modeling of Boston crime data. https://doi.org/10.20944/preprints202002.0108.v1
10. Vijaylakshmi C, Banderkar SR (2019) Design and analysis of machine learning algorithms for the reduction of crime rates in India. Procedia Comput Sci 172(3):122–127. https://doi.org/10.1016/j.procs.2020.05.018
11. Zhao X, Tang J (2018) Crime in urban areas: A data mining perspective. ACM SIGKDD explorations newsletter 20(1) New York, NY, United States
12. Das N (2023) Digital education as an integral part of a smart and intelligent city: a short review. Digital Learn Based Educ Transcending Phys Barr 81–96
13. Rudra Kumar M, Gunjan VK (2022) Machine learning based solutions for human resource systems management. In: ICCCE 2021: proceedings of the 4th international conference on communications and cyber physical engineering. Springer Nature Singapore, Singapore, pp 1239–1249
14. Kesarwani A, Maheshwari S, Sharma S, Rai BK (2022) Hand talk: Intelligent gesture based communication recognition & object identification for deaf and dumb AIP Conference Proceedings 2424, 080007 (2022). https://doi.org/10.1063/5.0076796. Published Online: 21 March 2022
15. Kumar A, Goyal A, Rai BK, Sharma S (2022) OCR based medical prescription and report analyzer AIP Conference Proceedings 2424, 070006 (2022). https://doi.org/10.1063/5.008 1176. Published Online: 21 March 2022
16. Usman M, Gunjan VK, Wajid M, Zubair M, Noor-e-alam Siddiquee K (2022) Speech as a biomarker for COVID-19 detection using machine learning. Computat Intell Neurosci 2022, Article ID 6093613:12 p. https://doi.org/10.1155/2022/6093613

17. Sharma S, Rai BK, Gupta M, Dinkar M (2023) DDPIS: Diabetes disease prediction by impro-
vising SVM. Inter J Reliable Qual E-Healthc (IJRQEH) 12(2):1–11.http://doi.org/10.4018/IJR
QEH.318090
18. https://www.datarobot.com/wiki/prediction/

An Empirical Comparison of Classification Machine Learning Models Using Medical Datasets

B. V. Saketha Rama, G. Suryanarayana, Mohd Dilshad Ansari, and Ruqqaiya Begum

Abstract Classification is a supervised learning model where the class labels are accurately identified for future samples. Medical data is an important source for understanding and improving health outcomes and classification algorithms are often used to analyze these data. Learning models give significant experiences into the situational needs of patients. Various hypotheses have been carried out on different datasets yet it is truly challenging to track down which model is suitable. Proposed work compares the performance of classification models like LR, DT, SVM, NB, KNN, and RF on various datasets. SVM classifier yields accuracy of 0.59 for the Diabetic dataset as it considers individual model opinion, while RF classifier surpassed them both with accuracy 0.9974 for the breast cancer Wisconsin dataset since it is an ensemble approach that takes majority opinions. These findings highlight the need for careful consideration of the choice of classification model when analyzing medical data and provide valuable insights for researchers and practitioners working with these data.

Keywords Supervised learning · Classification models · Empirical comparison · Medical datasets

B. V. Saketha Rama
Department of Computer Science Engineering, Indian Institute of Information Technology, Design and Manufacturing, Kancheepuram, Tamil Nadu, India

G. Suryanarayana
Department of Information Technology, Vardhaman College of Engineering, Hyderabad, Telangana State, India

M. D. Ansari (✉)
Guru Nanak University, Hyderabad, India
e-mail: m.dilshadcse@gmail.com

R. Begum
Department of Computer Science Engineering, Vardhaman College of Engineering, Hyderabad, Telangana, India

© The Author(s), under exclusive license to Springer Nature Singapore Pte Ltd. 2023   321
A. Kumar et al. (eds.), *Proceedings of the 4th International Conference on Data Science, Machine Learning and Applications*, Lecture Notes in Electrical Engineering 1038,
https://doi.org/10.1007/978-981-99-2058-7_29

1 Introduction

Classification techniques are mostly used in the field of medical research to predict the livelihood of a patients having various disease or condition based on their medical history and other relevant features. These techniques have the potential to dramatically increase the accuracy as well as efficiency of diagnosis, treatment, and prognosis making them an important tool in the practice of medicine. Below are some of the classification techniques, their merits, and demerits.

1.1 Logistic Regression

Binary classification challenges are handled by the nominal/ordinal machine learning technique known as logistic regression. It is frequently used in the medical sector to forecast the possibility of a specific result, such as the likelihood that a patient would contract a specific disease or the likelihood that they will respond to a specific therapy. The chance that a patient has a specific disease, for instance, might be predicted using a logistic regression model based on symptoms, test findings, and other characteristics.

One of the main advantages of using logistic regression in the medical field is to implement and can handle huge amount of data types, including continuous and categorical variables. It can also provide insights into the relationships between different features (e.g., symptoms and test results) and the likelihood of a particular outcome. However, logistic regression is a linear model and can only capture linear correlation between the input features and the output which might be a limitation when the data is not linearly separable or when there are nonlinear relationships in the data that are important to consider. Logistic regression can also be sensitive to the presence of outliers in the data which can affect the model's performance [1–4].

1.2 Decision Tree

A well-liked machine learning model called decision trees is utilized for both classification and regression problems. They function by building a tree-like structure, in which the leaf nodes indicate the predicted class or value and the inside nodes reflect judgments depending on the values of the input characteristics. Decision trees may be used for binary and multi-class classification tasks and can handle continuous and categorical information.

Decision trees have been used in the medical sector for a range of tasks, including determining the variables that affect the risk that a patient would contract a specific disease or forecasting the efficacy of a specific treatment for a specific patient. They have also been used to predict the likelihood of a patient being readmitted to the hospital. Decision trees are often chosen for these tasks because they are easy to interpret and implement and can often achieve good performance on many types of data. Decision trees may not generalize well to novel, untested data and might be prone to over fitting. They can also be computationally expensive to build and use which can be a drawback when working with larger datasets [5].

1.3 Support Vector Machine

The machine learning approach known as support vector machines (SVMs) is utilized for both classification and regression tasks. SVMs have been utilized in the medical profession for a number of purposes, including forecasting the chance that a patient would develop a certain disease, the likelihood that a patient will react to a specific therapy, and the risk that a patient will be readmitted to the hospital.

One of the main advantages of using SVMs in the medical field is that they can handle high-dimensional data and can find complex, nonlinear relationships in the data. They are also robust to noise and can handle large datasets efficiently. However, the choice of kernel and other hyperparameters, which might have an impact on the model's performance, is one SVM restriction. In addition, SVMs can be computationally expensive to train which can be a drawback when working with larger datasets. It is important for researchers to carefully evaluate the performance of SVM models and to consider alternative algorithms when appropriate [4–7].

1.4 Naive Bayes Classifier

Naive Bayes is an algorithm which is based on the concept of Bayes Theorem and is a probabilistic algorithm that makes predictions about the likelihood of an event based on prior knowledge and statistical data; Naive Bayes may be used in the medical industry to forecast a patient's chance of having a certain disease based on their symptoms or their likelihood of benefiting from a particular therapy, among other things.

The Naive Bayes algorithm's relative simplicity and ease of use are two of its key features. It also performs well when dealing with large amounts of data and can be used to make predictions in real time. The Naive Bayes algorithm's fundamental drawback is that it assumes that all characteristics are independent of one another, which may not always be the case in real-world scenarios. This can lead to less accurate predictions compared to other algorithms that do not make this assumption [1].

1.5 K-Nearest Neighbors

The machine learning technique K-nearest neighbors (KNN) is utilized for both classification and regression applications. It has been used in the medical industry for a range of purposes, including forecasting the risk that a patient would contract a certain illness or the efficacy of a specific treatment for a specific patient. A prediction is made using the class labels or values of the K data points in the training set that are closest to the new data point in KNN.

One of the main advantages of using KNN in the medical field is that it is simple to implement, doesn't involve a training phase, and can handle huge data types that includes continuous and categorical variables. KNN is also flexible and can be used for multi-class and binary classification tasks. However, one limitation of KNN is expensive to use, particularly when it is used with large datasets. The choice of K and the distance metric employed to gauge similarity between data points can both have an impact on how well KNN performs. It is important for researchers to carefully evaluate the performance of KNN models and to consider alternative algorithms when appropriate [8, 9].

1.6 Random Forests

An ensemble machine learning approach known as random forests is utilized for both classification and regression problems. They have been used in the medical industry for a range of purposes, including forecasting the risk that a patient would contract a certain illness or the efficacy of a specific treatment for a specific patient. The chance of a patient being readmitted to the hospital has also been predicted using them [10, 11].

One of the main advantages of using random forests in the medical field is that they can handle high-dimensional data and can find complex, nonlinear relationships in the data. They are also robust to noise and can handle large datasets efficiently. In addition, random forests can provide feature importance scores which can help researchers understand which features are most important for predicting the outcome. However, one limitation of random forests is that they can be difficult to interpret as the decision-making process is distributed across many different decision trees. They may also be sensitive to the selection of hyperparameters, which might impact the effectiveness of the model [12–14].

2   Related Work

Recent years have seen a significant amount of study on the use of classification algorithms to medical data. For example, Arwatki Chen et al. [15] have achieved their goal by predicting diabetes with a stable and high accuracy using various machine learning algorithms. The model's consistent accuracy was made possible by the use of several risk factors and cross-validation methods. The study was constrained by the fact that it was based on a single dataset, and more testing and validation on a bigger, more varied dataset may be required to fully assess the model's efficacy. The study could be expanded in the future to include more deep learning and deep learning methods and to test the model on a bigger dataset in order to increase precision and generalizability. The model's practical application and effectiveness in the early diagnosis of diabetes would also be further understood by using it in a real-world situation, such as a hospital or medical clinic.

Li et al. [10] included applications of machine learning classifiers like DT, LR, KNN, SVM, ANN, NB, employing feature selection methods including MRMR, Relief, LLBFS, and LASSO. The feature selection issue for heart disease diagnosis has also been addressed using the novel feature selection algorithm FCMIM. The models that use feature selection methods in addition to the LOSO cross-validation approach have given a good accuracy. Hence, they claimed to improve the performance of the diagnosis of cardiac problems by adding new feature selection methods and optimization approaches.

Li et al. [11] the preoperative identification and staging of pancreatic cancer, computed tomography (CT) images, and an ensemble learning-support vector machine (EL-SVM) were used. The Least Absolute Shrinkage and Selection Operator (LASSO) approach was employed for feature selection, and it achieved high accuracy at various points. The effectiveness of models with more feature selection and hyperparameter tuning techniques would be interesting to study.

Anthimopoulos et al. [2] have published a deep convolutional neural network (CNN) for categorizing lung computed tomography (CT) image patches into seven groups, including healthy tissue and six different interstitial lung illnesses (ILDs). The proposed network architecture, designed to capture the low-level textural features of the lung tissue, consisted of five convolutional layers and three dense layers. On a challenging dataset of 120 CT images from multiple hospitals and scanners, the proposed technique outperformed the state of the art and yielded encouraging results. Negative aspects of the recommended technique include its large number of parameters, delayed training period (often a few hours), and some variance in output for the same input due to the random initialization of the weights. In order to help in the differential diagnosis of ILDs, the authors intend to expand the approach to take into account three-dimensional data from multidetector CT volume scans.

Liu et al. [13], For the simultaneous classification of Alzheimer's disease and clinical score regression, a deep multi-task multi-channel learning (DM2L) architecture was built employing magnetic resonance imaging (MR) imaging data and demographic information (i.e., age, gender, and education of participants). The study found that the DM2L technique outperformed many cutting-edge algorithms in the tasks of illness classification and clinical score regression on four publically accessible datasets. Existing convolution neural networks trained on other sizable 3D medical image datasets were needed to fine-tune the proposed network. Other study limitations include discrepancies in data distributions between the training and testing data, independence of the proposed deep feature learning framework from the landmark identification procedure, and the need for more research. For further enhancing the performance of the DM2L approach, they have included a number of directions, such as researching model adaptation strategies, combining landmark detection and landmark-based classification/regression into a single deep learning framework, optimizing the convolution neural networks that have already been trained on other substantial 3D medical image datasets, and automatically learning weights for the tasks of disease classification and clinical score regression.

Tsanas et al. [17] used prolonged vowel phonations; a variety of traditional and cutting-edge speech signal processing methods were used to separate Parkinson's disease (PWP) patients from healthy controls. The accuracy of their categorization was improved from prior research' 93% accuracy using a subset of 22 features to the authors reported 99% accuracy using ten dysphonic measures. From the initial 132 features, four different feature selection algorithms found a small subset of 10 features that were informative for the binary classification task. Although RELIEF offered the subset with the lowest classification error, the FS methods still performed rather well. Signal-to-noise ratio (SNR) and mel-frequency cepstral coefficients (MFCCs) measurements were discovered to be consistently chosen by the FS algorithms, demonstrating the significance of these measurements for the evaluation of vocal pathology in PWP. For the purpose of mapping characteristics to the response, the scientists also examined the effectiveness of nonlinear random forests (RFs) and support vector machines (SVMs) and discovered that RF classifier works well than the SVM classifier. The authors did observe that the RF classifier was more susceptible to the FS method and training set selection. They also emphasized the necessity for more investigation into how PWP affects vocal tract articulatory dysfunction as well as the need of employing a sizable and varied dataset for vocal pathology evaluation.

Liu et al. [14] have discussed multimodal neuroimaging data, and a deep learning system has been suggested for detection of Alzheimer's disease (AD). They showed the system could discriminate between several phases of AD development by combining unsupervised feature representation with deep learning techniques. The framework was evaluated using data from the ADNI repository, and it was discovered to perform better than existing deep learning frameworks including the most advanced SVM-based approach for classifying AD. The authors claim that the approach might be extended to additional unlabeled data for feature engineering and could yet use more training data. The technique may be applied to more efficiently depict multimodal neuroimaging biomarkers.

Tao et al. [21] had conducted a study on machine learning methods using Magnetocardiography (MCG) data, and an efficient and precise approach for the automated diagnosis and localization of ischemic heart disease was created. Following the extraction of 164 features from the T wave segmentation and comparison of multiple classifiers, the SVM-XGBoost model gives the good results for IHD identification, with 94.03% accuracy and an AUC of 0.98. The XGBoost model successfully localized in ischemia in the left circumflex, left anterior descending, and right coronary arteries with accuracy values of 0.68, 0.74 and 0.65, respectively. This approach may broaden the use of MCG data in clinical settings by giving doctors a valuable tool for interpreting the data. Furthermore, the possibility of noninvasive ischemia localization is suggested by the link between magnetic field patterns and stenosis location. To increase the localization accuracy and validate the findings using bigger datasets, further effort is still required.

Arwatki Chen et al. [15] employed a variety of machine learning methods, and it was able to predict diabetes with a steady and high level of accuracy. The model's consistent accuracy was made possible by the inclusion of several risk variables and cross-validation methods. The study was constrained by the fact that it was based on a single dataset, and further testing and validation on a bigger, more varied dataset may be required to fully assess the model's efficacy. To increase accuracy and generalizability, machine learning and deep learning models are applied to bigger dataset. Moreover, applying the model in a genuine environment, such as a hospital or medical facility, would give additional understanding of the usefulness for diabetes detection.

Kumari et al. [9] have suggested ensemble voting classifier has great accuracy in predicting diabetes mellitus and breast cancer, with 97.02% and 79.04% accuracy, respectively. Modern techniques and basic classifiers such as logistic regression, AdaBoost, support vector machine, Naive Bayes, random forest, Bagging, XGBoost, CatBoost, and GradientBoost were surpassed by ensemble approach. To properly assess the efficacy and robustness of the suggested strategy, however, more testing on a larger and more varied dataset may be required. Future advancements in the suggested method's accuracy might come from using deep learning models and investigating various ensemble methodologies.

Ambrish et al. [6] used logistic regression technique on the UCI dataset, and a prediction of cardiovascular disease with a high accuracy of 87.10% was made. The model's performance was enhanced by pre-processing the data, which involved cleaning, identifying missing values, and conducting feature selection. With more training data, the model's accuracy also rose. The study was only able to use the UCI dataset, but future research might expand to include other datasets for more reliable findings. The application of additional machine learning algorithms or for greater performance in the prediction of cardiovascular illness might also be investigated in the future study.

Astani et al. [3] a system for categorizing 13 kinds of tomato diseases in both lab and field settings was put forth and put to the test. On the Taiwan database, which includes photos with a variety of difficulties like shadow, background clutter, noise, low image quality, many leaves, varied textures, and brightness fluctuations, the approach employed ensemble classification and obtained an accuracy of 95.98%. The suggested approach was also discovered to be quicker than deep learning models and was very accurate in classifying photographs with low image quality, shadows, and cluttered backgrounds. To completely evaluate the approach's efficacy, more testing on additional plant species and illnesses is required. The method was only tested on two databases, though.

Piao et al. [18], Feature subset-based ensemble technique is suggested for leveraging miRNA expression data to categorize various tumors. In comparison with other widely used ensemble approaches, the method was able to produce good results and greater prediction accuracy. The capacity of this technique to take feature relevance and redundancy into account while creating numerous feature subsets, leading to more independent and informative models for the classification problem, is one of its key accomplishments. Additionally, performance was enhanced by the integration of various base classifiers and the average posterior probability. The suggested approach does have some limitations, though, one of which is that it cannot be used for low-dimensional data since there are only so many subsets that can be formed. This might potentially affect the algorithm's capacity to produce a base no of classifiers.

Sambasivam and Opiyo [19] used photos from a dataset amassed in Uganda with the goal of developing a machine learning model to precisely diagnose illnesses affecting cassava leaves. The majority of the photos belonged to the Cassava Mosaic Disease and Cassava Brown Streak Virus Disease categories, making up the limited and severely unbalanced dataset. The authors achieved an accuracy of 93% by using methods like class weight, SMOTE, and focused loss with deep convolution neural networks to overcome this class imbalance. The suggested approach performed well in real-world situations and was able to categorize the underrepresented groups appropriately. However, one drawback of the study is that it was only tested on one dataset; hence, more testing on a larger and more varied dataset may be required to corroborate the findings.

Hameed et al. [7], this study's proposed intelligent digital diagnosis strategy for skin disorders used deep learning to obtain a high classification accuracy of 96.47%. This is a noteworthy accomplishment since prompt and efficient skin disease treatment depends on precise diagnosis. A restriction of earlier research that only concentrated on a small number of illnesses has been addressed by the use of the multiclass multi-level (MCML) classification method, which is inspired by the "split and conquer" approach. Further research might be done to include more diseases in the categorization algorithm as the study only took a small number of skin conditions into account. Better testing could be carried out on actual patient cases to further confirm the suggested algorithm's efficacy as it was only tested using photos gathered from various sources. The suggested approach may be used to create a mobile-enabled expert system for usage in remote locations with sparse access to diagnostic resources. The availability and accessibility of skin disease diagnosis and treatment in these locations may significantly increase as a result [18, 20, 21].

Overall, the use of classification techniques on medical data has the potential to significantly improve the accuracy and efficiency of diagnosis, treatment, and prognosis, making it an important area of research in the field of medicine. However, it is crucial to carefully weigh the advantages and disadvantages of various categorization algorithms and to assess how well they perform on pertinent medical datasets [16, 17, 22–27].

3 Proposed Work

Empirical analysis is done on medical datasets by using various classification algorithms. Essential pre-processing methods are done on the datasets, such as feature scaling and missing value imputation, before building the models. Decision trees (DT), logistic regression (LR), support vector machine (SVM), k-nearest neighbors (KNN), random forests (RF), and Naive Bayes classifier (NBA) were some of the classification models used in this study. Figure 1 provides an information about building a classification models.

Thyroid Disease dataset [DS-1] includes a total of 3772 occurrences and 29 characteristics. The Chronic Kidney Disease dataset [DS-2] with the exception of the target variable, it had 400 instances in total and 25 characteristics. Diabetes 130-US hospitals for the years 1999–2008 dataset [DS-3] has 100,000 instances and 55 characteristics. Breast Cancer Wisconsin (Diagnostic) dataset [DS-4] contains 569 occurrences and 31 characteristics and Pima Indians Diabetes dataset [DS-5]

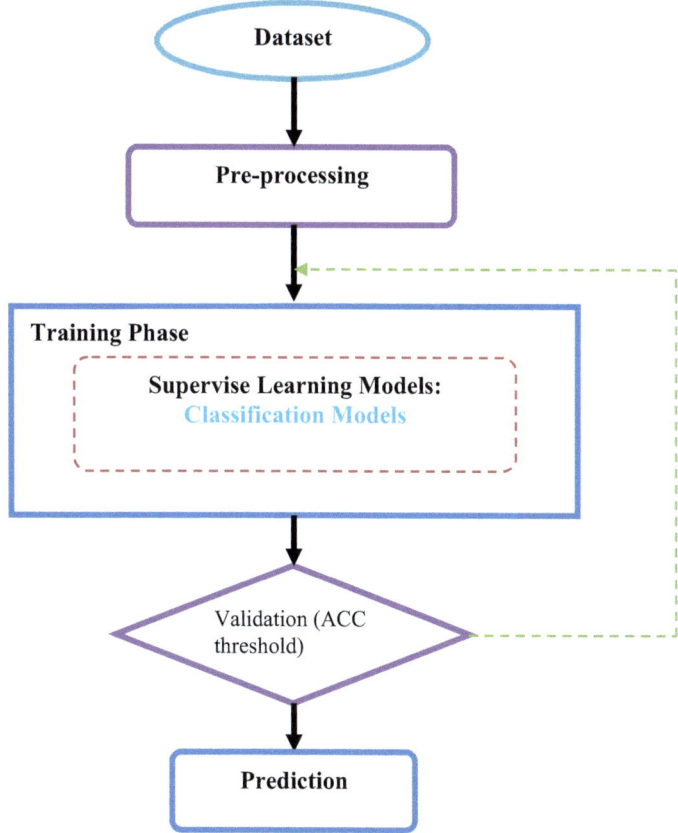

Fig. 1 Flow of the proposed work

Table 1 Accuracy of classification models

	LR	DT	SVM	GNB	KNN	RF
DS-1	0.940860	0.983870	0.9220430	0.674193	0.981182	0.981182
DS-2	0.875	0.975	0.6	0.958333	0.8	0.983333
DS-3	0.61	0.92	0.59	0.59	0.83	0.94
DS-4	0.989949	0.977386	0.987437	0.949748	0.962311	0.997487
DS-5	0.746753	0.720779	0.759740	0.740259	0.759740	0.779220

which consist of 700 occurrences and 8 characteristics. Table 1 represents accuracy of different classification models on various datasets.

4   Results and Discussion

Figure 2 shows classification models accuracy on various datasets. Out of all the datasets, logistic regression, support vector machine, and random forest with an average of 99% accuracy for Breast Cancer Wisconsin, decision tree, and KNN with an average 98% for Thyroid Disease dataset performed well, whereas decision tree, KNN, and random forest with an average of 75% for Pima Indians Diabetes, logistic regression, and GNB with an average of 60% for diabetes 130-US hospitals datasets.

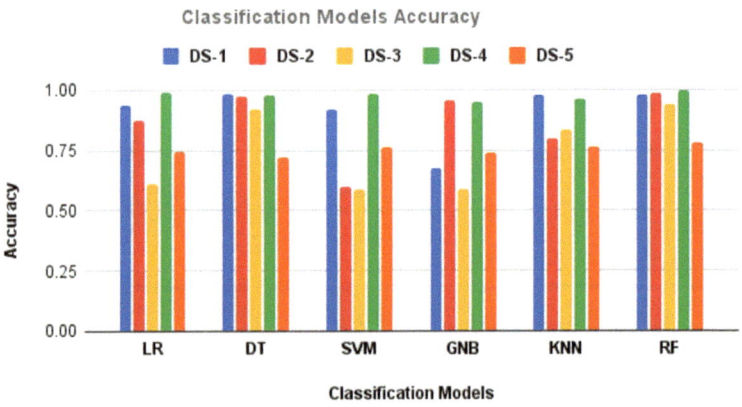

Fig. 2 Accuracy of classification models

5 Conclusion

This work is helpful for researchers and practitioners making the choice of classification model when analyzing medical data. Out of all classification models random forest with accuracy 99.74, and logistic regression with 98.99 for Breast Cancer Wisconsin dataset as random forest is an ensemble method and logistic regression is good for classification problems performed well, Gaussian Naïve Bayes with 0.59 for diabetes dataset and support vector machine with 0.60 for Chronic Kidney Disease dataset does not performed well. In the future, researchers can get good accuracy of classification models by considering feature selection methods.

References

1. Aksoy S, Koperski K, Tusk C, Marchisio G, Tilton JC (2005) Learning Bayesian classifiers for scene classification with a visual grammar. IEEE Trans Geosci Remote Sens 43(3):581–589
2. Anthimopoulos M, Christodoulidis S, Ebner L, Christe A, Mougiakakou S (2016) Lung pattern classification for interstitial lung diseases using a deep convolutional neural network. IEEE Trans Med Imaging 35(5):1207–1216
3. Astani M, Hasheminejad M, Vaghefi M (2022) A diverse ensemble classifier for tomato disease recognition. Comput Electron Agric 198:107054
4. Barakat N, Bradley AP, Barakat MNH (2010) Intelligible support vector machines for diagnosis of diabetes mellitus. IEEE Trans Inf Technol Biomed 14(4):1114–1120
5. Fayn J (2010) A classification tree approach for cardiac ischemia detection using spatiotemporal information from three standard ecg leads. IEEE Trans Biomed Eng 58(1):95–102
6. Ambrish G, Ganesh B, Ganesh A, Srinivas C, Dhanraj, Mensinkal K (2022) Logistic regression technique for prediction of cardiovascular disease. Glob Trans Proc 3(1):127–130. Int Conf Intell Eng Approach (ICIEA-2022)
7. Hameed N, Shabut AM, Ghosh MK, Hossain MA (2020) Multi-class multi-level classification algorithm for skin lesions classification using machine learning techniques. Expert Syst Appl 141:112961
8. Hossain E, Hossain MF, Rahaman MA (2019) A color and texture based approach for the detection and classification of plant leaf disease using knn classifier. In: 2019 international conference on electrical, computer and communication engineering (ECCE). IEEE, pp 1–6
9. Kumari S, Kumar D, Mittal M (2021) An ensemble approach for clas-sification and prediction of diabetes mellitus using soft voting classifier. Int J Cognitive Comput Eng 2:40–46
10. Li JP, Ul Haq A, Ud Din S, Khan J, Khan A, Saboor A (2020) Heart disease identification method using machine learning classification in e-healthcare. IEEE Access 8:107562–107582
11. Li M, Nie X, Reheman Y, Huang P, Zhang S, Yuan Y, Chen C, Yan Z, Chen C, Lv X et al (2020) Computer-aided diagnosis and staging of pancreatic cancer based on ct images. IEEE Access 8:141705–141718
12. Lindner C, Thiagarajah S, Wilkinson JM, Wallis GA, Cootes TF, arcOGEN Consortium et al (2013) Fully automatic segmentation of the proximal femur using random forest regression voting. IEEE Trans Med Imag 32(8):1462–1472
13. Liu M, Zhang J, Adeli E, Shen D (2019) Joint classification and regression via deep multi-task multi-channel learning for alzheimer's disease diagnosis. IEEE Trans Biomed Eng 66(5):1195–1206
14. Liu S, Liu S, Cai W, Che H, Pujol S, Kikinis R, Feng D, Fulham J, ADNI (2015) Multimodal neuroimaging feature learning for multi-class diagnosis of Alzheimer's disease. IEEE Trans Biomed Eng 62(4):1132–1140

15. Lyngdoh AC, Choudhury NA, Moulik S (2021) Diabetes disease prediction using machine learning algorithms. In: 2020 IEEE-EMBS conference on biomedical engineering and sciences (IECBES), pp 517–521
16. Gunjan VK, Kumar S, Ansari MD, Vijayalata Y (2022) Prediction of agriculture yields using machine learning algorithms. In: Proceedings of the 2nd international conference on recent trends in machine learning, IoT, smart cities and applications: ICMISC 2021. Springer, Singapore, pp 17–26
17. Tsanas A, Little MA, McSharry PE, Spielman J, Ramig L-R (2012) Novel speech signal processing algorithms for high-accuracy classifica-tion of parkinson's disease. IEEE Trans Biomed Eng 59(5):1264–1271
18. Piao Y, Piao M, Ryu KH (2017) Multiclass cancer classification using a feature subset-based ensemble from microrna expression profiles. Comput Biol Med 80:39–44
19. Sambasivam G, Opiyo GD (2021) A predictive machine learning application in agriculture: Cassava disease detection and classification with imbalanced dataset using convolutional neural networks. Egypt Inf J 22(1):27–34
20. Springer DB, Tarassenko L, Clifford GD (2015) Logistic regression-hsmm-based heart sound segmentation. IEEE Trans Biomed Eng 63(4):822–832
21. Tao R, Zhang S, Huang X, Tao M, Ma J, Ma S, Zhang C, Zhang T, Tang F, Jianping L, Shen C, Xie X (2019) Magnetocardiography-based ischemic heart disease detection and localization using machine learning methods. IEEE Trans Biomed Eng 66(6):1658–1667
22. Kumar S, Gunjan VK, Ansari MD, Pathak R (2022) Credit card fraud detection using support vector machine. In: Proceedings of the 2nd international conference on recent trends in machine learning, IoT, smart cities and applications: ICMISC 2021. Springer, Singapore, pp 27–37
23. Gaddam DKR, Ansari MD, Vuppala S, Gunjan VK, Sati MM (2022) A performance comparison of optimization algorithms on a generated dataset. In: ICDSMLA 2020: proceedings of the 2nd international conference on data science, machine learning and applications. Springer, Singapore, pp 1407–1415
24. Narayana GS, Ansari MD, Gunjan VK (2022) Instantaneous approach for evaluating the initial centers in the agricultural databases using K-means clustering algorithm. J Mob Multimedia 43–60
25. Kumar S, Ansari MD, Gunjan VK, Solanki VK (2020) On classification of BMD images using machine learning (ANN) algorithm. In: ICDSMLA 2019: proceedings of the 1st international conference on data science, machine learning and applications. Springer, Singapore, pp 1590–1599
26. Gunjan VK, Prasad PS, Pathak R, Kumar A (2020) Machine learning methods for extraction and classification for biometric authentication. In: ICDSMLA 2019: proceedings of the 1st international conference on data science, machine learning and applications. Springer, Singapore, pp 1984–1988
27. Kumar MR, Gunjan VK (2020) Review of machine learning models for credit scoring analysis. Ingeniería Solidaria 16(1)

Risk Prediction of Chronic Kidney Disease Using Machine Learning Algorithms

A. S. Chaithra, D. K. Chandana, S. M. Chetana, and N. Greeshma

Abstract Chronic kidney disease (CKD), which increases a patient's chance of developing end stage renal disease, is a significant public health concern. In recent years, dialysis has had a significant financial impact on the national health insurance system, and the cost is steadily rising. Using machine learning methods, we will create a system that could identify diseases at an early stage.

Keywords Kidney disease · Glomerular filtration rate (GRF) · End stage renal disease (ESRD) · Attributes · Logistic regression classifying · Decision tree classifying · Support vector machine classifying · Random forest classifying

1 Introduction

The word "chronic" means the slow decline of kidney cells over a period of time. The CKD leads to major kidney failure due to heavy fluid buildup in body. Chronic kidney disease is a very serious issue faced by many people in the recent years [1]. The main reason is it is very difficult to identify this disease until it reaches advanced stage. Therefore, we try to identify this disease at early stage with our prediction system with accuracy.

A. S. Chaithra (✉) · D. K. Chandana · S. M. Chetana · N. Greeshma
Don Bosco Institute of Technology, Bangalore, India
e-mail: aschaithra@gmail.com

© The Author(s), under exclusive license to Springer Nature Singapore Pte Ltd. 2023   333
A. Kumar et al. (eds.), *Proceedings of the 4th International Conference on Data Science, Machine Learning and Applications*, Lecture Notes in Electrical Engineering 1038, https://doi.org/10.1007/978-981-99-2058-7_30

2 Problem Statement

2.1 Existing System

The following are some earlier predictive methods to diagnose CKD in the field of using machine learning algorithm. There are few drawbacks of the existing system such as the implementation still lacks in accuracy of result in some cases, more optimization is needed. Also, database extension is required in order to reach more accuracy. This process of detecting chronic kidney disease is expensive, because chronic kidney disease is invasive, costly so that many patients reached at last stages without treatments [2]. These are few problems faced by the existing systems which we are trying to solve it in the proposed system.

2.2 Proposed System

This work focusses on investigating machine learning methods and techniques that combines with feature selection techniques for effective chronic kidney disease detection with regards to detection accuracy [3]. This study analyses many prediction models for correctly occurring occurrences while using various classification algorithms and approaches offered by the WEKA tool. Prediction values for early chronic kidney disease diagnosis can be obtained using the identified classification technique [4]. Advantages of the proposed system are that it includes high processing speed and high classification accuracy. Large datasets can be classified. The model may be trained with increasing amount of data and may produce data faster, more accurate answers without requiring human interaction. Most of the time, it can assist doctor in quickly identifying the symptoms and taking appropriate action to lessen them in early stage. Early prediction helps patients to undergo proper treatment and reduce the consequences. We can lessen the flow of water products through our bodies by making early predictions. It also removes hazardous substances from our blood in addition to excess fluid.

3 System Design

This project is involved into three parts:

- **Data preprocessing**: missing values must be handled in this method based on their distribution to attain appropriate accuracy. The mechanism which causes the data to be missing will determine the potential bias brought on by missing data. By using the same number of estimators and full instances while running

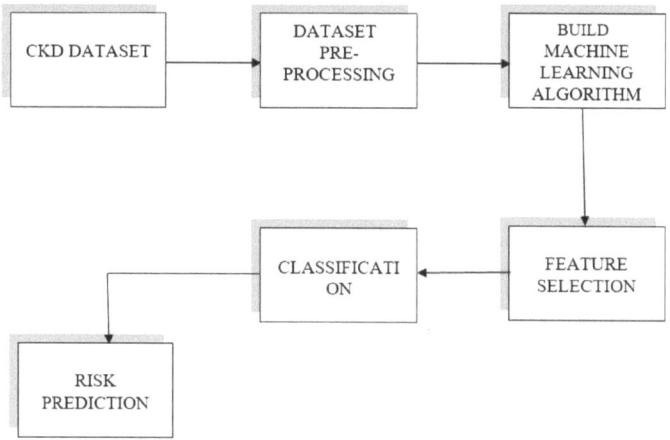

Fig. 1 Block diagram for proposed system

the algorithm, which resulted in the least amount of mean and standard deviation change, the dataset's original distribution was preserved.

- **Training and tuning the model**: models for classifying data are regarded as training. The dataset was split into three parts: 70% of training data, 15% of cross validation data, and 15% of test data. Grid search for the training dataset and hyper parameter tuning using a genetic algorithm are used to optimize the model. Using python and the keras framework, the implementation and assessment were carried out.
- **Model selection**: in order to clearly demonstrate each algorithm's bias toward particular traits, the standard deviation of the feature importance of each method was measured once the relevance of selected features had been determined.
- **Attribute used**: here we are using dataset from gov.com and the few attributes we used in the proposed system are (Fig. 1; Table 1).

4   Methodology

The algorithms we have used in our proposed system are:

- **Decision tree**: it falls under supervised learning technique. It uses both classification and regression to solve problems. Any Boolean function on discrete attribute can be represented using decision tree.
- **Logistic regression**: it falls under supervised learning technique. It predicts the probability of categorical dependent variable. The output value will be between 0 and 1. The base form of logistic regression is Binary classifier. In the logistic regression algorithm formula, there is a linear module that is integrated to sigmoid function.

Table 1 Attributes set

Attribute name	Attribute code
Age	Age
Albumin	Al
Anemia	Ane
Appetite	Appet
Bacteria	Ba
Blood glucose random	Bgr
Blood pressure	Bp
Blood urea	Bu
Class	–
Coronary artery disease	Cad
Diabetes mellitus	Dm
Hemoglobin	Hemo
Hypertension	Htn
Packed cell volume	Pcv
Pedal edema	Pe
Potassium	Pot
Puss cell	Pc
Puss cell clumps	Pcc
Red blood cells	Rbc
Red blood cell count	Rbcc
Serum creatinine	Sc
Specific gravity	Sg
Sodium	Sod
Sugar	Su
White blood cell count	Wbcc

- **Random forest**: Bierman's bagging ensemble approach is based on a machine learning technique known as "decision trees". Decision trees are considered to be the "weak learners" in a random forest. By choosing a random characteristic, random forest imposes the diversity of each tree individually. They then cast their votes for class that is most prevalent after producing a lot of trees. The runtime of the random forest approach is significantly shorter, and it is robust against overfitting and can handle unbalanced data.
- **Support vector machine**: it falls under supervised leaning algorithm. It employs both classification and regression problems, favoring the former. Goal of SVM algorithm is to create hyperplanes in N-dimensional space that classifies the new data point in a correct category and to identify a plane with the greatest separation between the data points for the two groups. SVM is an algorithm that

selects extreme locations that aid in building hyperplanes. These extreme points are referred to as support vectors.

5   Results

The metrics listed below provide data on the caliber of the findings from thus investigation. This is helped by a confusion matrix, which provides information on the effectiveness of the classifier.

Confusion matrix	CKD	NOT CKD
CKD	TP	FN
NOT CKD	FP	TN

- **Confusion Matrix Precision**: Precision, or positive predictive value in this instance, is the proportion of patients who actually had CKD to those who had CKD predicted (true positive and false positive).

$$\text{"Precision} = TP/TP + FP\text{"}$$

- **Recall**: It refers the sensitivity and measures the proportion of actual chronic kidney disease patients who were actually recognized to all CKD patients.

$$\text{"Recall} = TP/TP + FN\text{"}$$

- **F-Score**: It evaluates the test's accuracy. It is the recall harmonic mean and precision.

$$\text{"F-Score} = 2 * ((recall * precision)/(recall + precision))\text{"}$$

- **Accuracy**: It is the proportion of output cases that were successfully predicted to all the other cases in the data collection.

$$\text{"Accuracy} = (TP + TN)/(TP + FP + TN + FN)\text{"}$$

Algorithm result comparison:

	Accuracy	Precision	F1-score	Recall
Decision tree	1.00	0–1.00 1–1.00	0–1.00 1–1.00	0–1.00 1–1.00
Random forest	1.00	0–1.00 1–1.00	0–1.00 1–1.00	0–1.00 1–1.00

(continued)

(continued)

	Accuracy	Precision	F1-score	Recall
Logistic regression	0.98	0–1.00 1–0.95	0–0.99 1–0.98	0–0.97 1–1.00
SVM	0.98	0–1.00 1–0.95	0–0.99 1–0.98	0–0.97 1–1.00

Acknowledgements Don Bosco Institute of technology gave the planned system its whole support during the development. We express our appreciation to our guide. Together with the team members Chandana D. K., Chetana S. M., and Greeshma N., Mrs. Chaithra A. S., Assistant Professor, Department of Information Science and Engineering, DBIT, generously contributed additional help to make this study effective. We appreciate the department and management's help in providing the concept for this system and the abundance of resources.

References

1. Almutary H, Bonner A, Douglas C (2013) Chronic kidney disease in Saudi Arabia: a nursing perspective. Middle East J Nur 7(6):17–26
2. Al-Sayyari A, Shaheen F (2011) End stage chronic kidney disease in Saudi Arabia. A rapidly changing scene. Saudi Med J 32:339–346
3. Magoulas GD, Prentza A, Machine learning in medical applications. Springer, Berlin
4. Jerlin Rubin L (2001) UCI machine learning repository: Chronic_Kidney_Disease Data Set, pp 300–307

Sandalwood Tree Deduction Using Deep Learning

K. R. Nataraj, A. V. Mohan Kumar, P. Archana, C. A. Chandana, and S. Chaitanya

Abstract Deep learning with image processing provides better results. Image processing are used for spike disease detection, oil extraction, age detection, and so on by using CNN and image processing techniques such as preprocessing the image, extracting the feature are used to detect spike disease. Spike disease detection, oil extraction, and age detection have a huge scope in agricultural domains. Visually observing the rings in tree bark is tedious job henceforth, counting the concentric rings in tree barks to estimation the age of a tree is done using deep learning and image processing. Just by considering the tree log size we cannot predict the amount of oil extraction, so we use CNN technique to predict approximate amount of oil extracted (Muzamal et al. in Crowd counting with respect to age and gender by using faster R-CNN based detection. IEEE [1]). This paper provides an overview of different techniques that are used for spike disease detection, age detection, and Oil extraction of sandalwood tree.

Keywords Spike disease detection · Age detection · Oil extraction · Deep learning (DL) · CNN network · Image processing · Feature extraction

1 Introduction

Santalum Album L. known as precious tree that contribute value to Indian culture. It is the second costliest wood in the world. Sandalwood oil is used in cosmetics (makeup kits) like perfumes and soaps [2]. The monopoly of sandalwood trade done by the Government of Karnataka, Tamil Nadu, Kerala and its consequences have shown in severe exploiting this tree the vulnerable category in the IUCN Red List. Detailed oriented research has shown that sandalwood showcasing considerable hormonal

K. R. Nataraj (✉) · A. V. M. Kumar · P. Archana · C. A. Chandana · S. Chaitanya
Don Bosco Institute of Technology, Bangalore, Karnataka, India
e-mail: director.research@dbit.co.in

A. V. M. Kumar
e-mail: avmohan@dbit.co.in

diversity for the different traits and features. However, the information persisting to the heartwood oil content is highly lack in quantity mainly because of not being available of the sandalwood tree plantations carrying out the research in future about these two important traits are burdensome in natural population has gone down swiftly.

2 Problem Statement

The aim of this project is to predict the age of sandalwood tree age using rings of tree trunk. To predict the disease of the tree through leaves. To predict the oil extraction in sandalwood tree using woods, roots of the sandalwood tree. We use deep learning model in this project through this model we can predict age, disease, and oil extraction. We use convolutional neural networks. Here we take sample of leaves, wood, and rings of branch of tree. Deep learning algorithm are adopted to monitor the characteristics of leaf, tree log across section of tree trunk. Upon this deep learning we can accurately predict the spike disease, age of tree, and amount of oil extracted. A normal human cannot be monitoring this features of tree accurately. Therefore, and artificial Perceptron tells which tree is affected by spike disease, could predict trees age by looking at the concentric circles in cross-section of trunk and could predict quantity of oil that been extracted based on sandalwood log size.

3 Aims and Objectives

The target of this project is to predict age in sandalwood tree age using rings of tree trunk. To predict the disease of the tree through leaves. To predict the oil extraction in sandalwood tree using woods, roots of the sandalwood tree. We use deep learning model in this project through this model we can predict age, disease, and oil extraction. We use convolutional neural networks. Here we take sample of leaves, wood, and rings of branch of tree. We use convolutional neural networks. Here we take sample of leaves, wood, and rings of branch of tree. Deep learning algorithm are adopted to monitor the characteristics of leaf, tree log, and cross-section of tree trunk. Upon this deep learning we can accurately predict the spike disease, age of tree, and amount of oil extracted. A normal human cannot monitoring this features of tree accurately. Therefore, and artificial Perceptron tells which tree is affected by spike disease, could predict trees age by looking at the concentric circles in cross-section of trunk, and could predict quantity of oil that been extracted based on sandalwood log size.

4 Motivation

Sandalum Album L. is mainly known as the east Indian sandalwood and its own oil the east Indian sandalwood oil. The mainwood that constitutes the middle part of the tree is value for its good smell. The outside part of wood has no smell. The mainwood is described as a contracting, durability in yellow or brown in a one sight, with an oily texture and is an impressive material for carving complex designs. The carved sculptures of gods and mythological numbers do have a high demand in the merchandise. There is no wonder that sandalwood is the second most being precious wood in the world next to the Africa's blackwood. A vast number of variety of articles such as boxes, cabinet panels, jewel cases, combs, picture frames, hand fans, pen holders, card cases, letter opener, and bookmarks are made from sandalwood. The Vidhana Soudha which houses legislative chambers of state of Karnataka in Bengaluru has a complex as a carved, exposing sandalwood door leading to the Cabinet Sandalwood is sacred and is used in religious festivities and as an important ingredient in 'homa' as Sanskrit word that refers to any program in which making givings into a devoted fire. The wood is white or yellow and not fragmented, that is used in preparing aromas in devoted sticks.

5 Existing System

The existing method for leaf disease detection is simple eye beaming conservative study by experts through which identifying and detecting of leaf diseases is being done. For doing that, a large team of well-versed people as well as continuous looking after of leaf is needed, which is worth very high when we do with larger farms. At the same time, in some of the countries, farmers will not have proper necessary things so that they can contact to experts. Because of which visiting well-versed people even worth more as well as consumes the time too. In such cases, the most wanted technique shows to be useful in checking large disease. Automatic detection of the disease by just seeing the indications on the plant leaves makes it easier as well as inexpensive. This also supports vision in the machines to provide image-based automatic processing control, inspection, and robot guidance. Leaf disease is identified in visual way that is more grueling task and at the same time, little accuracy can be done only in narrowing areas. Where in case of automatic detection technique is used it will take minute efforts, tiny time and becomes more accurate [3].

The execution still lacks in accuracy of result in some cases. More steps are needed. Priority information is needed for segmenting. Extension of database is needed in order to reach the high amount of accuracy. Tiny diseases have been covered so far. So, task needs to be worked to cover more diseases. The possible causes that can lead to errors in classifications can be as follows: disease indications vary from one leaf to another, features perfection is needed, more training samples are needed to cover more cases and to predict the disease more accurately.

6 Proposed System

To resolve this problem, the image of the diseased plant is taken as input and the device is trained to classify the disease of the plants by extracting features of the leaf and predict the output. Apart from image classification into Bacterial, Viral, and Fungal types of plant disease using CNN and data augmentation, we would like to convert the model into a Web App. The basic design of the Web App would allow a person to upload an image of the plant and the model built would classify the image into appropriate disease type.

Block Diagram

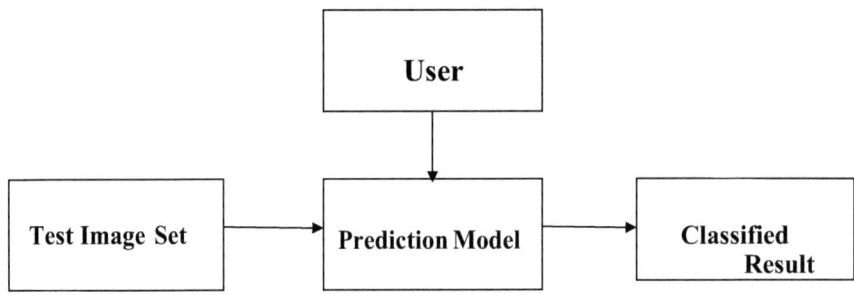

Data Flow Diagram

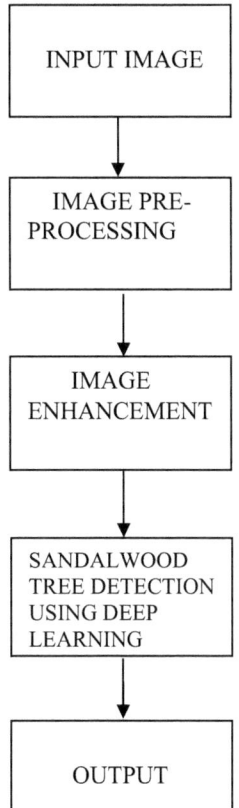

7 Methodology

Methodology is the specific procedures used to detect process, analyze information. Here we are predicting three subtopics, i.e., the spike disease usually found in variety of sandalwood trees, age of a tree using concentric rings of wood, and amount of oil is extracted by taking sample of tree log.

7.1 Sandalwood Spike Disease Detection

Basically here we use labeled dataset of sandalwood leaves to detect the spike disease and the dataset are trained and tested. Convolutional, pooling layer are used for extracting the features, whereas the fully connected layer are used for classification. Images are divided into two parts that is, training and testing, the ratio of train test

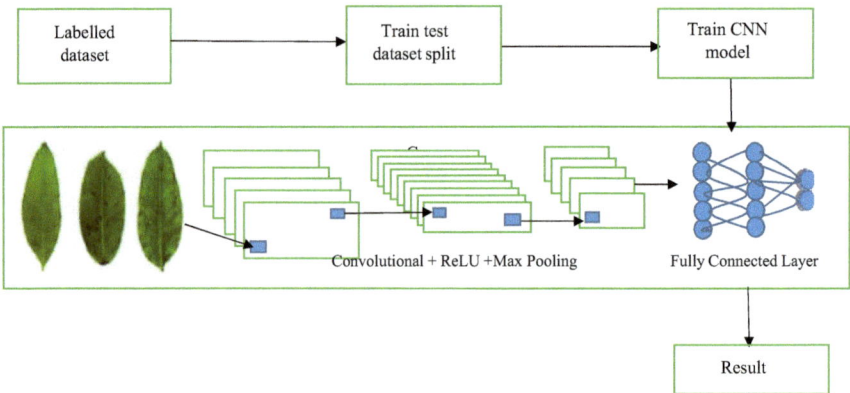

Fig. 1 Block diagram of spike disease detection

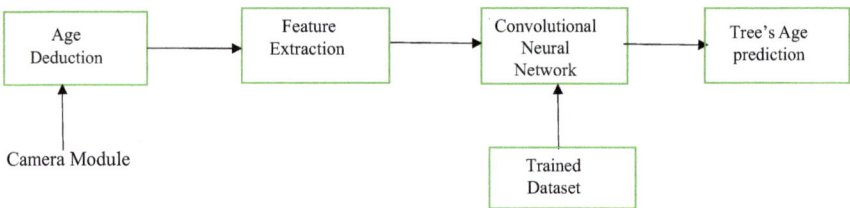

Fig. 2 Block diagram of age deduction

split is 80–20 (80% of the whole dataset used for the training and 20% for testing). Trained datasets go through convolutional layer where they are classified to predict the output (Fig. 1).

7.2 Sandalwood Age Detection

Here we will detect the age of sandalwood tree using concentric rings of wood sample. In this process through camera concentric rings of wood, we are going to deduct the age by extracting the features, and then through CNN network we are going to train the dataset (Fig. 2).

7.3 Sandalwood Oil Extraction

Here we are determining amount of oil extraction using tree log sample. Here we take diameter of tree log to calculate the output. Basically it takes the input from

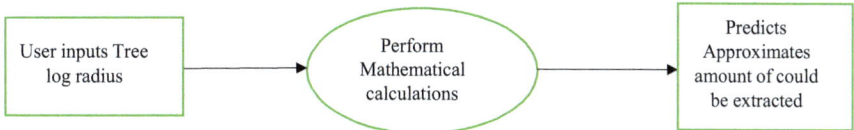

Fig. 3 Block diagram of oil extraction

user, i.e., radius and by user formula we predict how much oil could be extracted from specified tree log (Fig. 3).

8 Software Requirements Functional Requirements

- Device must do less computations on its own.
- Devices must be capable detect the disease of the plant leaf [3].
- Device should be able to compare with training data for detection and prediction analysis.

Non-functional Requirements:

- The web app should allow the user to give image as input.
- The image should be detected with its appropriate disease and nutrition.

Hardware Requirements:

- Processor: Intel Core 2.30 GHz.
- Hard Disk: 250 GB.
- Monitor: 15 VGA Color.
- RAM: 4 GB/8 GB.

Software Requirements:

- MATLAB Tool.
- Image Processing Toolbox.
- Python language.
- NVIDIA GPUs.

9 System Design and Implementation

System design is a process of determining the components like data, modules, architecture, and interface for a system to accomplish specified requirements. System development is the process of developing, designing systems, creating or altering system which satisfies the specific needs and requirements of an organization. System design is one of the most important stage of software development process. The main purpose of the system design is to provide information and data useful and necessary for implementation of the system elements.

Object-oriented analysis and design methods are most generally used methods for computer systems design. The Unified Modeling Language (UML) is one of the standard language used in object-oriented analysis and design system. The purpose of UML is to modeling software system and to design non-software system.

Systems implementation is the process of determining how the information system should be in physical system design, realization of an application, or execution of a plan, idea, model, design, specification, standard, algorithm, or policy, ensuring that the information system is operational and used, ensuring that the information system meets quality standard (i.e., quality assurance).

Implementation is one of the most important phases of the Software Development Life Cycle (SDLC). The SDLC highlights different stages (phrases or steps) of the development process. The life cycle approach is used so users can see and understand what activities are involved within a given step. It is also used to let them know that at any time, steps can be repeated or a previous step can be reworked when needing to modify or improve the system.

10 Conclusion

Hence, we have finished categories of image, extracting the features, and training data. The whole development of algorithm is done using [MATLAB] tool. We have used several toolboxes like statistics and deep learning toolboxes, convolutional neural network, and image processing toolkits. The outcomes are training, data in form of image, text, and video. Image can be classified using CNN network. This algorithm can be implemented with training data and classifying of given image dataset. The input test image is compared with the trained data for detecting and predicting the analysis. By utilizing all the above following method we find sandalwood spike disease by using sample leaves, detect the age of tree by using concentric circles of wood and calculate the amount of oil that can be extracted from a tree by using log of wood.

References

1. Muzamal JH, Tariq Z, Khan UG (2019) Crowd counting with respect to age and gender by using faster R-CNN based detection. IEEE
2. Kotwal K, Mostaani Z, Marcel S (2019) Detection of age-induced makeup attacks on face recognition systems using multi-layer deep features. IEEE
3. Patel S, Jaliya UK Dr, Patel P (2020) A survey on plant leaf disease detection. IEEE
4. Zhang N, Zhao H, Liu Y (2020) Deep learning for automatic recognition of oil production related objects based on high-resolution remote sensing imagery. IEEE, China

Aspect-Based Sentiment Analysis: A Survey of Deep Learning Methods

Ugranada Channabasava, Golu Kumar Ram, Bhanu Bhakta Jaishi,
Chitrasen Raj, and Keshav Kumar Shandliya

Abstract The process of analyzing, processing, concluding, and inferring the sentiment of subjective texts is known as sentiment analysis (Shandilya et al. in Aspect-based sentiment analysis survey of deep-learning. IEEE [1]). Sentiment analysis is used by businesses to better understand their customers public opinion polling, market research, and brand evaluation reputation, comprehension of customer experiences, and social media research the media's influence. Depending on the various aspect requirements granularity is classified as positive, negative and neutral of the sentiment analysis. This article provides an overview of recently proposed methods for dealing with a sentiment analysis problem based on aspects (Liu et al. in Aspect-based sentiment analysis-a survey of deep learning methods. IEEE [2]). There are currently three popular approaches: deep learning, lexicon-based, and traditional machine learning methods.

Keywords Sentiment analysis based on aspects · Sentiment analysis based on natural language processing · Web scraping · Opinion mining

1 Introduction

Many people read online customer reviews and ratings. According to studies, consumers trust online reviews or comments from strangers before purchasing a product or service. In this field, numerous statistical surveys and studies have been conducted. According to a study conducted in, 39% of customers read about eight reviews, while only 12% read 16 or more reviews before purchasing a product; 98% of customers admit that customer reviews of previous buyers influence their purchasing decision. According to statistics, potential buyers are willing to spend 31% more on a product or service that has received positive feedback.

U. Channabasava (✉) · G. K. Ram · B. B. Jaishi · C. Raj · K. K. Shandliya
Don Bosco Institute of Technology, Bangalore, India
e-mail: channasan11@gmail.com

A. Kumar et al. (eds.), *Proceedings of the 4th International Conference on Data Science, Machine Learning and Applications*, Lecture Notes in Electrical Engineering 1038,
https://doi.org/10.1007/978-981-99-2058-7_32

Many reviews are lengthy, making it difficult for a potential customer to read them and decide whether or not to purchase the product. The large number of reviews also makes it difficult for product manufacturers to track customer sentiments and opinions about their products and services.

As a result, creating a review summary is required. Reviews are described using sentiment analysis [3]. Sentiment analysis employs the natural concept of natural language processing to extract subjective information required for source materials. The main task is to determine whether the stated opinion is positive or negative [4].

Because customers rarely express their opinions in simple terms, judging an opinion stated can be a difficult task. Some perspectives are comparative, while others are direct. By simply condensing these ratings into two more general categories positive or negative sentimental analysis helps shoppers visualize customer satisfaction while making purchases [5]. Feedback is largely used to help customers make online purchases and learn about current product market trends, which helps retailers create market strategies.

2 Problem Statement

- A word of opinion that is regarded as positive in one circumstance may be regarded as negative in another.
- We can significantly increase the precision and capability of sentiment analysis with the use of machine learning.

2.1 Existing System

At different granularities, enough work has been done in the field of sentiment analysis. Some works at the document level classify the entire review based on the reviewer's subjective judgment. In certain sentence-level studies, the focus is on determining the polarity of a sentence (e.g., positive, neutral, or negative) using semantic data gleaned from the sentences' textual content. Additionally, several recent researches also include sentiment analysis at the phrase level, with the major emphasis being on phrases, which are collections of words that frequently have a unique idiomatic meaning. The topic of sentiment analysis at the aspect level, however, is still developing and needs additional study. Sentiment analysis has been used in a variety of industries, including the travel and entertainment sectors. While another article employs Perceptron neural networks, the work employed a combination of machine learning characteristics and lexical features. Additionally, research has been done on the data derived from social media, such as Twitter's mapping of social media attitudes using observations and quantifiable data. The study made the case that tracking customer opinion online may serve as dynamic feedback for any firm. The study classified the moods of Twitter tweets into three classifications:

positive, negative, and neutral using a tree kernel-based model. This can also be used to track how the general public feels about a specific incident, piece of news, etc.

Method	Year of proposal	Classification	Text level	Prediction accuracy	Pros	Cons
OPINE	2005	Unsupervised rule-based approach	Word	87%	Domain independent	Difficulty in availing OPINE system, thus rare to get applied in real life
Sentiment analysis: Adjectives and adverbs are better than adjectives alone	2006	Linguistic approach	Document	Pearson correlation of 0.47	Adjectives are given more priority (adjectives expresses human sentiments better than adverbs alone)	None
Opinion digger	2010	Unsupervised machine learning method	Sentence	51%	Rates product at aspect level	Requires rating guidelines to rate. Works only on known data
Sentiment classification using lexical contextual sentence structure	2011	Rule-based approach	Sentence	86%	Said to be domain independent	Depends solely on wordNet
Interdependent latent Dirichlet allocation	2011	Probabilistic graphical model	Document	73%	Faster in comparing and correlating sentiment and rating	Correlation between identified clusters and feature or ratings are not explicit always [6]
A joint model of feature mining and sentiment analysis for product review rating	2011	Machine learning	Document	71% (in 3 categories) 46.9% (in 5 categories)	Automatic calculation of feature vector	Use of WordNet

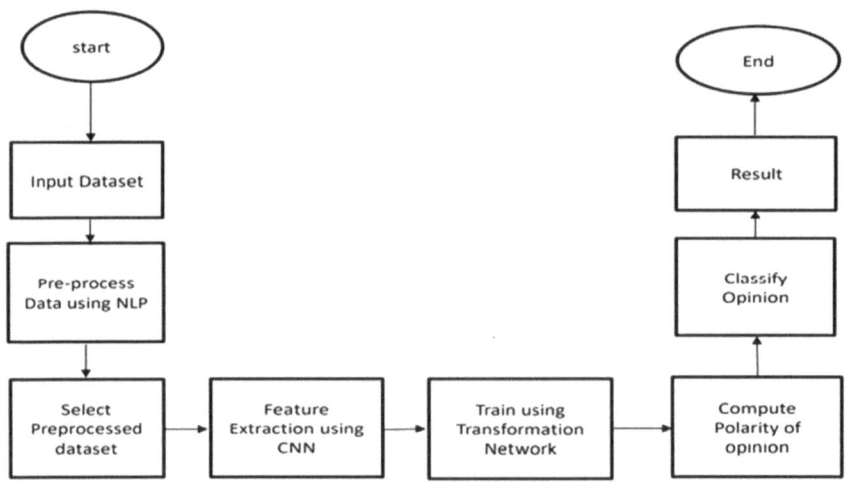

Fig. 1 Flowchart

2.2 Proposed System

Figure 1 says that, the architecture of the proposed system, the main goal is to be the process the data using an NLP and then used VADER analysis to get the priority of user opinion.

3 Literature Survey

Sl. No.	Title	Author	Methodology	Limitations
Paper-1	Survey of Deep Learning Techniques for Aspect-Based Sentiment Analysis	Ishani Chatterjee, Haoyue Liu	ABSA is treated as a multiclassification problem by traditional machine learning and deep learning techniques	Data preprocessing is underrated process. People focus more on methodology and give less attention to preprocessing of data
Paper-2	A Sentiment Analysis Survey	Preeti Routray, Smita Prava Mishra	Here, various aspects of text document sentiment analysis are reviewed	Need to improve the quality of system such as accuracy
Paper-3	Deep Learning Sentiment Analysis	Shilpa P C, Rissa Shereen, Vinod P	Twitter message sentiment analysis system. The tweets that we take into account for the analysis are a mix of various and emotions	Further analysis is required to obtain personality of the user from their tweets

(continued)

(continued)

Sl. No.	Title	Author	Methodology	Limitations
Paper-4	Survey on Sentiment Analysis and Opinion Mining	G. Vinodhini, RM. Chandrasekaran	In this study, issues in the subject of sentiment analysis are discussed along with methodologies and methods	Major obstacles include the use of several languages, opinions based on features, and phrase complexity
Paper-5	Sentiment Analysis Algorithm and Application	Walaa Medhat, Ahmed Hassan, Hoda Korashy	This study primarily focuses on providing a concise overview of SA techniques and the connected topics	More work is needed for sentiment analysis to analyze a context-based SA
Paper-6	Deep Learning and Machine Learning for Sentiment Analysis	Yogesh Chnadra, Antoreep Jana	Different techniques for sentiment analysis have been considered Sentiment analysis is done using machine learning classifiers	One of the difficulties with sentiment analysis is accuracy

4 Methodology

The implementation of the project consist of four steps that can be defined below [7–12]:

1. Data Preprocessing
2. Filtering
3. Compute Polarity of Opinion
4. Classify Opinion.

Data Preprocessing

The review contains word that are not required in the classification model. It can consist of hyperlinks, emoji special characters, double quotation, punctuation, extra white space. Data preprocessing in defined for the removal of such words. Stopwords such as 'is', 'are', 'the', which do not contain any meaning are filtered out by using inbuilt python module. Streaming and lemmatization are also done using NLP to normalize the text for further preprocessing using the model [13–16].

Data Scraping

It is done to extract data from the preprocessed data. In this project we have used this process to extract the features that are required for the analysis of the sentence. Once the data scraping is done then the extracted feature is used for the classification of the polarity of the user opinion.

Compute Polarity of Opinions

Once the data is done then the extracted data is used by the VADER analysis tool to compute the polarity of the sentence. A list of features/words is used; these words have been labeled as either positive or negative.

Classify Opinions

The statement is categorized as positive or negative depending on the compound score after the polarity calculation [17, 18]. The compound score totalizes ratings with values ranging from − 1 (negative) to + 1 (positive).

The sentiment is favorable (complex score ≥ 0.05). Sentiment of Neutrality: (− 0.05, compound score 0.05). Unfavorable Attitude: (compound score = − 0.05)

5   Results

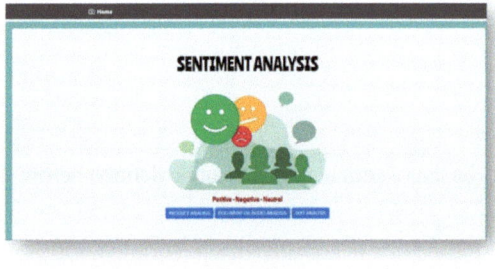

Snapshot 1: Home Page
- It contains 3 options –
    1. Product Analysis
    2. Document Analysis
    3. Text Analysis

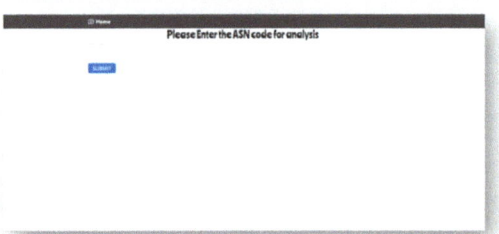

Snapshot 2: PRODUCT ANALYSIS
- Enter Product URL

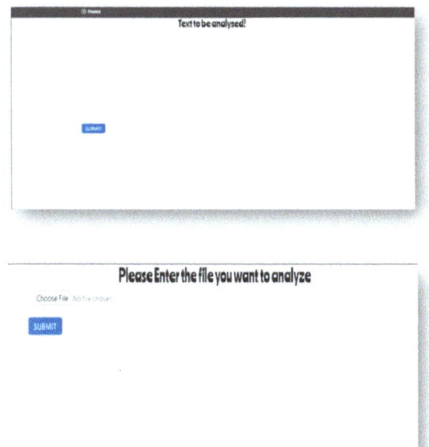

Snapshot 3: Enter Text for Text analysis

Snapshot 4: Upload .PDF or .TXT for document analysis.

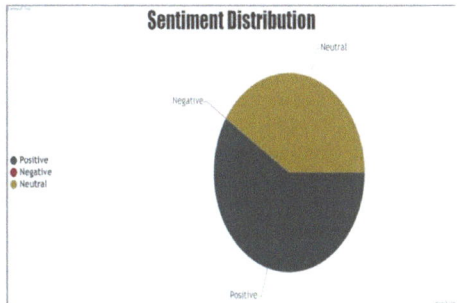

Snapshot 5: Sentiment Distribution using pie chart.

Acknowledgements The development of the proposed system was extremely supported by Don Bosco Institute of Technology. We present gratitude to our guide Prof. U. Channabasava, Assistant Professor, Department of Information Science & Engineering, DBIT co-heartedly provided extra support to make this project successful with team members—Golu Kumar Ram, Bhanu Bhakta Jaishi, Keshav Kumar Shandilya We thank the department and management for providing the idea in coming up with this system & extremity in resources.

References

1. Raj CS, Shandilya KK, Jaishi BB, Ram GK (2022) Aspect-based sentiment analysis survey of deep-learning. IEEE
2. Liu H, Chatterjee I, Zhou MC, Lu XS, Abusorrah A (2021) Aspect-based sentiment analysis-a survey of deep learning methods. IEEE
3. Routray P, Swain CK, Mishra SP (2021) A survey on sentiment analysis
4. Shilpa PC, Shereen R, Jacob S, Vinod P (2020) Sentiment analysis using deep learning

5. Medhat W, Hassan A, Korashy H (2019) Sentiment analysis algorithms and applications: a survey

6. Nisha Jebaseeli A, Kirubakaran E (2020) A survey on sentiment analysis of (product) reviews. Int J Comp Appl 47(11):36–39

7. Wang M, Shi H (2020) Research on sentiment analysis technology and polarity computation of sentiment words. In: 2016 IEEE international conference on progress in informatics and computing (PIC), vol 1. IEEE

8. Saleh K (Apr 11, 2018) The importance of online customer reviews [Infographic]. Invesp. Accessed: 5 Nov 2020. [Online]. Available: https://www.invespcro.com/blog/the-importance-of-online-customerreviews-infographic/

9. Qualtrics (Apr 10, 2019) 20 online review stats to know in 2019. Accessed: 6 Dec 2019. [Online]. Available: https://www.qualtrics.com/blog/online-review-stats/

10. Harrag F, Alsalman A, Alqahtani A (2019) Prediction of reviews rating: a survey of methods, techniques and hybrid architectures. J Digit Inf Manage 17(3):164

11. He R, Lee WS, Ng HT, Dahlmeier D (2019) An interactive multitask learning network for end-to-end aspect-based sentiment analysis. In: Proceedings 57th annual meeting association for computational linguistics, pp 504–515

12. Akhtar MS, Garg T, Ekbal A (2020) Multi-task learning for aspect term extraction and aspect sentiment classification. Neurocomputing 398:247–1193

13. Carley KM (2019) Syntax-aware aspect level sentiment classification with graph attention networks. In: 9th international joint conference on empirical methods in natural language processing, pp 5469–5477, Nov 2019

14. Zhang B, Xu X, Li X, Chen X, Ye Y, Wang Z (2019) Sentiment analysis through critic learning for optimizing convolutional neural networks with rules. Neurocomputing 356:21–30

15. Do HH, Prasad PWC, Maag A, Alsadoon A (2019) Deep learning for aspect-based sentiment analysis: a comparative review. Expert Syst Appl 118:272–299

16. Song M, Park H, Shik Shin K (2019) Attention-based long short-term memory network using sentiment lexicon embedding for aspect-level sentiment analysis in Korean. Inf Process Manag 56(3):637–653

17. Ma X, Zeng J, Peng L, Fortino G, Zhang Y (2019) Modeling multi-aspects within one opinionated sentence simultaneously for aspect-level sentiment analysis. Futur Gener Comput Syst 93:304–311

18. Tang F, Fu L, Yao B, Xu W (2019) Aspect based fine-grained sentiment analysis for online reviews. Inf Sci (Ny) 488:190–204

Phishing Attack Detection Using Machine Learning

K. R. Nataraj, D. K. Yashaswini, R. Hema, Nayana S. Pawar, and S. Yashaswi

Abstract Phishing, an online scam in which people are tricked into sensitive personal and account information, is a serious threat both to consumers and institutions doing business on the web. Statistics show a continued increase in phishing attacks. Not only that, but these scams are becoming more and more elaborate and therefore more difficult to detect. Thus, there is a need to employ intelligent algorithms to solve this problem. One of the most successful methods for detecting these malicious activities is machine learning. This is because most phishing attacks have some common characteristics which can be identified by machine learning methods. In this article, we select important phishing URL, and website content-based features are extracted. We used machine learning algorithms like Gradient Boosting Algorithm to give greater accuracy (97%). We compare the performance of other related techniques with our scheme. The proposed scheme performs better compared to other selected related work. This shows that our approach can be used for real-time applications in detecting phishing URLs.

Keywords Phishing attack · Machine learning · Phishing detection · Uniform resource locators (URLs)

1 Introduction

A phishing website is a common social engineering method that mimics trustful uniform resource locators (URLs) and webpages. Through such attacks, the phisher targets naïve online users by tricking them into revealing confidential information, with the purpose of using it fraudulently. Phishing is committed so that the criminal may obtain sensitive and valuable information about a consumer, usually with

K. R. Nataraj (✉) · D. K. Yashaswini · R. Hema · N. S. Pawar · S. Yashaswi
Don Bosco Institute of Technology, Bangalore, India
e-mail: director.research@dbit.co.in

D. K. Yashaswini
e-mail: yashaswinidk5@dbit.co.in

© The Author(s), under exclusive license to Springer Nature Singapore Pte Ltd. 2023   355
A. Kumar et al. (eds.), *Proceedings of the 4th International Conference on Data Science, Machine Learning and Applications*, Lecture Notes in Electrical Engineering 1038, https://doi.org/10.1007/978-981-99-2058-7_33

the goal of fraudulently obtaining access to the consumer's bank or other financial accounts. Often 'phishers' will sell credit card or account numbers to other criminals, turning a very high profit for a relatively small technological investment [1]. A complete phishing attack involves three roles of phishers. Firstly, mailers send out a large number of fraudulent emails (usually through botnets), which direct users to fraudulent websites. Secondly, collectors set up fraudulent websites (usually hosted on compromised machines), which actively prompt users to provide confidential information. Finally, cashers use the confidential information to achieve a payout. Monetary exchanges often occur between those phishers.

When a user clicks on a malicious URL, malware is downloaded, and an attempt is made to steal login credentials, whereas phishing websites deceive the user into entering credit card information while acting as a genuine website. Attackers examine authentic websites first, looking for traits that separate them from the original. The attackers then attempt to develop their own phishing websites that are nearly identical to the original. Some clear signals can be left out for users and developers to recognize and protect themselves against phishing websites, even if some functionality cannot be recreated. In order to avoid getting phished, users should have awareness of phishing websites and have a blacklist phishing websites which requires the knowledge of website being detected as phishing. With the technologies evolving day by day, there have been drawbacks associated with it. The fraudulent activities are happening on a very large scale which needs to be stopped. It has been estimated that the loss due to cybercrimes has projected to reach $6 trillion by 2021 which is very huge. To deal with this, an automated system that uses sophisticated techniques to identify phishing URLs is required. One of the most critical stages in combating online fraud is detecting fraudulent URLs. As a result, tools that assist in recognizing phishing URLs are in high demand to make the internet a safer place [2].

2 Literature Survey

Phishing is a terrifying threat in the world of Internet security. In this assault, the user enters personal information on a fake website that appears to be authentic. We conducted a survey on visual similarity-based phishing detection techniques. This survey gives you a better understanding of phishing websites, different solutions, and the future of phishing detection. Several ways for phishing detection are mentioned in this work; nevertheless, most of the approaches still have problems such as accuracy, protection against new phishing websites, failure to notice embedded objects, and so on. To detect phishing assaults, these methods use several aspects of a webpage, such as text similarity, font color, font size, and graphics on the page. Although text-based similarity techniques are quick, they are unable to detect phishing attacks when the text is replaced with an image. Image processing-based techniques have a high accuracy rate, despite their complexity and length. Furthermore, the majority of the job is done offline. These include data collection and profile creation steps that must be completed first. A comparative table is made for quick comparison of the benefits of the various available ways established by various individuals and published. There is no single strategy that is sufficient for phishing detection. The accurate detection of phishing websites is an open challenge for further research and development [3].

Sl. No.	Paper title, author	Year	Methodology/ algorithms used	Findings/results obtained
1	Efficient detection of phishing attacks with hybrid neural network-Xiaoqing Zhang, Dongee Shi, Hongpo Zhang, Wei Liu, Runzchi Li	2018	Random Forest Naïve Bayes algorithm	This shows that we adopt an AE to cover the shortage of classic convolution filter, which helps to build correlation among the whole features
2	Detection of phishing websites using hybrid model-Ch. Chakradhara Rao, A. V. Ramana	2018	Decision Tree, IBK, Naïve Bayes, and Bayes Net algorithms	Experiments have been done to measure the accuracy of all the algorithm at the beginning, since the accuracy measure of Naïve Bayes algorithm is very low when compared to other algorithm such as Random Forest, IBK, and Decision Tree
3	Phishing attack detection using feature selection techniques-Aniruddha Narendra Joshi, Thanuja R Pattanaashetti	2019	Random Forest algorithm and Relief F algorithm	The combination of the algorithms for feature selection approach showed very good accuracy
4	Detecting Phishing sites using hybrid algorithm (IJERT)-Poonam Kumari, Spruthi MN, Bhavya MU	2020	Linear model with logistic regression method Random Forest method Decision Tree classifier	Higher accuracy predicting result is logistic regression or other model by comparing the best accuracy
5	A machine learning approach for phishing and its detection techniques, Dhananjay Merat, Anurag Patil, Sourabh Gavsane, Vivekanand Jadhav, Prof. Himanshu Joshi	2020	Support Vector Machines and Naïve Bayes Classifier	The different methods that achieved highest accuracy have been tested by selecting minimum number of features and by reducing the dimensionality of feature subset

(continued)

(continued)

Sl. No.	Paper title, author	Year	Methodology/ algorithms used	Findings/results obtained
6	Machine learning approach based on hybrid features for detection of Phishing URL's-(IEE2021) Sushruthi Mishra, Awishkar Ghimire	2021	Data collection Feature Engineering Data Preprocessing Classification algorithm Performance measures	The results are promising and can help us to detect the phishing URLs with great accuracy
7	Hybrid model of phishing email detection: A combination of technical and non-technical anti-phishing approaches-Melad Mohamed Al-Daeef, Nurlida Basir, and Madihah Mohd Saudi	2018	Proposed anti-phishing models	Separated evaluation processes have been conducted to individually evaluate each of the modules from which the system is consisting of
8	A hybrid approach for phishing website detection using machine learning (VIVA-TECH INTERNATIONAL FOR RESEARCH AND INNOVATION)-Faiz Shaikh, Prof Saniket Kudoo	2021	Random Forest algorithm and TF-IDF approach	Query will be calculated and compared with the original query of the program
9	A machine learning approach for phishing and its detection techniques (INTERNATIONAL RESEARCH JOURNAL OF ENGINEERING AND TECHNOLOGY)-Dhananjay Merat, Anurag Patil	2020	Tokenization Word Embedding Classification Testing	Provides a good accuracy and detects whether the email is phished or safe

3 Objectives

The primary objectives are

- For generation of an effective Phishing Detection system that effectively manages phishing attacks and protects user data.
- Overcoming the problem of increasing Phishing attacks and poor detection measures.
- Our system provides a feature of phishing detection that helps users to protect their secured data.
- Prevent cyber-attacks.
- Avoid users from malicious sites.

3.1 Existing System

In the existing model, algorithms such as logistic regression, SVM, KNN, and Naïve Bayes algorithms are used. And the classifiers can classify the text content and image content. Also the performance of different classifiers is based on classification ratio, F-score, and precision values. To detect phishing sites, the existing system employs Classifiers, Fusion Algorithms, and Bayesian Models. Text and visual material can be classified by the classifiers. Text classifiers are used to categories text material, whereas Image classifiers are used to categories image content. The threshold value is calculated using a Bayesian model, the Fusion Algorithm uses the results of both classifiers to determine whether or not the site is phishing [4]. Correct classification ratio, F-score, Matthews' correlation coefficient, false negative ratio, and false alarm ratio are used to evaluate the performance of different classifiers. The developer will be the only decider of the threshold value. As a result, issues such as false positive and false negative occur. False positive means the likelihood of the webpage being a phishing webpage is higher than the threshold value, yet it is not a phishing webpage. False negative indicates that the likelihood of a phishing webpage is lower than the threshold number, yet the webpage is still a phishing webpage. As a result, security levels are lowered. The current technology can only handle one type of phishing assault. If it was a phishing site, the existing system would just issue a warning to the user.

Disadvantages of Existing System: The Accuracy of face recognition under low lighting and low-resolution condition is relatively less.

3.2 Proposed Approach

Fake websites are becoming increasingly widespread, and there are a variety of attacks behind them. Phishing is identified as a key security issue in these types of attacks, and new innovative ideas are arising with it every second; therefore, preventive mechanisms must be equally effective. As a result, the security in these circumstances must be extremely good and should not be easily tractable due to ease of implementation. Most programs today are only as secure as their underlying operating system. Since the design and technology of middleware has continually improved, detecting them has become a difficult task. As a result, determining whether or not a computer connected to the Internet is trustworthy, and safe is almost difficult. In the proposed system, we use Gradient Boosting Classifier. Gradient Boosting Algorithm is used for URL classification which consist of various levels that are used for classification task to detect whether the website is safe to use or not [5].

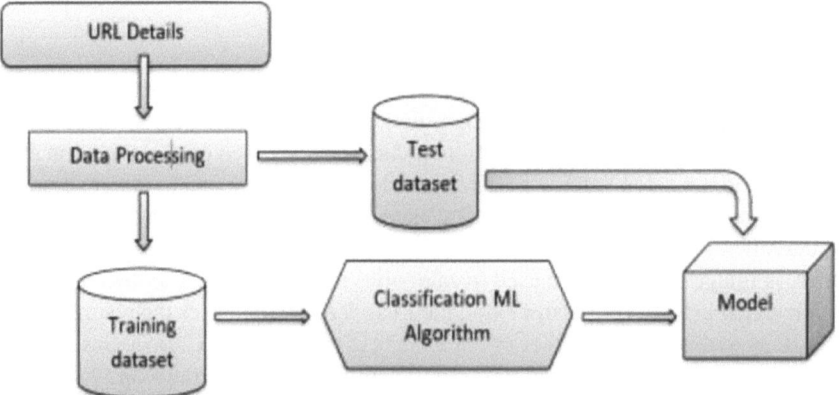

Fig. 1 Architecture of proposed model

3.3 Advantage of Proposed System

- Gives more accuracy nearly 97%.
- The main advantage of Gradient Boosting Classifier is that it combines many weak learning models together to create a strong predictive model.
- Thus, the model is effective in classifying complex dataset (Fig. 1).

3.4 Block Diagram

See Fig. 2.

4 Methodology

1. **Data Collection**

The dataset is borrowed from Kaggle, https://www.kaggle.com/eswarchandt/phishing-website-detector. A collection of website URLs for 11000+ websites. Each sample has 30 website parameters and a class label identifying it as a phishing website or not (1 or − 1). The overview of this dataset is, it has 11054 samples with 32 features [6].

2. **Feature Extraction**

The URLs associated with phishing have certain peculiar characteristics that set them apart from the benign ones. Different features like 'Index', 'UsingIP', 'LongURL', 'ShortURL', 'Symbol@', 'Redirecting//', 'Prefix Suffix-', 'SubDomains', 'HTTPS',

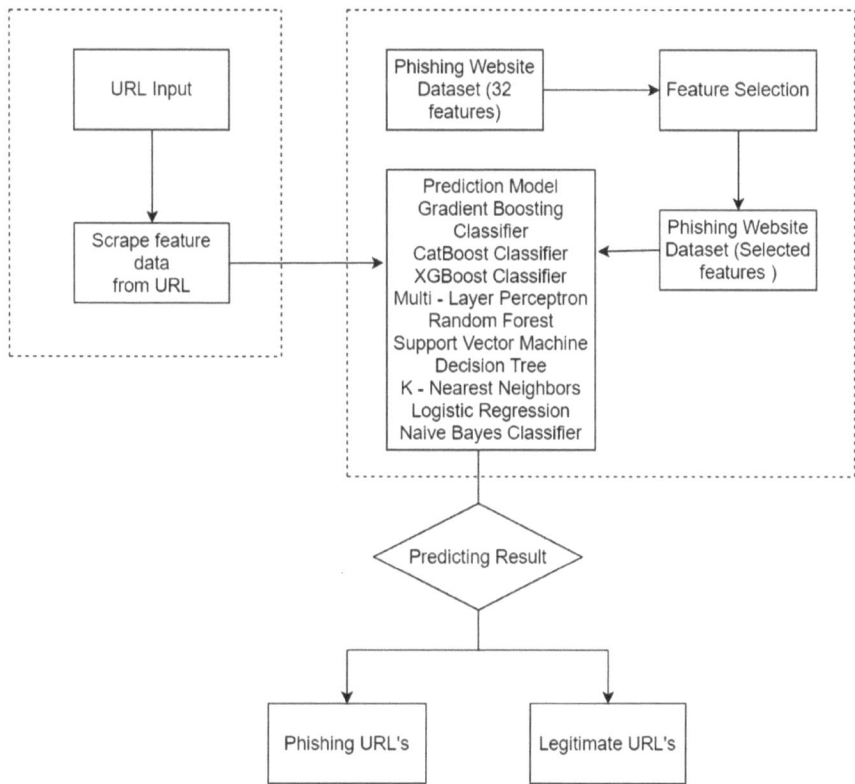

Fig. 2 Proposed framework for detecting phishing URLs

'DomainRegLen', 'Favicon', 'NonStdPort', 'HTTPSDomainURL', 'RequestURL', 'AnchorURL', 'LinksInScriptTags', 'ServerFormHandler', 'InfoEmail', 'AbnormalURL', 'WebsiteForwarding', 'StatusBarCust', 'DisableRightClick', 'UsingPopupWindow', 'IframeRedirection', 'AgeofDomain', 'DNSRecording', 'WebsiteTraffic', 'PageRank', 'GoogleIndex', 'LinksPointingToPage', 'StatsReport', 'class', and dtype. There are 11054 instances and 31 features in dataset. Out of which 30 are independent features, whereas 1 is dependent feature. Each feature is in datatype, so there is no need to use Label Encoder. There is no outlier present in dataset, and there is no missing value in dataset [7].

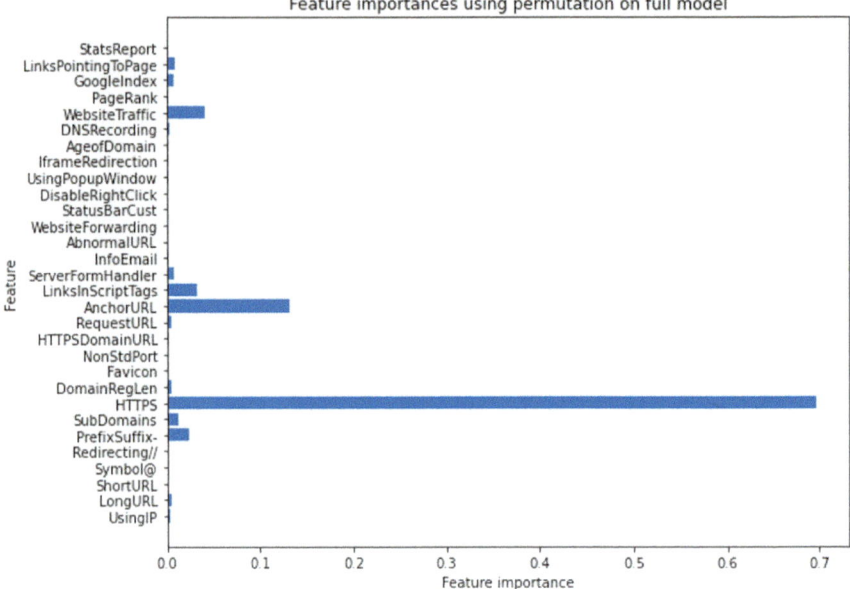

5 Machine Learning Algorithms

Machine learning classification models such as Gradient Boosting Classifier, CatBoost Classifier, XGBoost Classifier, Multi-Layer Perceptron, Random Forest, Support Vector Machine, Decision Tree, K-Nearest Neighbors, logistic regression, and Naïve Bayes Classifier have been used to detect phishing websites [8].

5.1 Logistic Regression

Logistic regression predicts the output of a categorical dependent variable. Therefore, the outcome must be categorical or discrete value. Logistic regression is much similar to the Linear Regression except that how they are used. Linear Regression is used for solving regression problems, whereas logistic regression is used for solving the classification problems.

5.2   K-Nearest Neighbors Classifier

K-Nearest Neighbor is one of the simplest machine learning algorithms based on supervised learning technique. KNN algorithm assumes the similarity between the new case/data and available cases and put the new case into the category that is most similar to the available categories [7].

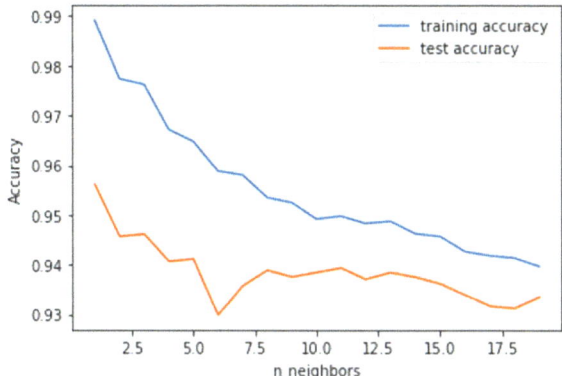

5.3   Support Vector Machine Classifier

Support Vector Machine or SVM is one of the most popular supervised learning algorithms, which is used for classification as well as regression problems. The goal of the SVM algorithm is to create the best line or decision boundary that can segregate n-dimensional space into classes so that we can easily put the new data point in the correct category in the future [9].

5.4   Naïve Bayes Classifier

Naïve Bayes algorithm is a supervised learning algorithm, which is based on Bayes theorem and used for solving classification problems. It is mainly used in text, image classification that includes a high-dimensional training dataset. Naïve Bayes Classifier is one of the simple and most effective classification algorithms which helps in building the fast machine learning models that can make quick predictions [10].

5.5   Decision Tree Classifier

Decision Tree is a supervised learning technique that can be used for both classification and regression problems, but mostly it is preferred for solving classification problems. It is a tree-structured classifier, where internal nodes represent the features of a dataset, branches represent the decision rules, and each leaf node represents the outcome.

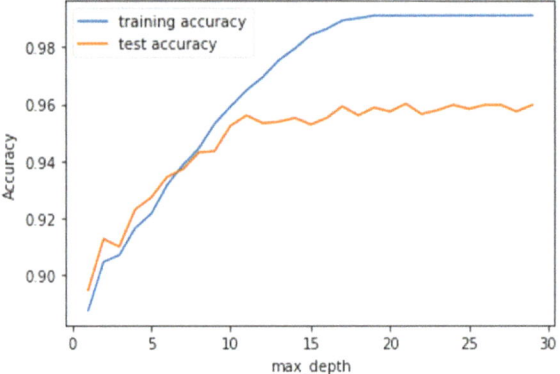

5.6   Random Forest Classifier

Random Forest is a popular machine learning algorithm that belongs to the supervised learning technique. It can be used for both classification and regression problems in ML. It is based on the concept of ensemble learning, which is a process of combining multiple classifiers to solve a complex problem and to improve the performance of the model.

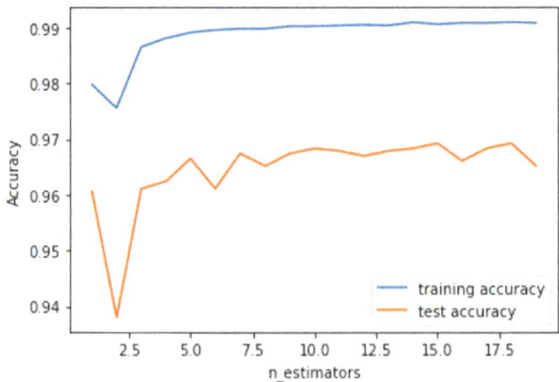

5.7 Gradient Boosting Classifier

Gradient Boosting Classifiers are a group of machine learning algorithms that combine many weak learning models together to create a strong predictive model. Decision Trees are usually used when doing Gradient Boosting. Boosting algorithms play a crucial role in dealing with bias variance trade-off. Unlike bagging algorithms, which only controls for high variance in a model, boosting controls both the aspects (bias and variance) and is considered to be more effective [10].

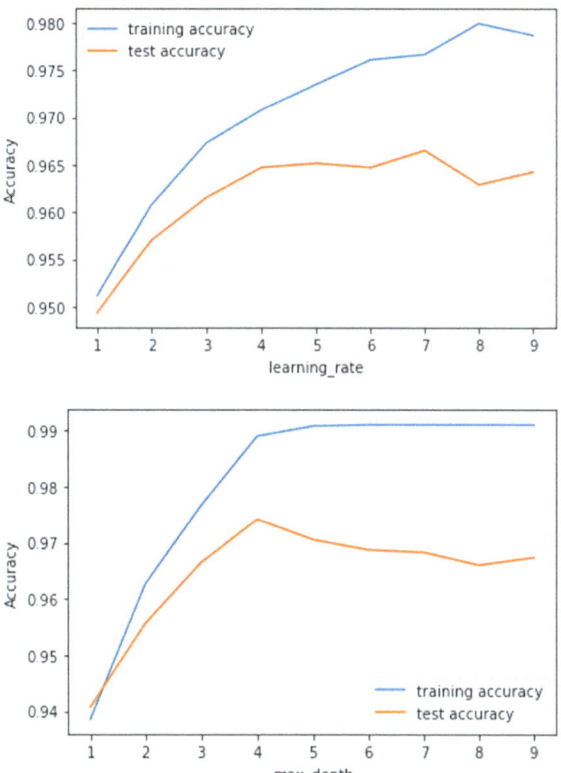

5.8 CatBoost Classifier

CatBoost is a recently open-sourced machine learning algorithm from Yandex. It can easily integrate with deep learning frameworks like Google's Tensor Flow and Apple's Core ML. It can work with diverse data types to help solve a wide range of problems that businesses [11].

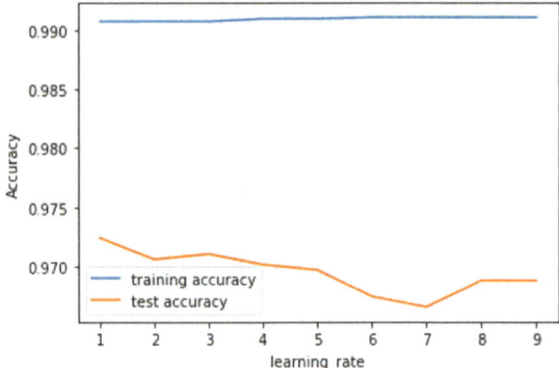

5.9 Extreme Gradient Boosting (XGBoost) XGB Classifier

XGBoost is an implementation of Gradient Boosted Decision Trees designed for speed and performance that is dominative competitive machine learning. In this post, you will discover how you can install and create your first XGBoost model in Python [12].

5.10 Multi-layer Perceptron Classifier

MLP classifier stands for Multi-Layer Perceptron classifier which in the name itself connects to a Neural Network. Unlike other classification algorithms such as Support Vectors or Naïve Bayes Classifier, MLP classifier relies on an underlying Neural Network to perform the task of classification [13].

6 Performance Evaluation

One of the most important needs for any phishing detection system is for it to have a high detection accuracy. If the detection system incorrectly classifies phishing websites as legitimate, user safety is jeopardized. Users will also protest if legal websites are labeled as phishing sites by the detection system. The following evaluation metrics were used to assess the detection accuracy of the phishing detection system.

- True Positive (tp)—The number of phishing websites identified as phishing.
- True Negative (tn)—The number of legitimate websites identified as legitimate.
- False Positive (fp)—The number of legitimate websites misidentified as phishing.

- False Negative (fn)—The number of phishing websites misidentified as legitimate.
- Accuracy: Accuracy is the measurement used to determine which model is best at identifying relationships and patterns between variables in a dataset based on the input, or training, data [14].

$$\text{Accuracy} = (tp + tn)/(tp + tn + fp + fn)$$

- Precision (*p*) measures the rate of phishing websites which are identified correctly as the websites detected as phishing. To be precise, it measures the degree to which the blocked websites are in fact phishing.

$$p = tp/(tp + fp)$$

- Recall (*r*) measures the rate of phishing websites which the phishing detection system identifies correctly as phishing websites.

$$r = tp/(tp + fn)$$

- F-Score (f1-Score) is the weighted harmonic mean of precision and recall. This project uses the f1-Score as an index for testing accuracy.

$$\text{f1-score} = 2 * (p * r)/(p + r)$$

Results

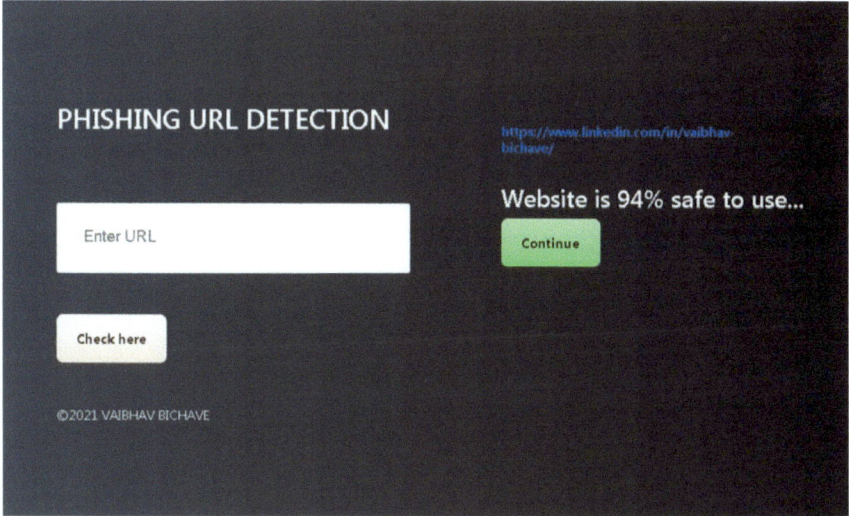

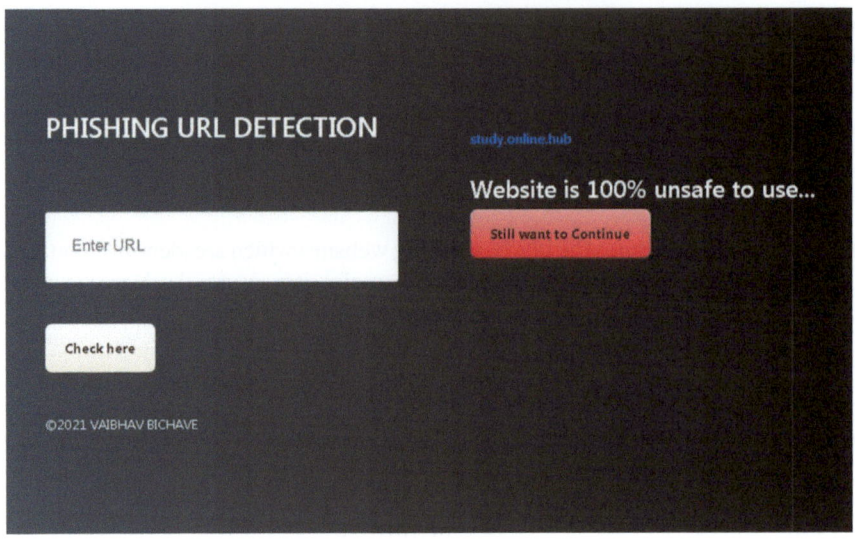

	ML Model	Accuracy	f1_score	Recall	Precision
0	Gradient Boosting Classifier	0.974	0.977	0.994	0.986
1	CatBoost Classifier	0.972	0.975	0.994	0.989
2	XGBoost Classifier	0.969	0.973	0.993	0.984
3	Multi-layer Perceptron	0.969	0.973	0.995	0.981
4	Random Forest	0.967	0.971	0.993	0.990
5	Support Vector Machine	0.964	0.968	0.980	0.965
6	Decision Tree	0.960	0.964	0.991	0.993
7	K-Nearest Neighbors	0.956	0.961	0.991	0.989
8	Logistic Regression	0.934	0.941	0.943	0.927
9	Naive Bayes Classifier	0.605	0.454	0.292	0.997

7   Conclusion

At present generation attackers are more in networks, phishing has become major security problem, causing many losses by hacking the legal data that are used by the user. Phishers set their own fake websites which exactly looks like the original website including applying DNS server name, setting up web server and creating web pages similar to destination website. The final conclusion on the phishing dataset is that some feature like 'HTTPS', 'AnchorURL', 'WebsiteTraffic' have more importance to classify URL is phishing URL or not. Gradient Boosting Classifier currently classifies URL upto 97.4% respective classes and hence reduces the change of malicious attachments.

Future Enhancement

Future work could include the use of a variety of other factors that are useful to detecting phishing URLs. To improve accuracy, more network-based features can be employed in conjunction with length-based features. The paper's algorithms are run with default hyperparameters. Hyperparameter tuning can be used to make the algorithms more robust in the future. Only machine learning algorithms are used in this study, whereas deep learning methods can be studied further. Furthermore, phishing has become more complex, with attackers employing a variety of innovative strategies. As a result, the tests should be conducted using newer datasets that account for more complex threats.

In addition, it is possible to deploy the website on the cloud for ease of use. A Google Chrome extension can be created to save time and can automatically check every website. The administrator has control over the home page and can make changes as needed. We can add a comment box to the front page so that users can leave admin remarks. In the future, we can introduce more sorts of phishing schemes.

References

1. Krishnan DM, Subramaniyaswamy V (2015) Phishing website detection system based on enhanced itree classifier. ARPN J Eng Appl Sci 10(14):5688–5699
2. Tahir AUH, Asghar S, Zafar A (2016) A hybrid model to detect phishing sites using supervised learning algorithms. In: 2016 International conference on computational science and computational intelligence (CSCI), 15–17 Dec 2016
3. Phishing | What is phishing? Phishing.org, 2018. [Online]. Available: http://www.phishing.org/what-is-phishing. Accessed 15 Oct 2018
4. Mei Y (2008) Anti-phishing system: detecting phishing e-mail. School of mathematics and systems engineering
5. Yadav S, Bohra B (2015) A review on recent phishing attacks in internet. In international conference on green computing and Internet of Things 2015. IEEE, pp 1312–1315
6. Hsu CH, Wang P, Pu S (2011) Identify fixed-path phishing attack by STC. In: Proceedings of the 8th annual collaboration, electronic messaging, anti-abuse and spam conference (CEAS'11), ACM, Perth, Australia, September 2011, pp 172–175
7. Jagatic TN, Johnson NA, Jakobsson M, Menczer F (2007) Social phishing. Commun ACM 50(10):94–100
8. Jingguo W, Herath T, Rui C, Vishwanath A, Rao HR (2012) Phishing susceptibility: an investigation into the processing of a targeted spear phishing e-mail. IEEE Trans Prof Commun 55(4):345–362
9. http://www.phishing.org/history-of-phishing. Accessed 10 May 2022
10. Sheng S, Holbrook M, Kumaraguru P, Cranor LF, Downs J (2010) Who falls for phish? a demographic analysis of phishing susceptibility and effectiveness of interventions. In: Proceedings of the 28th international conference on human factors in computing systems, CHI '10. ACM, New York, NY, USA, 2010, pp 373–382
11. https://dictionary.cambridge.org/dictionary/english/phishing. Accessed 8 May 2022

12. Mohammad R, Thabtah F, McCluskey L (2015) Phishing websites dataset. Available: https://archive.ics.uci.edu/ml/datasets/Phishing+Websites. Accessed January 2016
13. Khonji M, Iraqi Y, Jones A (2013) Phishing detection: a literature survey. IEEE
14. Parmar B (2012) Protecting against spear-phishing. Comp Fraud Secur 2012(1):8–11
15. APWG Q1-Q3 Report, 2015. http://docs.apwg.org/reports/apwgtrendsreportq1-q32015.pdf
16. https://www.cisco.com/c/en_in/products/security/email-security/what-isphishing.html

Smart Farming Assistance with Plant Disease Detection Using Deep Learning

A. S. Chaithra, A. J. Chirag, S. Karan, and R. Jayaram Krishna

Abstract In the field of agriculture, huge proportions of crops are lost due to the imprecise selection of the crop to be cultivated in a particular portion of land (Kumar et al. in Recommendation system for crop identification and pest control technique in agriculture [1]). With our model acting as a catalyst to this hindrance, we attempt to identify and predict the pre-eminent and much suited crop to the farmers and diagnose the pest that may be the cause of the affect as well as advocate the most appropriate and effective pest control techniques. Also we have incorporated plant disease and diagnosis model which depicts if the crop is healthy or the type of disease it has caught along with the remedy to be followed to treat the plant. For crop recommendation, we have used random forest algorithm, which when tested provided the highest accuracy as compared to other algorithms, and for plant disease detection, we have used ReLU architecture of the convolution neural network for higher precision in detecting the plant disease by uploading the image of affected area of the plant leaf.

Keywords Random forest · Rectified linear unit (ReLU) · Convolution neural network (CNN)

1 Introduction

India is an agricultural country with the second-highest land area of more than 1.6 million square-km under cultivation. Agriculture has been the backbone of our country. It is very much essential for the growth of our country. There is no such universal system to assist farmers in agriculture. India is an agriculture-based developing country. In spite of having lot of digital data, they are not able to access real time to the factual information such as the crop yield data in particular soil and crop disease detection techniques, pesticides to be used, weather conditions and pest

A. S. Chaithra (✉) · A. J. Chirag · S. Karan · R. Jayaram Krishna
Don Bosco Institute of Technology, Bangalore, India
e-mail: aschaithra@gmail.com

© The Author(s), under exclusive license to Springer Nature Singapore Pte Ltd. 2023   371
A. Kumar et al. (eds.), *Proceedings of the 4th International Conference on Data Science, Machine Learning and Applications*, Lecture Notes in Electrical Engineering 1038, https://doi.org/10.1007/978-981-99-2058-7_34

management. To overcome these problems, technological solution is needed which can help the farmers. The productivity of farming is not only depending on natural resources but it also depends on input provided to the system. The main difficult task for farmers is information access and management for the quantity of data and the complication of processes in precision farming. The data for farming like crop life cycle detail, seeds, crop selection, crop processes weather, pesticides, fertilizer, etc., are accessible from a lot of different sources like newspaper, printed media, audio, mobile, TV, Internet, visual aids, etc. [2]. But the structures and formats of data are different. So it is extremely hard for farmer to get exact information and to know variety of information which have distributed from diverse sources. Sometime several manual steps are essential to handing out data for translating data from one format to another format.

2   Problem Statement

2.1   Existing System

In the existing system, farmers are not connected with any technology and analysis. In traditional system, farmer uses "trial and error" method. Farmer experiments on land with different crops, water availability, etc., and after many such "tries", farmer probably gets the best crop suitable in particular land. In the existing model, RGB images are converted to grayscale images using color conversion and classification according to KNN and ANN. Various enhancement techniques are used to improve image quality [3, 4]. They are combined with different methods of image preprocessing in favor of better feature extraction.

Disadvantages of existing system

- High chance of time and money loss.
- Particularly when growing new crops, farmers may face the risks of either market failure or production problems.
- KNN and ANN are not suitable for large datasets of image classification. With ANN, concrete data points must be provided, and in KNN, the cost of calculating the distance between the new point and each existing points is huge.

2.2   Proposed System

In this system, we propose to build a platform which can help farmers to get best crop on the basis of the highest requirement predicted after analyzing all necessary attribute of soil like crops that are suitable for their area according to the soil type, average rainfall, temperature, location, etc. Pudumalar et al. [5] proposed the use of technology to provide result to farmers for recommendation of best crops. It also focuses on farmers to be able to identify crop diagnosis by uploading the image to the application. Also the farmer by using our project will get information easily even though he is an amateur at using technology. Thus, the farmer gets maximum profit and knowledge which in turn reduces digital divide. In the proposed system, we use CNN algorithm for image classification which consists of various levels that are used for forecasting to detect disease in plant leaves, and with the help of CNN, the maximum accuracy can be achieved if the data are huge. It also, after identification of the disease, suggests the name of pesticide to be used.

Advantages of proposed system

- Random forest gives more accuracy
- The main advantage of CNN is that it automatically detects the important features. This is why CNN would be an ideal solution to computer vision and image classification problems.
- Once disease is classified, the remedy to treat the disease is also provided.

3 System Design

The idea for implementing this system design is to bridge the gap between the farmer and technology through which he could increase his yield. The proposed project is built with the most trending programming language also used in major industries—Python. The system is provided with two main modules.

3.1 Crop Recommendation

This module deals with scattered datasets which are cleaned and processed using the data preprocessing step discussed above [6]. The dataset is trained using classification models like SVM, random forest, XG boost algorithm and decision tree, and by calculating accuracy, the algorithm with the highest accuracy is preferred for suitable crop prediction (Fig. 1).

3.2 Performance Evaluation of the Algorithms Used

When the trained data were fed to the algorithms, the following accuracies were determined. We can clearly see that random forest and XG boost have the highest accuracy among them, and here we propose to make use of random forest algorithm (Fig. 2).

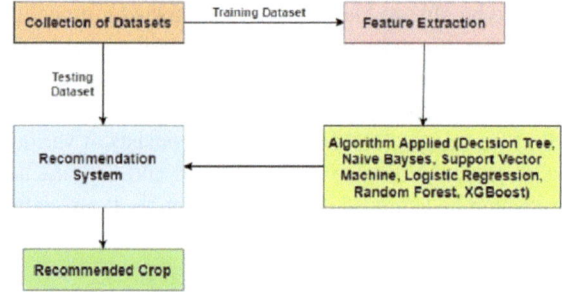

Fig. 1 Dataflow diagram of crop recommendation

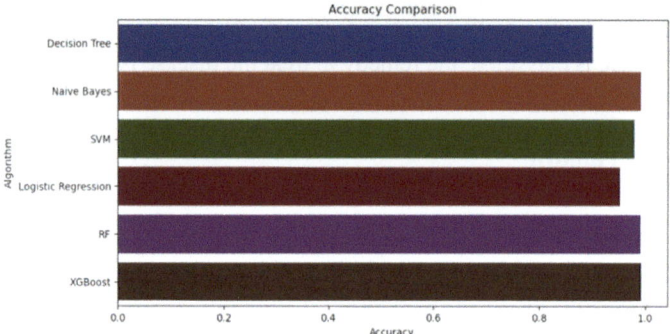

Fig. 2 Accuracy comparison of different algorithms

3.3 *Plant Disease Detection*

The database is preprocessed such as image reshaping, resizing and conversion to an array form. Similar processing is also done on the test image. A database consisting of about 32,000 different plant species is obtained, out of which any image can be used as a test image for the software. The train database is used to train the model (CNN) so that it can identify the test image and the disease it has. CNN has different layers that are Dense, Dropout, Activation, Flatten, Convolution2D, and MaxPooling2D. After the model is trained successfully, the software can identify the disease if the plant species is contained in the database. After successful training and preprocessing, comparison of the test image and trained model takes place to predict the disease (Fig. 3).

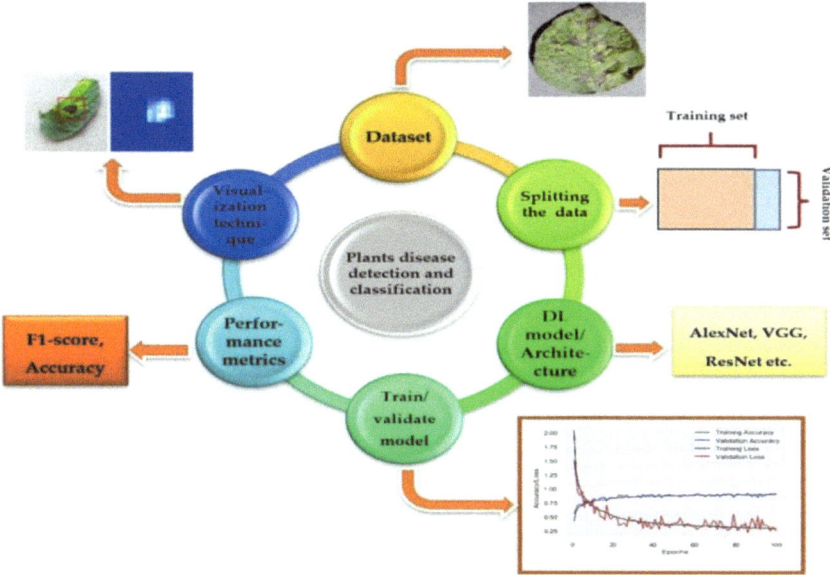

Fig. 3 Dataflow diagram of plant disease detection

3.4 Validation Accuracy and Loss

Validation accuracy is how model is able to classify the images with the validation dataset. Validation loss is a metric used to assess the performance of a deep learning model on the validation set. The validation set is a portion of the dataset set aside to validate the performance of the model. In the proposed model, the graph shows evidently that the accuracy is high (Fig. 4).

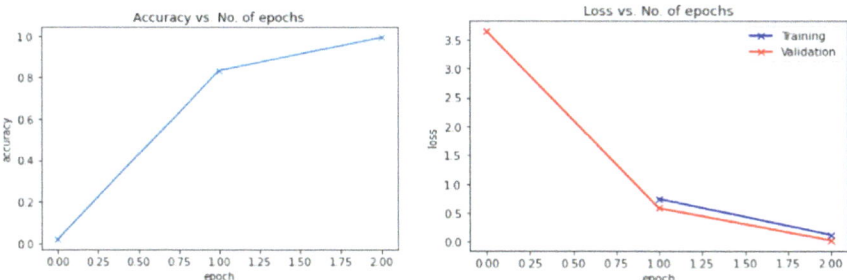

Fig. 4 Validation accuracy and validation loss

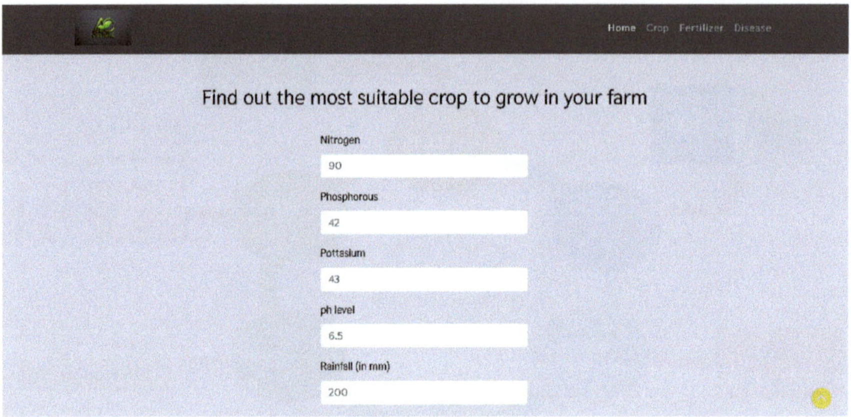

Fig. 5 Crop recommendation page

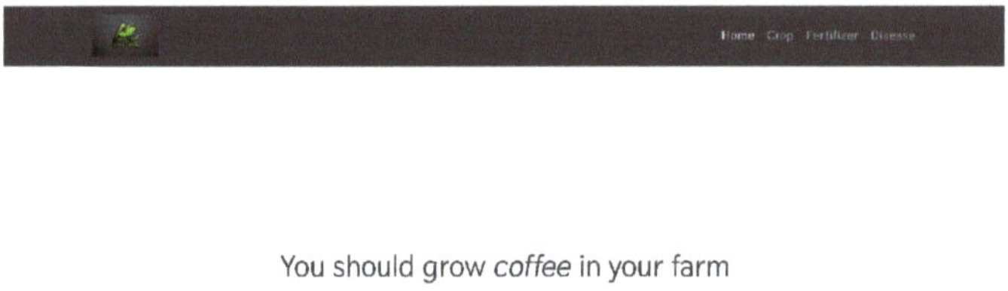

You should grow *coffee* in your farm

Fig. 6 Result of crop recommendation system

4 Results

The proposed system hands the role of crop recommendation and plant disease detection [7]. Suitable crop to grow at the right time and plant leaf disease can be determined. All these modules are available for single system general purpose. Below are results obtained from two modules for different input (Figs. 5, 6, 7 and 8).

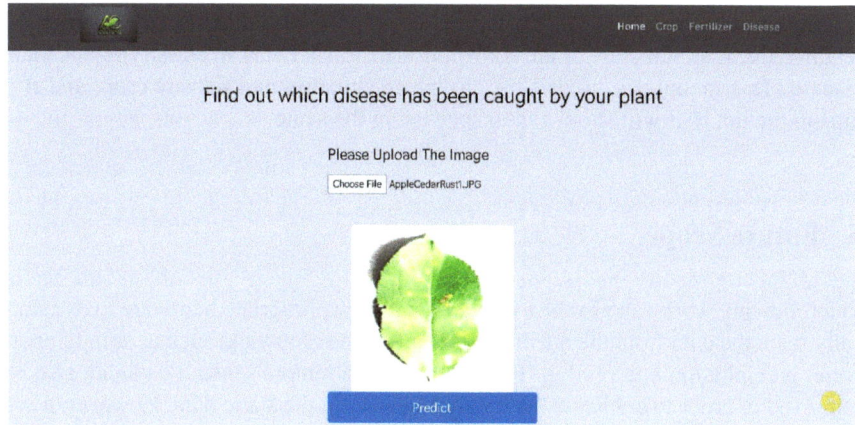

Fig. 7 Plant disease detection page

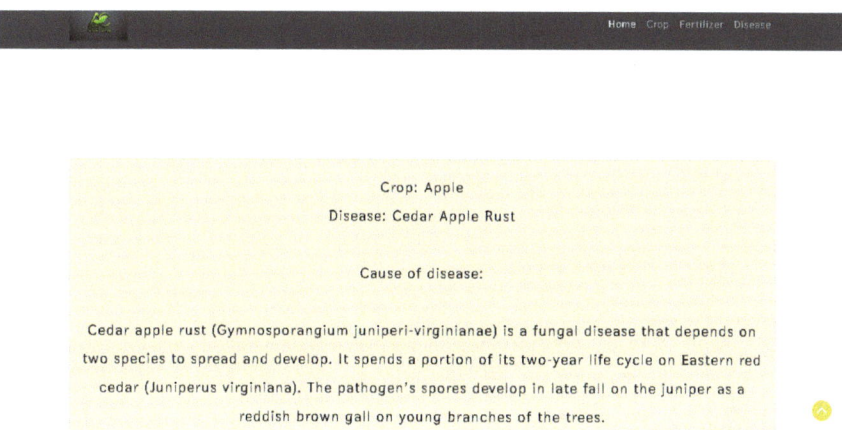

Fig. 8 Result of plant disease detection system

5 Conclusion

Agriculture which is an important part of our economy is essential to ensure that even the smallest investment which is done in the agriculture sector should be taken care of. So it is essential to check if the right crop has been chosen for a land holding which matches its requirements to benefit the nation and farmer in particular [8]. In the current study, we collected multiple datasets and performed appropriate feature engineering to build a single source of data that accounts for all of the essential features to help model correctness. It was clearly seen in the graphical evaluation that random forest not just outperformed other classical machine learning algorithms in terms of accuracy which was found to be 99%, but also was able to predict the result with minimum error. Similarly, we acquired an accuracy of 99.2% from the rectified linear architecture of the convolution neural network in classifying the plant disease. This model can predict for any given situation and for any crop, and if it fails to predict it, it will show a message stating the same.

6 Future Scope

The following work can also be extended by using appropriate hardware to dynamically fetch the data from the attributes that affect the crop yield such as soil, temperature, precipitation and rainfall. Forecasting using remote sensing data can also be improved in order to eradicate the hassle of handling the static data. However, these satellite images can be used in association with the in-land data to provide a new dimension to the following work.

Acknowledgements The development of the proposed system was extremely supported by Don Bosco Institute of Technology. We present gratitude to our guide Mrs. Chaithra A.S., Assistant Professor, Department of Information Science and Engineering, DBIT, co-heartedly provided extra support to make this project successful with team members—A.J. Chirag, Karan S. and Jayaram Krishna R. We thank the department and management to provide the idea in coming up with this system and extremity in resources.

References

1. Kumar A, Sarkar S, Pradhan C (2019) Recommendation system for crop identification and pest control technique in agriculture. In: 2019 International conference on communication and signal processing (ICCSP). https://doi.org/10.1109/iccsp.2019.8698099
2. Paul M, Vishwakarma SK, Verma A (2015) Analysis of soil behaviour and prediction of crop yield using data mining approach. In: 2015 International conference on computational intelligence and communication networks (CICN). https://doi.org/10.1109/cicn.2015.156K. Elissa, An overview of decision theory. unpublished. (Unplublished manuscript)
3. Jagan Mohan K, Balasubramanian M, Palanivel S (2016) Detection and recognition from paddy plant leaf images. Int J Comp Appl 144
4. Nemishte D, Patil R, More S, Sayali
5. Pudumalar S, Ramanujam E, Rajashree RH, Kavya C, Kiruthika T, Nisha J (2017) Crop recommendation system for precision agriculture. In: Eighth international conference on advanced computing (ICoAC).https://doi.org/10.1109/icoac.2017.7951740
6. Doshi Z, Nadkarni S, Agrawal R, Shah N (2018) AgroConsultant: intelligent crop recommendation system using machine learning algorithms. In: 2018 Fourth international conference on computing communication control and automation (ICCUBEA). https://doi.org/10.1109/iccubea.2018.8697349

7. Raja SKS, Rishi R, Sundaresan E, Srijit V (2017) Demand based crop recommender system for farmers. In: 2017 IEEE technological innovations in ICT for agriculture and rural development (TIAR). https://doi.org/10.1109/tiar.2017.8273714

8. Kumar R, Singh MP, Kumar P, Singh JP (2015) Crop selection method to maximize crop yield rate using machine learning technique. In: 2015 International conference on smart technologies and management for computing, communication, controls, energy and materials (ICSTM)

Malpractice Detection System for Online Examination Using AI

K. R. Nataraj, Domnic Xavier, T. C. Vishrutha, S. Kavya, and M. Hemalatha

Abstract A boom in remote learning has recently occurred. However, there has not been a good answer for academic exams. While some colleges have remote proctoring in place, where a human proctor continuously monitors student activity, others have collected assignments that students can copy and paste from the Internet. There must be a solution found if the way we live is to become the new norm. In this essay, we have put out a plan for creating an integrated AI-based system that can aid in discouraging exam cheating. The system detects fraud and records the evidence. This technology will be both affordable and secure. A boom in remote learning has recently occurred. However, there has not been a good answer for academic exams. In some universities, students can copy and paste collected assignments.

Keywords Malpractice detection · Convolutional neural networks (CNN) · TensorFlow · OpenCV · Transfer learning · Head poses (single shot detector) · Object detection

1 Introduction

Academics now conduct their work online. This is a significant barrier from both the perspective of learning and exam preparation. One of the most difficult problems to be overcome is how to conduct exams honestly. In the last six years, the number of Internet users in India has almost doubled. As a result, many students were able to continue their study, which was great for academics. It is challenging to manually proctor an online exam when it is being taken remotely since multiple students cannot be watched at once. A teacher can see pupils physically and with all of their senses during manually proctored exams at the centres. They may easily ensure that the event runs smoothly by being aware of the sounds and actions of the pupils exams taken online [1].

K. R. Nataraj (✉) · D. Xavier · T. C. Vishrutha · S. Kavya · M. Hemalatha
Don Bosco Institute of Technology, Bengaluru, Karnataka, India
e-mail: director.research@dbit.co.in

© The Author(s), under exclusive license to Springer Nature Singapore Pte Ltd. 2023   381
A. Kumar et al. (eds.), *Proceedings of the 4th International Conference on Data Science, Machine Learning and Applications*, Lecture Notes in Electrical Engineering 1038,
https://doi.org/10.1007/978-981-99-2058-7_35

1.1 Need For Prediction

Our approach guarantees a fair examination environment free of unethical behaviour. System monitors the movements of students in distant locations and can assist in the identification and avoidance of cheating throughout the online exam procedure. You can conduct your entire exam without using any paper. By switching to our technology, we can significantly reduce the cost of managing and administering exams. Our solution is compatible with the conventional exam method.

2 Machine Learning Techniques

Machine learning encompasses problems in which the relationship between the input and output parameters is undetermined. These systems use algorithms to perceive patterns in datasets, which might contain structured or unstructured textual data, numeric data, or even some media like audio files, images, and videos. Machine learning algorithms are computationally intensive, requiring specialized infrastructure to run at large scale [2].

2.1 OpenCV

This research focuses on the development of sign language translator application using OpenCV tool, and Python could be a Python linking library designed to unravel issues with PC vision. Python could be a general artificial language started by Guido van Rossum, which quickly became very talked-about, primarily thanks to its simplicity and readability of code. It permits the computer user to precise ideas in fewer code lines while not decreasing readability. OpenCV application areas include Toolkits 2D and 3D Recognition of the ego phenomenon recognition of leadership robot mobile [3].

Advantages:

- OpenCV has more than sufficient image processing functions and is available free of cost.
- It also entails a vast number of algorithms which can be incorporated into various parts of the program and can perform various tasks like removing red eyes, extracting 3D models of objects, following eye movements, and many more.
- It is highly portable and can be executed on any system that supports the C language while consuming low RAM.

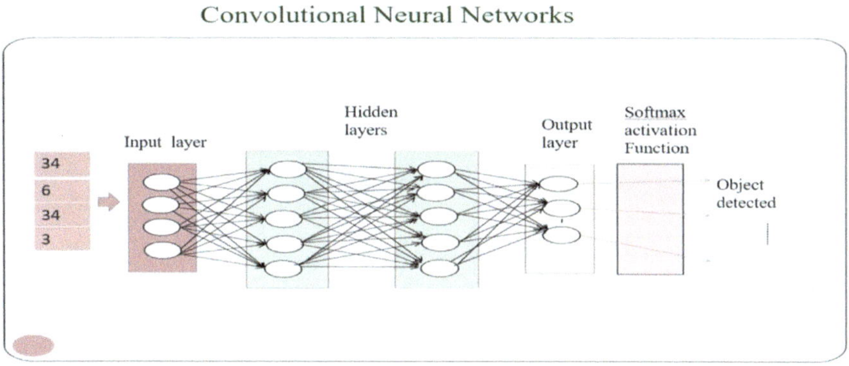

Convolutional Neural Networks

Fig. 1 Layers of CNN

2.2 Convolutional Neural Network (CNN)

Convolutional neural network (CNN) is a deep learning algorithm which can process an input image, assign importance (learnable weights and biases) to various aspects/ objects in the image, and be able to differentiate one from the other based on the user's command or importance assigned. The region of interest which, in our case is the hand gesture which has been conveyed, is segmented. The features extracted are the binary pixels of the images which are the smallest units. We incorporate CNN for training and to classifying the images [4] (Fig. 1).

Advantages:

- The main advantage of CNN lies in automatic detection the important features without any human intervention.
- It implements special convolution and pooling operations and performs parameter sharing.
- The usage of CNN is motivated by the fact that they can capture and are able to learn relevant features from an image/video which necessarily comes down to feature learning.

3   Objectives

In this project, we are going to take an online exam and do the following:

- Detecting mobile phones near students and taking appropriate action.
- Candidate verification and attendance management.
- In an online exam, the system will offer a single portal for logging in, accessing the question paper, a chat window for communicating with the examiner, and uploading the answer sheet using a scanner that is integrated into the portal.

Convolutional Neural Networks

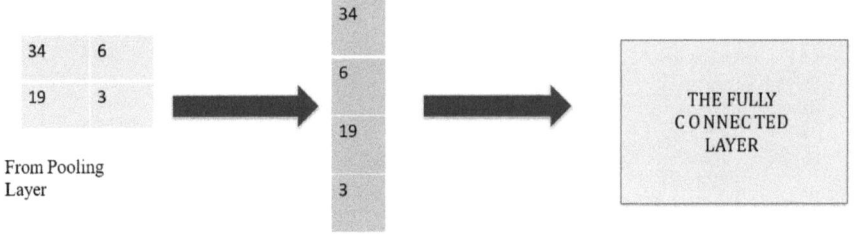

Fig. 2 CNN workflow

- This prevents candidates from opening or accessing any other applications while taking the exam online on a desktop or mobile device (Fig. 2).

4 Module Specification

It is the way to improve the structure design by breaking down the system into modules and solving it as independent task. By doing so, the complexity is reduced and the modules can be tested independently. The number of modules for this model is two, namely hazed data collection set and dataset de-hazing module [5].

- **Image Preprocessing Module**

  Name of the Module: Image Preprocessing
  Actors: User, System
  Use Cases: Captured image, captured image, generate RGB matrix, grey to binary image
  Functionality: The main functionality of this module is to convert the RGB image to binary format for faster process.

- **Segmentation Module**

  Name of the Module: Segmentation
  Actors: System
  Use Cases: Binary image, thresholding, segmented matrix
  Functionality: This module is to obtain segmented matrix from binary image by performing masking.

- **Feature Extraction Module**

  Name of the Module: Feature extraction
  Actors: System
  Use Cases: Segmented matrix, apply GLCM, generate statistical values

Functionality: The main functionality of this module is to apply principle component analysis and obtain statistical values.

- **Classification Module**

  Name of the Module: Classification Actors: System
  Use Cases: Statistical values, obtain feature vector
  Functionality: The main functionality of this module is to calculate minimum distance.

- **Convolution Matrix**

  Name of the Module: Convolution
  Actors: System
  Use Cases: Feature vectors matched with dataset, calculating minimum distance value
  Functionality: The main functionality of this module is to calculate minimum distance value to recognize the corresponding image.

- **Image Preprocessing**

  Before images are utilized for model training and inference, preprocessing is the process of formatting them. This comprises, but is not limited to, collecting, resizing, orienting, and color [6, 7]. As a result, in some circumstances a transformation that could be an augmentation may be better as a pretreatment step. A camera input image is recorded and can be utilized as an input image to identify characters or to save as a dataset for training. Any supported format given by the device is used to capture and save the image. Every image must undergo preprocessing in order to improve the efficiency of image processing. Images that are captured are in RGB format. The collected images have extremely high pixel values and complexity as pictures [8] (Fig. 3).

5 Proposed Work

In this research, we present a web-based system that uses voice recognition and artificial intelligence to detect and assess cheating by students during online exams.

- **Registration**: When students register for the first time on a portal, they provide personal information, an ID card, and a photo that is recorded in the database and used to authenticate them before the exam.
- **Face Recognition**: Using face recognition, the student is identified, and if the face matches the saved face image, the student is validated and permitted to deliver the exam. The student's computer must have a webcam installed, or the front camera if the student is providing the exam on a smartphone.
- **Many face detection**: If multiple faces are detected in the frame, it will also be marked as malpractice in the database.

Fig. 3 Conversion of RGB to grey scale

- **Head posture detection**: In MCQ-based tests, when using a pen and paper is not required, students' head positions will be examined, and if it appears that they are looking away from the screen, a record will also be saved.
- **Mobile phone detection**: If a student is observed using a mobile device, it will be recorded as malpractice in the database (Fig. 4).

Fig. 4 Block diagram for proposed system

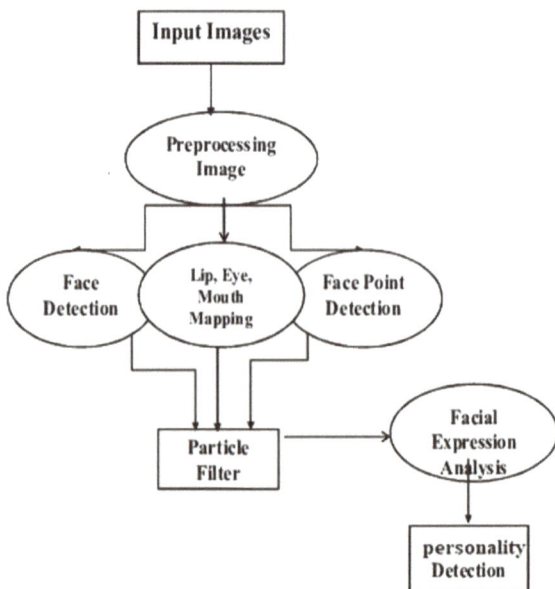

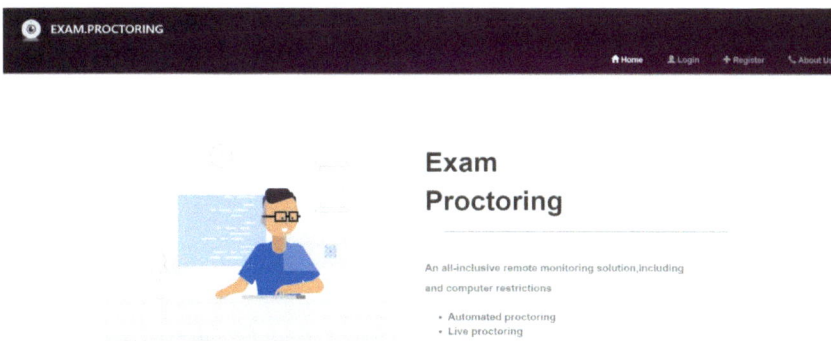

Fig. 5 Dashboard of exam proctoring

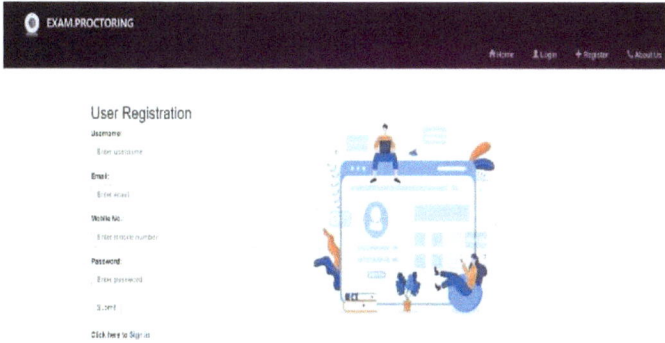

Fig. 6 User registration

6 Experimental Results

This project can be enhanced by building it as a web or a mobile application for users to conveniently access it and can be enhanced in such a way that we can identify most gestures on live video calls or broadcasts, and we will try to recognize signs which include motion. This model has proven to be much faster than any other former methods like support vector machine, hidden Markov model, backpropagation algorithm, and CNN method (Figs. 5, 6, 7 and 8).

7 Conclusion and Future Scope

This project's primary goal is to close the communication gap between hearing-impaired people and those around them who do not understand sign language. This system was created to categorize static signs and numbers, and it will be taught in

Fig. 7 User login

Fig. 8 Proctoring screen

such a way that it can recognize dynamic hand movements or gestures that arrive in a persistent succession of images with a specific amount of time between each action. We have looked into many machine learning methods that can be used to recognize and interpret sign language. The project was created primarily using the TensorFlow model, convolutional neural network (CNN), OpenCV, transfer learning, and single shot detector (SSD), and it was created to demonstrate the highest level of accuracy in identifying.

Acknowledgements The satisfaction and euphoria that accompany the successful completion of any task would be incomplete without the mention of the people who made it possible, whose constant guidance and encouragement crowned the efforts with success. I would like to express my gratitude to our project guide and mentor **Prof. Hemalatha M.**, who gave us the opportunity to carry out this wonderful research followed by the hands-on project titled "**Malpractice Detection System for online examination using AI**". It has also helped us carrying out loads of research work and hone our skills accordingly. At last, I would like to extend my heartfelt thanks to my teammates without which this project would not have been successful. Finally, I would like to thank our professors who were adept with the topics of our research and have been quick to answer our queries.

References

1. Asha KH, Manjunathswamy BE Dr (2021) Analysis of biometric modalities using iris face multimodal framework. J Des Eng, A Scopus Indexes J 2021(9):15153–15172
2. Asha KH, Manjunathswamy BE Dr (2002) Challenge responsive multi model biometric authentication resilient to presentation attack. Int J Int Eng Syst A Scopus Indexes J 15(2)
3. https://caveon.com/wp-content/uploads/2013/03/Online-Proctoring-Systems-Compared-Mar-13-2013.pdf
4. Moten J Jr., Fitterer A, Elise, Brazier, Jonathan. Examining online college cyber cheating methods and prevention measures
5. https://esc.fnwi.uva.nl/thesis/centraal/files/f2025073908.pdf
6. Alessio HM, Malay N, Maurer K, Bailer AJ, Rubin B. Examining the effect of proctoring on online test scores
7. https://towardsdatascience.com/automating-online-proctoring-using-ai-e429086743c8
8. Miller A, Shoptaugh C, Wooldridge J (2011) Reasons not to cheat, academic-integrity responsibility, and frequency of cheating. J Exp Educ 79:69–184

Deep Learning Approach in Alzheimer's Phase Classification and Analytical Prediction in COVID-19 Sequencing

B. E. Manjunathswamy, S. Anusha, R. Athirakrishnan, R. Bumika, and R. Divyashree

Abstract Alzheimer's disease (AD) is a unique neurodegenerative sickness. Despite the way that the most broadly perceived results of Alzheimer's ailment consolidate precarious thinking, judgment, and route, feebleness to perform conventional tasks will similarly cause changes in character and lead. As the disorder progresses, a person with Alzheimer's ailment will experience serious amnesia and lose the ability to regular capacity. This difficult sickness is by and large typical in the more seasoned. An all-around learning approach has been taken on. Neuroimaging procedures, for instance, MRI and PET result are used for the assurance of AD. For the best results, various neuroimaging techniques are used with an all-around learning computation for Alzheimer's disease. In this adventure, we will use MRI and PET breadth and convolution mind association (CNN) to isolate the image into ordinary mental (NC), delicate mental weakness (MCI), and Alzheimer's affliction (AD).

Keywords Alzheimer's · Amnesia · Neuroimaging

1 Introduction

Alzheimer's contamination is an issue of the neural connection provoking hippocampus cramps, fits of the cerebral cortex and development of the ventricles, and at last leads to mental degradation not set in stone to have Alzheimer's strength experience issues regulating everyday presence. It impacts the prosperity of patients locally. The Alzheimer's association surveys that just about 6 million Americans experience the evil impacts of the affliction and is the sixth driving justification behind death in the USA. The evaluated cost of AD was $277 billion in the USA in 2018. The affiliation surveys that early and exact examinations can save to $7.9 trillion in clinical and care costs throughout the accompanying several numerous years. Early acknowledgment of Alzheimer's ailment will aid early treatment, which with

B. E. Manjunathswamy (✉) · S. Anusha · R. Athirakrishnan · R. Bumika · R. Divyashree
Computer Science Engineering, Don Bosco Institute of Technology, Bengaluru, India
e-mail: manjube2412@gmail.com

© The Author(s), under exclusive license to Springer Nature Singapore Pte Ltd. 2023   391
A. Kumar et al. (eds.), *Proceedings of the 4th International Conference on Data Science, Machine Learning and Applications*, Lecture Notes in Electrical Engineering 1038,
https://doi.org/10.1007/978-981-99-2058-7_36

canning prevent the spread of incidental effects. There is no prescriptive drug that will stop the movement of sentiments. For a successful finding of Alzheimer's infection, a couple of tests are expected, for instance, a minor mental evaluation, physical and mental tests, and a patient history profile are moreover required. Self-finding of Alzheimer's disease is dreary and leaned to human botch, so it is really smart to use PC benefits, for instance, speed and accuracy to make an assurance of Alzheimer's [1].

In convolutional neural networks, there are four essential sorts of layers that complete the fundamental jobs of such associations: convolution, pooling, and normalization. Because of the convolution layer, the embedded picture is presented to various channels to disengage the features contained in those areas. The blend layer is used to reduce the size of the analyzed data, in like manner decreasing the association repugnance for the researches of the separated bundle. The fundamental strategy used in this layer is the top blend, where the most outrageous worth is picked in the split window and the scale, where its worth is limited. ReLED layer (fixed layout unit layers) for data tying down builds network capacity to ponder atypical issues.

CNN develops various layers at constant levels, yet the last association in such a system to streamline the results is in the last layer—the totally planned layer. This layer brings the last angle, taking into account different capacities. The intriguing piece of CNN over the old cerebrum network is that the amount of layers is much higher. Significance of cerebrum network designing is described as the length of quite far among info and dynamic neurons. There is no specific limit to the amount of layers, which licenses one to call a "significant" network, yet it is made sure to imply an association with more than two mystery layers. CNN is used to decrease counts diverged from a customary mind association. Convolution makes working out significantly less complex without losing data full scale. They are really capable at managing picture request. They apply comparative information to all areas of photography.

Alzheimer's sickness (AD) is an irredeemable condition that causes perpetually neural connections to fail horrendously. Considering the general debilitating of the frontal cortex, it could occur during old age. Advancement is the justification for around 60–70% occurrences of dementia. Fleeting loss of memory is the most notable early result of the sickness. Language and lead issues, confusion, perspective swings, loss of hankering, and feebleness to control one's certainty are reflected in the ailment. As a singular's condition declines, he habitually feels up close and personal and mental issues from family and neighborhood. As the ailment propels, dynamic work is lost and the patient eventually prompts end. Though the earnestness of the disease could move, the run of the mill future after assurance is under nine years [2].

All over the planet, 29.5 million still up in the air to have Alzheimer's sickness in 2015. At age 65, we conventionally start with people, yet 4–5% of cases start before this age. Due to the dementia cause, 1.9 million people passed on in 2015. Advancement is maybe of the most broadly perceived money-related affliction made in countries. In India, some sort of dementia impacts various million people.

2 Objective

- Cultivate programming to recognize Alzheimer's in MRI checks and separate between different levels of Alzheimer's reality.
- Give supportive programming that experts can use to reduce the time spent at the outset periods of assurance.
- Give a quick strategy for discarding Alzheimer's as a possible illness for patients with ordinary results of Alzheimer's and other clinical issue.

3 Problem Statement

Alzheimer's disorder (AD) is a condition that impacts the dim frontal cortex which achieves loss of memory, thinking, language capacities, and inconvenience in low-level activities. Alzheimer's is the most broadly perceived sort of dementia. Degenerative brain development and hippocampal degeneration are seen as solid areas for an of AD. Research is in the works to break down these neurodegenerative issues at the outset stages. Early investigation helps patients with looking for the most benefits from treatment before there is outrageous melancholy. The work presented in this paper dissects picture imaging assist with magnetic resonance imaging (MRI) results to assess the likelihood of early distinguishing proof. Picture dealing with methodologies, for instance, pre-strength processing, blend of K-strategies, for instance, creating regional computation to remove white matter and faint have an effect on figure out frontal cortex volume for express conditions. To figure mind volume, we consider three frontal cortex planes, for instance, the center plane, the coronal plane, and the sagittal plane by finding the entrancing locale.

4 Related Work

The last period of the disease can continue onward for 1–3 periods. At this stage, memory and mental capacities continue to rot and the patient could need the support of a clock as they lose the ability to answer the environment. It is indispensable to break down the disease quite a while before the irreversible nerve hurts. Current non-automated procedures like mental tests, mini-mental state examination (MMSE), and clinical dementia rating (CDR) and thinking strategies, for instance, magnetic resonance imaging (MRI), positron emission tomography (PET), and single-photon emission computed tomography (SPECT) are used to follow surprising psyche changes [3].

Troubles/Limits

Plaques and tangles are a piece of the critical results of the disease. As the amount of plaques and tangles creates, strong neurons begin to work progressively and consistently lose their ability to convey and in the end fail miserably provoking absolute breakdown of frontal cortex tissue.

The death of neurons especially in the hippocampus subdues the ability to acquire new encounters. The hippocampus is the main region in the affected frontal cortex.

Challenges: The test

The most generally perceived incidental effects consolidate inability to examine well with others, a more noticeable penchant to sickness, misinformed thinking, misleading ideas, transient mental deterioration, and vision issues. Another report suggests that around 50 million people in general have Alzheimer's ailment. The disease addresses a remarkable test to scientists and experts today as it is only from time to time seen until patients show up at the last works of the ailment considering the way that their psychological incidental effects are every now and again suggested as developing. But assuming better treatment is given, the bet of the contamination will continue to grow. Along these lines, more settled people are at more serious bet of getting the sickness [4].

Alzheimer's disease is a debilitating and irreversible frontal cortex disorder. As expected, someone in the not entirely set in stone to have Alzheimer's ailment. The result is deadly, as it prompts passing. In this way, it is essential to contract the sickness early. The principal wellspring of dementia is Alzheimer's contamination. Dementia achieves a decline in thinking and abilities to adjust, which impacts people's ability to work openly. The patient will neglect to recollect the latest events to start with stages. If the sickness continues on, they will dynamically neglect to recall all of the events. Diagnosing the ailment at the earliest open door is huge [5].

5 Proposed System

Suggested Techniques and Strategies

Feature yield, feature lessening, and division are the three essential stages where standard AI methodologies. These classes are then planned into standard CNN. By using CNN, there is convincing explanation need to play out the component removal process up close and personal. The heaps of its most vital layers go probably as specific parts, and their characteristics are updated by examined. CNN offers preferred execution over various classification.

Through and through scrutinizing

Through and through learning is a sort of AI where the model tracks down how to isolate issues directly from a given informational collection, which may be pictures. Through and through learning is normally wrapped up by mind network plan. As the amount of layers constructs, the association will expand. Appeared differently in relation to normal cerebrum networks containing 2 or 3 layers, significant mind associations can have different layers. Through and through learning models are particularly suitable for helpful information, for instance, PC vision, facial affirmation, typical language dealing with, voice affirmation, and virtual amusement assessment. It has conveyed equivalent results once in a while and better than human trained professionals.

Accuracy improves all-around scrutinizing than other AI instruments. This raised level of precision has been made possible in three areas of advancement.

(i) Easy permission to free tremendous enlightening assortments (like ImageNet, Caltec101, MNIST), gave permission to stamped colossal educational records.
(ii) High PC power with the help of tip top execution GPUs. It will chip away at the planning of the tremendous course of action of data expected for all-around learning and diminishing getting ready time.
(iii) Advanced cerebrum network models pre-arranged. With the help of a trade learning technique, pre-arranged significant cerebrum network models (like AlexNet, VGG-16, Resnet-50) can be retrained to perform new disengagement works and model affirmation.

DNN plans and move learning models

Adequate resources for through and through learning model execution gave as follows:

(i) Data containing the movement of magnetic resonance images (fMRI) from the ADNI.
(ii) An outstandingly capable GPU maintained by CUDA—Python stage.
(iii) Advanced mind network models are pre-arranged like AlexNet, VGG-16, VGG-19, and GoogleNet.

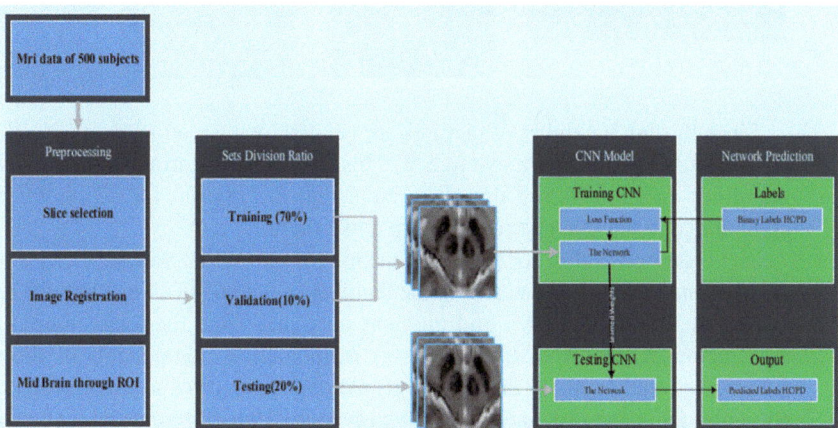

Method

The proposed system perceives MR pictures as embedded, which are in the end named PD or HC. The model contains an amount of eight tremendous layers. The solicitation for these layers is according to the accompanying, convolution 1, max-pool 1, convolution 2, max-pool 2, convolution 3, 1 layer, 2 layer, and leave layer. The piece size of all convolutional layers (Convt 1, Convt 2, Convt 3) and max-pooling layers (max-pool 1, max-pool 2) are 3 × 3 making 32 component maps. CNN scrutinizes these component maps, allowing CNN to perceive PD and HC in MR pictures. Routinely, on CNN, convolutional layers are followed to harden layers. In the proposed model, two multi-layer blend Max-pool 1 and Max-pool 2 come after Convt 1 and Convt 2 layers with stage 1 and padding 0.

Undeniable level processing strategies: Adding an image is a technique for changing existing data to make additional data for the model readiness process. All things are considered; it is a course of normally extending the open data to set up a more significant learning model.

In this photo, the image on the left is only the essential picture, and the large number of different pictures is delivered utilizing the chief planning picture.

Since this huge number of pictures are created by the readiness data itself, we do not need to assemble them eye to eye. This overhauls the arrangement test without going out and assembling this data. Note that the imprint for all photos will be comparable to the main picture used to make them.

The proposed ability for the distinguishing proof of AD was performed on MRI checks. The surface, region, and shape features are taken out using the gray-level co-occurrence matrix (GLCM) and moment invariants from the hippocampus picked as a region of interest. GLCM releases mathematical components of the second solicitation mathematical and moment invariants describes a lot of plans used to recognize the shape. Advancement is then described into different classes considering factors got from ROI. It uses an artificial neural network (ANN) arranged using the error back propagation (EBP) computation.

Advantage

The surface, region, and genuine characteristics are removed using the gray-level co-occurrence matrix (GLCM) and moment invariants from the hippocampus picked as the region of interest.

Ibe AD is then arranged into different orders considering components conveyed in ROI, using an artificial neural network (ANN) arranged using the error back propagation (EBP) estimation.

6 Architecture

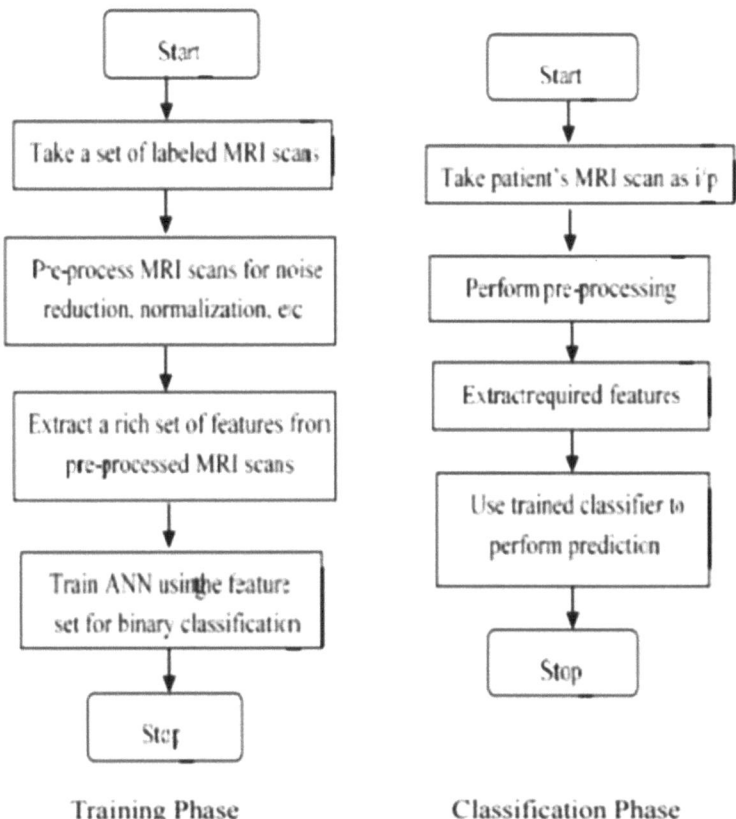

Training Phase   Classification Phase

7   System Design

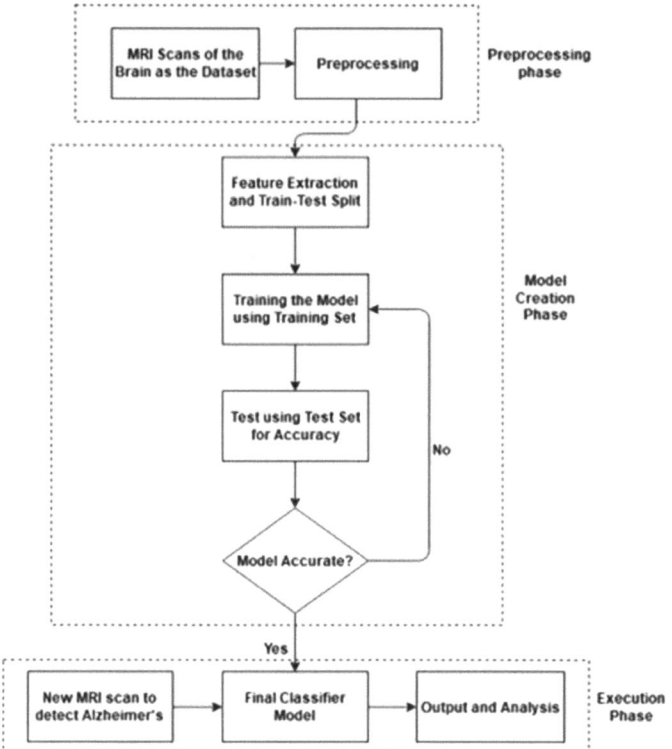

Dataflow outline

- DFD is in like manner implied as an air pocket frame. A fundamental graphical structure can be used to address the system to the extent that in structure input, the various cycles were performed on this data.
- Data stream diagram (DFD) is one of the primary exhibiting gadgets and used to show system parts. These parts are the structure cycle, process data, outside system working system, and information that travels through the system.
- DFD shows how information streams in the structure and the status quo changed as per a movement of changes. A graphical system shows the movement of information and changes being utilized as data goes from commitment to yield.
- DFD is generally called an air pocket chart. DFD can be used to address the structure at any level of separation. DFD may be apportioned into levels that address the movement of information and execution information.

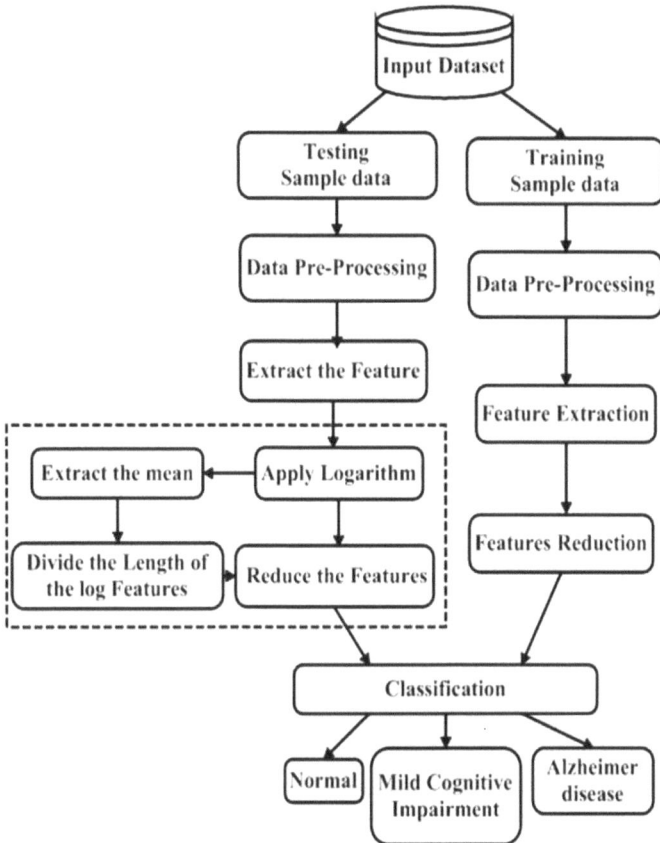

Progression frame

Progressive diagram shows the correspondence of a thing coordinated in consecutive solicitation. It shows the articles and classes related to the situation and the course of action of messages that are exchanged between the parts expected to complete the job of the situation. Back–to-back charts are habitually associated with the fulfillment of events in a reasonable point of view on the fundamental structure.

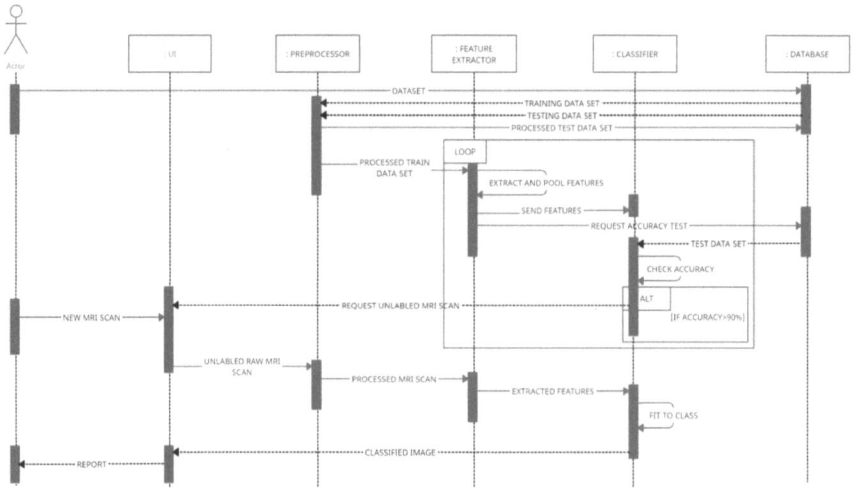

8   Conclusion

We have cultivated Alzheimer's expressive programming for MRI analyses and isolated between different levels of Alzheimer's reality to early help experts. The motorized AI contraption for expecting Alzheimer's sickness uses a through and through learning computation that has been successfully arranged and used. The show standards of CNN models are similarly assessed. Through and through assessments show a high accuracy of 80–90% in predicting Alzheimer's sickness.

References

1. Ali H, Al-nuaimi et al (2016) Changes in the EEG amplitude as a biomarker for early detection of Alzheimer's disease. In: 2016 30th annual international conference of the IEEE engineering in medicine and biology society (EMBC)
2. Jiang J et al (2015) A computed aided diagnosis tool for Alzheimer's Disease based on 11C-PiB PET imaging technique. In: IEEE international conference on information and automation Lijiang, China
3. Chaves R et al (2021) FDG and PIB biomarker PET analysis for the Alzheimer's Disease detection using association rules. In: 2012 IEEE nuclear science symposium and medical imaging conference record (NSS/MIC)
4. Garali I et al (2015) Region-Based brain selection and classification on PET images for Alzheimer's Disease computer aided diagnosis. In: 2015 IEEE international conference on image processing (ICIP)
5. Al-Nuaimi AH et al (2015) Tsallis entropy as a biomarker for detection of Alzheimer's Disease. In: 2015 37th annual international conference of the IEEE engineering in medicine and biology society (EMBC)

Voice-Controlled Robot for Surveillance

**A. N. Maheswarappa, R. Gowtham, D. K. Harsha, C. Karthik,
and M. Nagaraj**

Abstract The project model was designed in a way that the Raspberry Pi robotic is disciplined by user voice command. The communication between the robot and the Android application is established using Bluetooth technology. The robotic is disciplined by user speech commands. The motion of the robotic is simplified by using two or more DC motor connected with microprocessor or Raspberry Pi. The command from the Android application is transformed into digitise signal by radio frequency module. At the destination end, the data gets decoded and is stored into the microprocessor, which guides the H Bridge to drive the DC motors for the necessary work. The main ambition of this project is to perform the specific task by listening to the user command. The movement of the robot can be monitor in PC (capable of night vision) through camera which is attached to Raspberry Pi module port.

Keywords Android app · Bluetooth · Raspberry Pi · Voice command

1 Introduction

An IoT is gaining plenty of popularity, and we can control any device in the house, and we can operate it casually. Our goal is to do the robotic vehicle which can be disciplined by the user voice command, and these types of system are also called as robot control with voice command [1]. The robot is casually controlled by an Android application; for the user-required aim in this paradigm, an Android application and a microcontroller are utilised.

The Android app and the robot car communicate with each other via a Bluetooth module. Examining the user's verbal command and taking appropriate action is the primary motivation. The most frequent orders used to move the robot are forward, backward, right, and stop.

A. N. Maheswarappa · R. Gowtham (✉) · D. K. Harsha · C. Karthik · M. Nagaraj
Department of ECE, Don Bosco Institute of Technology, Bengaluru, India
e-mail: Gowthamr1802@gmail.com

A. Kumar et al. (eds.), *Proceedings of the 4th International Conference on Data Science,
Machine Learning and Applications*, Lecture Notes in Electrical Engineering 1038,
https://doi.org/10.1007/978-981-99-2058-7_37

If any object is sensed while moving, it notifies the user with images which is shared to the user GMAIL ID and then stops [2]. This project paper tell of quit and uncomplicated hardware for robot implementation and speech observation. The speech command is converted to text, and with the help of the Bluetooth, we are communicating the robot.

With the aid of an Android application that converts spoken commands into Google Text and transmits them through Bluetooth, user spoken commands have been carried out. When the voice order is heard, the robotic vehicle moves and should search for the object using an attached ultrasonic sensor. The robot will move left, right, forward, and backward; we can view the place with our PC with the help of Pi camera which shares the live video. The four-wheeled robotic vehicle is controlled by an Android application. If the robot enters into the dark place, the LED turns on with the help of LDR sensor.

The main ambition of this project is that human cannot go in the tunnel area which is more than 20 m long, it is highly risk. So that robotic vehicles are developed according to the application with buck of user voice command.

The user Android application Voicebot is paired to Bluetooth module (HC-05) which is directly connected to Raspberry Pi pin 10. Voice command robotic vehicles are very helpful in places where people cannot travel, such as in the event of a fire or an area that is polluted. For the purpose of accepting voice commands from users, the Raspberry Pi is directly linked to the Bluetooth module HC-05.

2 Literature Survey

In 2003, global speculation in contemporary robots increased by 19%. Orders for robots increased by 18% in 2004 to the greatest level ever recorded. Overall growth is expected to be around 7% each year between 2004 and 2007. During 600,000 family robots will be in use millions of times over the following several years. Various studies have been conducted in order to create this idea by various researchers. However, they have diverse innovations implemented and serve a different purpose.

The technology and applications of several of these articles are listed below.

Voice Control Robot Using an Android App Soniya Zope and others: The goal of this research is to develop less complex robot equipment designs while still providing wonderful computational Android stages. This article describes how to control a robot using a portable using Bluetooth connection, as well as certain key features of Bluetooth technology, as well as other aspects of the portable and robot. It provides an examination of robots controlled by smartphones by using an Android application to move the robot forward, backward, left, and right [1].

Voice Control Robot Using Android Application is the title of the research study. The project is described by Anup Kumar et al. and is intended to control a robot remotely using physical controls and vocal instructions. The control unit for sensing the signals provided by any Android application is computed using an ATmega32 microcontroller and a Bluetooth device. Through Bluetooth, control commands are transmitted. The speed of the motor may be controlled, and the Bluetooth can sense and send information to a smartphone. Additionally, it transmits data on its orientation and how far it is from the closest obstacle [2].

"Arduino based voice-controlled robot" is the title of the paper. In the International Journal of Engineering Innovation & Research, Pittsburgh. K. Kannan et al. employ an Android app to recognise speech that has been converted to text. The robot automobile is operated using this text once it has been further analysed. The robot vehicle receives this test through Bluetooth technology, which the microcontroller then processes to instruct the robot accordingly. We conducted several experiments on robot control mechanisms using this approach [3].

The robot is a locomotive robot vehicle in the paper "Arduino based voice-controlled robot" Author: K. Kannan et al. The user may control the robot's locomotion by issuing precise voice instructions. The voice module processes the speech that was picked up by the microphone. The speech module sends a command message to the robot's microcontroller when a command for it is recognised. The microcontroller examines and evaluates the message before acting accordingly. The goal of this project is to equip the robot with a voice recognition system and an AI hearing sensor so that it can communicate with users by using spoken natural language (NL) [4].

The method for controlling a robotic vehicle via voice input is presented in the paper named "Voice Recognition System for A Voice Controlled Robot with A Real Time Obstacle Detection and Avoidance." An Android smartphone serves as the voice recognition platform, and it communicates with the robot through Bluetooth. With this approach, data flow is continuous, and recognition is exact. The robot also has the capacity to recognise impediments and alert the user. Applications like assistive robots for persons with impairments or industrial applications like labour robots would benefit from our suggested approach [5].

"Robot Controlled using voice command" is the title of the article and includes how a robot functions when voice input is provided. Author: Dr. Srma Yavuz et al. This paper elaborates on the vocal commands that a user gives to a robot to interact with them. It describes in detail how everything works and advises using an Android smartphone and voice commands to operate a robot [6].

The robot will move in line with the vocal orders supplied by the user in the paper "Remote Voice Controlled Robot," by Ron Oommen Thomas et al. This robot's design was created using a Raspberry Pi 3, and an Android phone is used to operate it. Geared motors operating at 60 RPM may move this in both forward and backward directions. Additionally, this robot makes quick turns to the left and right. The Android application will identify a voice command provided by the user. This robot follows the commands from the Android phone that is interacting with it over Bluetooth. A camera is used to command the robot and watch live video feed [7].

A study on voice command in industrial automation is provided by Swetha. As she noted in her paper, experiments with voice recognition in a loud setting were undertaken. The signal was impacted by noise from many sources, including as electrical motors, machine tools, and human operators, as opposed to the laboratory setting. The noise made by an induction machine is used as an example. Finally, a few voice recognition simulation results in a noisy setting are shown. The robot will get the proper control signal from the LabVIEW. When compared to previous sound-based robot control systems, the research is made simpler by using LabVIEW for the interpretation of speech signals [8].

According to Ankush Sharma's research, the speech recognition system is the process of turning an audio signal acquired by a microphone into a collection of words. By employing LabVIEW, a virtual instrument technology platform, a type of voice recognition system is used in this work. The object control system is used in the robotics and artificial intelligence fields to regulate robot movement. Particularly, this is really beneficial to our work [9].

A set of vocal instructions is functionally comparable to a set of switches used to control a wheelchair, according to Richard C in his study. Therefore, speech control of a wheelchair may be simple and utilised by those who are physically disabled in place of buttons. This research led to the biggest advancement in the field of speech recognition and paved the path for voice automation [10].

Sathya Chandran A brand-new video-based coal mine rescue robot was demonstrated by C and Anjaly. Going to the mine without understanding, the environmental status is extremely risky in the event of a coal mine catastrophe since the situation will not be known. The rescuers' initial task is to identify mine scenarios by taking into account a variety of factors [11].

Robot controlled via Bluetooth on a smart phone. Authors: Reeteshverma and Arpit Sharma. In this study, a robot has been built that can be controlled by an Android smartphone application. It uses Bluetooth to deliver control orders, and some of the features include managing the engine's speed and recognising and exchanging information with a phone regarding the robot's bearing and distance from the next obstruction [12].

Bluetooth robot controlled by Android and an 8051 microcontroller. Ritika Pahuja and Narender Kumar, authors. Typically, a robot is an electromechanical device that is controlled by electrical and PC programming. Robots have been used extensively for manufacturing purposes and are present on production lines all around the world. This article develops the remote controls for the Android application that move the robot using them. Additionally, Bluetooth connection is used to connect the Android and controller. Although UART convention Robot Controlled Car Using Wi-Fi Module, the controller is interfaced to the Bluetooth module [13].

S. R. Madkar, Vipul Mehta, Nitin Bhuwania, and Maitri Parida are the authors. This paper discusses how to operate a robot-driven car using a Wi-Fi module via an Android smartphone application. The devices may also be managed by sending a standard SMS even if you do not have an Android phone. Effective modifications may be made to this assignment to include a hidden agent camera that streams records to the client over Wi-Fi. For the project, solar-powered cells will be used instead of the typical lithium-ion battery [14] (Table 1).

Table 1 Summary

Year	Title	Authors name	Implementation
2017	Voice control robot using android application	Zopie [1]	The goal of this research is to develop less complex robot equipment designs while still providing wonderful computational Android stages. This article describes how to control a robot using a portable using Bluetooth connection, as well as certain key features of Bluetooth technology, as well as other aspects of the portable and robot. It provides an examination of robots controlled by smartphones by using an Android application to move the robot forward, backward, left, and right
2017	" Arduino based voice control robot"	Kannan [3]	Using an Android app to recognise speech that has been converted to text. The robot automobile is operated using this text once it has been further analysed. The robot vehicle receives this test through Bluetooth technology, which the microcontroller then processes to instruct the robot accordingly. We conducted several experiments on robot control mechanisms using this approach
2019	" Speech recognization using lab view"	Sharma [9]	By employing LabVIEW, a virtual instrument technology platform, a type of voice recognition system is used in this work. The object control system is used in the robotics and artificial intelligence fields to regulate robot movement. Particularly, this is really beneficial to our work
2017	Design and implementation of a avoice command robot	Richard [10]	A set of vocal instructions is functionally comparable to a set of switches used to control a wheelchair, according to Richard C in his study. The speech control of a wheelchair may be simple and utilised by those who are physically disabled in place of buttons. This research led to the biggest advancement in the field of speech recognition and paved the path for voice automation
2014	Android phone controlled robot using bluetooth	Arpit Sharma, Reeteshverma, Surabh Gupta	In this study, a robot has been built that can be controlled by an Android smartphone application. It uses Bluetooth to deliver control orders, and some of the features include managing the engine's speed and recognising and exchanging information with a phone regarding the robot's bearing and distance from the next obstruction

3   Methodology

First of all, all speech instructions are transformed into text in order to drive a robotic vehicle utilising user voice commands. All the implementation takes place on an Android application and that command is being transmitted to Bluetooth module which is connected to the Raspberry Pi.

For data transmission and reception, this Bluetooth serves as a middleman between the Raspberry Pi and an Android application. When the command is received, it adjusts the robot's motion accordingly. And we are using Pi camera, Wherever the robot goes, we will be able to see that place in laptop. If any obstacles or object is detected, then the robot will stop without my permission, automatically Raspberry Pi sends the images of obstacle to the user GMAIL ID (Fig. 1).

- A direct current motor normally has just two wires: one is positive, and the other one is negative. The two wires of two DC motors are connected to the H Bridge driver. H Bridge receives the command from Raspberry Pi, and it controls the DC motor for make the robot to move in user command direction.
- H Bridges can be found in a variety of applications, where it is desirable to have control over the direction of current flow. Controlling the direction of electric motor can turn is a common example of this. This is done by allowing current to flow in one way and then reversing that direction in the H Bridge, causing the motor to turn in the opposite direction.
- Ultrasonic sensor module is used for obstacle or object detection. It is connected to Raspberry Pi, and it detects the obstacles in the straight-line path. If the obstacle is detected, it sends the signal to the Raspberry Pi, the rover will stop automatically, we are using Pi camera, it captures the obstacle, and it sends to the user email id with the help of Raspberry Pi and Python code.
- Raspberry Pi 3 model B+: Raspberry Pi is a 64-bit quadcore processor arm cortex, and it is 900 MHz. The first generation of Raspberry Pi is Broadcom BCM2835. It is used in first modern-generation smartphone, which includes 700 MHz and video core graphics processing unit. It has two caches: one is level1, and another is level 2. Level 2 is used by the GPU. The Raspberry Pi 3 B+ has a built-in Wi-Fi module and CSI camera port. The Raspberry Pi also consists of 2 Ethernet ports and 4 USB ports, and it has 40 GPIO pins.
- LDR sensor: A component that enables their usage in light sensor circuits is an LDR. LDRs, or light dependent resistors, are sometimes referred to as photoresistors. The resistance of an LDR is high in the dark and low at light levels, and little current can pass through it. This resistance reduces as the light intensity increases.

4   Components and Materials

Designing was done by the following part available in the market.

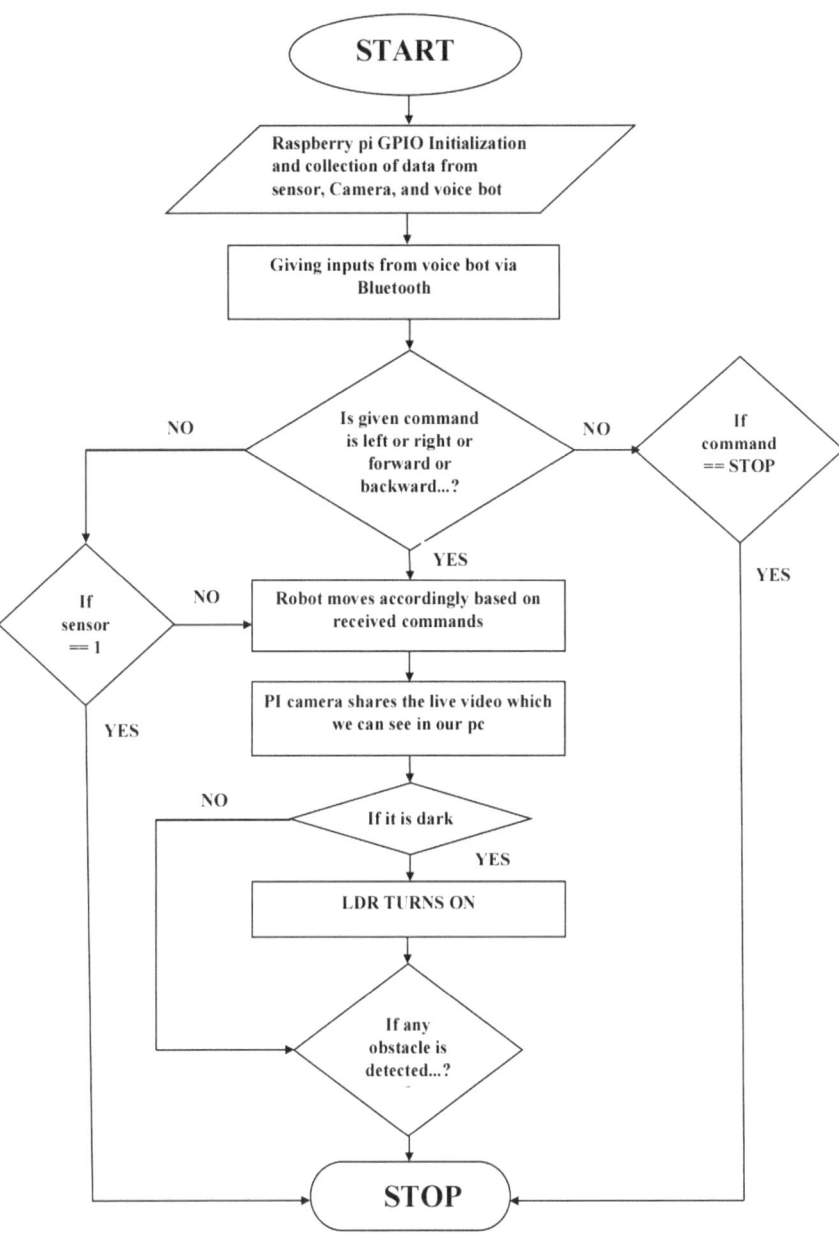

Fig. 1 Proposed system

Fig. 2 Framework

4.1 Chassis

Chassis is a metal which is used to place all the components as shown in Fig. 2. It is also called as framework.

4.2 DC Motor

A rotatory motor called a DC motor uses direct current to produce mechanical work. And this is directly connected to the H Bridge driver, and wheel is connected to the DC motor (Fig. 3).

Fig. 3 DC motor

Fig. 4 H Bridge motor driver

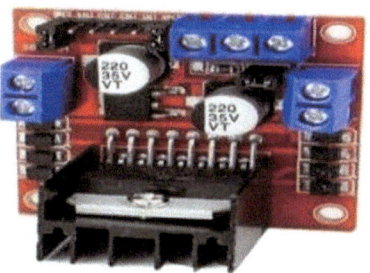

4.3 Wheels

Wheel is used to move the robotic vehicle forward and backward, which is connected to the DC motor.

4.4 L298D Motor Driver

The DC motor is managed by the H Bridge driver. It has the capability that it switches the polarity of a voltage applied to a load. With the help of the H Bridge, we can run the DC motor in clockwise and anticlockwise directions (Fig. 4).

4.5 PI Camera

In this project, we used Pi camera which is directly attached to the Raspberry Pi CSI camera port. Pi camera is used to share the pictures and live video to our GMAIL and PC.

4.6 Raspberry Pi 3 Model+

Raspberry Pi is a 64-bit quadcore processor arm cortex, and it is 900 MHz. The first generation of Raspberry Pi is Broadcom BCM2835. It is used in first modern-generation smartphone, which includes 700 MHz and video core graphics processing unit. It has two caches: one is level1, and another is level 2. Level 2 is used by the GPU. The Raspberry Pi 3 B+ model consists of built-in Wi-Fi module and CSI camera port. The Raspberry Pi also consists of 2 Ethernet ports and 4 USB ports, and it has 40 GPIO pins.

Once the setup is over, and your preferred operating system is installed, you need to run the Raspberry Pi in a desktop computer. The Raspberry Pi camera is directly

Fig. 5 Raspberry Pi

attached to the CSI port which is placed in Raspberry Pi board. We need to add micro-SD card to dumb the programme. We can add email which alerts user ID if the motion is detected (Fig. 5).

5   Implementation

Block Diagram (Fig. 6).

Fig. 6 Block diagram

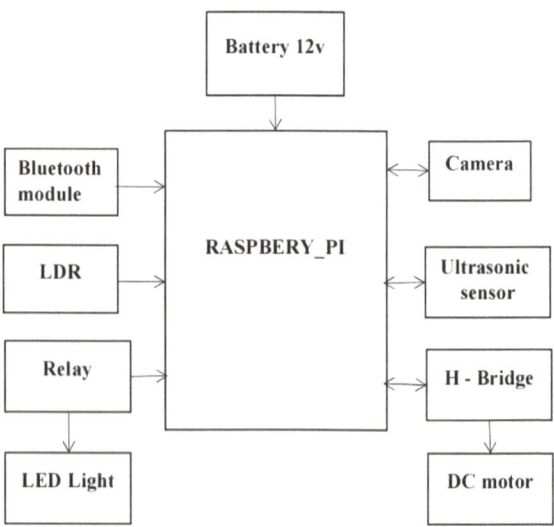

In this model, smartphone is used for to control the robot using voice command with the help of Android application called as voice bot which is installed in smartphone. Using the Google speech recognition, the voice command is converted to text command. That text command is passed to microprocessor via Bluetooth. We are using Bluetooth module, this module connected the Raspberry Pi, and this Bluetooth should connect to your voice bot application. The Bluetooth module HC 05 will serve as the transmitter (TX), which will deliver the code to the decode circuit after receiving it from a smartphone (Rx).

Following points are shown that how to communicate the robot using voice command:

1. Download "Voicebot" app from google.
2. Ensure that HC 05 Bluetooth is paired to your Android application.
3. Then click connect button.
4. Then choose the HC 05 option in the Voicebot application in search list.
5. After pairing, click on microphone icon.
6. Tell "Forward" if you want robot to move in the forward.
7. Tell "Backward" if you want robot to move in backward.
8. Tell "Left" if you want to turn robot towards left.
9. And tell "Right" if u want the robot to turn in the right.
10. Say "stop" to stop the robot.
11. Press disconnect button after use.

6   Results and Discussions

Figures 7, 8, and 9 show the moving robot and images which can be viewed in our PC.

Figure 7 shows that the robot is in moving condition with the help of voice command and Bluetooth module. We are given the voice command using Voicebot mobile application as shown in Fig. 8.

Figure 10 shows that the robot is moving based on the voice command via Bluetooth. Wherever the robot goes, we will be able to see that place in our laptop. In

Fig. 7 Robot is in moving condition

Fig. 8 Voicebot application

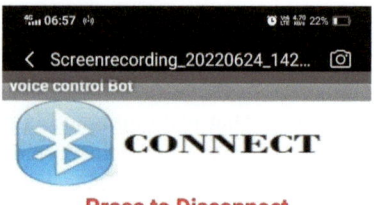

Fig. 7, we have placed the bottle in front of robot, so that bottle can be able to see in our laptop as shown in Fig. 10.

And also we implemented smart lighting system. If the robot enters into the darker place, the LED turns on with the help of LDR.

Table 2 describes the expected output and actual output used in the proposed system.

Fig. 9 Implementation

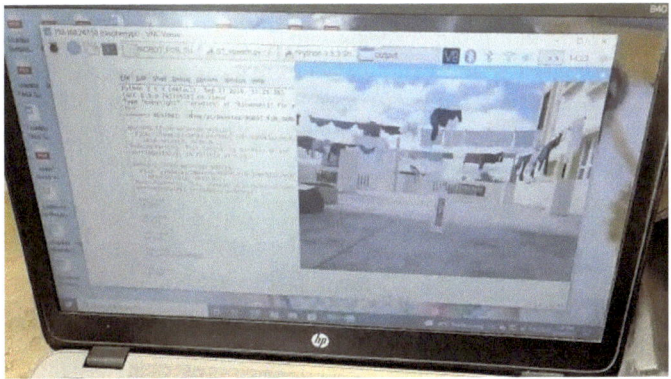

Fig. 10 Robot shared images

Table 2 Test cases

Test case no.	Functions with parameters under test	Expected result	Actual result	Result	Comment
1	Motion detection (ultrasonic sensor)	When object is detected, the robot is stopped automatically	When motion detected, the robot is stopped	Expected output matched the actual output	Pass
2	Camera module (Raspberry Pi camera)	Capture image, sharing live video, and send image to email	Captured image sent to mail and shared live video	Expected output matched the actual output	Pass
3	When command is sent	H Bridge controls the DC motor to move/stop	DC motor runs	Expected output matched the actual output	Pass

7 Conclusion

In this project, we used a Raspberry Pi to create a robotic surveillance truck. The surveillance robot works as a security monitoring device which replaces the human security. The outcome of this project deals with the recorded evidences of images and videos when a criminal activity occurs and alerts to the remote user host immediately.

The border area is so large, so a very large amount of manpower is required, so this kind of security system can effectively provide more security. In future, the system can be further extended using face recognition, so that the system will work more efficiently.

References

1. Zope S et al (2017) Voice control robot using android application. Imperial J Interdisc Res (IJIR) 3(2)
2. Kumar A et al, Voice control robot using android application. IJIRT 1(11)
3. Kannan PK et al (2015) Arduino based voice-controlled robot. Int Res J Eng Technol (IRJET) 02(01)
4. Kannan K et.al (2015) Arduino based voice-controlled robot. Int Res J Eng Technol (IRJET) 02(01)
5. Ramdhave S et al, Voice recognition robot control using android device. In: Proceedings of IRFIEE forum international conference, ISBN: 978-93-85973-95-6 (Kumar A et al, Voice controlled robot. IJIRT 1(11))
6. Yavuz S et al, Robot control with voice command. Yildiz technical university faculty of electric and electronics department of computer engineering senior project
7. Thomas RO et al (2014) Remote control of robotic arm using raspberry pi. Int J Emerg Technol Comput Sci Electron (IJETCSE) 8(1)

8. Patil S, Abhigna A (2018) Voice controlled robot using LabVIEW. In: 2018 International conference on design innovations for 3Cs compute communicate control (ICDI3C). IEEE, pp 80–83
9. Sharma A, Srinivas Perala PD, Objects control through speech recognition using LabVIEW
10. Richard C, Ahmed IU, Osman SB, Newaz B, Rasheduzzaman M, Reza ST (2017) Design and implementation of a voice command robot. In: Proceedings of the international conference on computer, communication, chemical, materials and electronic engineering, Rajshahi, Bangladesh, pp 26–27
11. Sarath Chandran C, Anjaly K (2014) Real time video-controlled traction for surveillance robots in coal mine. Int J Innovative Res Electr Electron Instrum Control Eng 2(1)
12. Sharma A, Verma R, Gupta S, Bhatia SK (2014) Android phone-controlled robot using Bluetooth. IJEEE 7:443–448
13. Pahuja R, Kumar N (2018) Android phone controlled Bluetooth robot using 8051 microcontroller. IJSER 2(7):14–17
14. Madkar SR, Mehta V, Bhuwania N, Parida M (2015) Robot-controlled vehicle utilizing Wi-Fi modulethrough android application of an android Smart Phone. IJEEE 3:19–23

Automated Proctoring System to Detect Anomalous Behavior in E-learning During Times of Crisis

B. E. Manjunathswamy, Ranjeet Rathod, Padmaraj D. Managave, Karthik S. Sagar, and R. Rakshith Gowda

Abstract A boom in remote learning has happened recently. However, there has not been a good answer for academic exams. Some institutions use online copy-and-paste assignments, while others use remote proctoring, where a human proctor maintains an eye on students' activities. There must be a solution as if way we live is to become the new norm. In this essay, we have put out a plan for creating an integrated AI-based system that can aid in discouraging exam cheating. The method recognizes fraud and records the evidence. This technology will be both affordable and secure.

Keywords AI proctoring · CNN algorithm · Face detection · Mobile detection · CNN algorithms · Haar cascade algorithm

1 Introduction

Academics now conduct their work online. This is a significant problem from both a learning and an assessment perspective. A significant problem that needs to be solved is how to conduct exams honestly. In India, there are now roughly twice as many internet users as there were six years ago. As a result, many students were able to retain their studies, which was convenient for the academics. This also permitted online tests, introducing online proctoring as it applies to academic settings. Web-based administration references to a computerized style of monitoring made possible by state-of-the-art monitoring software. A proctored exam enables the examiners to supervise remotely. To ensure the integrity of the exam, they make use of video, audio, and numerous anti-cheating measures. The remote exam manual online proctoring is a challenging undertaking since multiple students cannot be proctored at once. An instructor can physically observe pupils using all of their senses during manually proctored exams at the remote centers. They may easily ensure that the event runs

B. E. Manjunathswamy (✉) · R. Rathod · P. D. Managave · K. S. Sagar · R. R. Gowda
Don Bosco Institute of Technology, Bengaluru, India
e-mail: manjube2412@gmail.com

© The Author(s), under exclusive license to Springer Nature Singapore Pte Ltd. 2023   417
A. Kumar et al. (eds.), *Proceedings of the 4th International Conference on Data Science, Machine Learning and Applications*, Lecture Notes in Electrical Engineering 1038, https://doi.org/10.1007/978-981-99-2058-7_38

smoothly by being aware of the sounds and actions of the pupils. By paying attention to the noises and movements of the students, they can simply make sure that the event goes as planned. Understudies, the workforce, and academic foundations all face advantages and disadvantages as a result of the increased use of online assessment. Geographical locations and time zones are no longer obstacles for students to administer exams because tests can now be emailed practically everywhere on Earth with a web connection and safe software. Therefore, the goal is to develop an AI system that would watch over kids through camera and microphone so that teachers may watch over several pupils at once. Additionally, the system must maintain a log of potential malpractices. If there are any suspicions, the logs of misconduct can be utilized to manually check the student. Additionally, the system should retain. In the event that tests need to be interrupted, you can keep track of them of a power outage and that students may re-login and continue their work from the point at which the exam was closed.

2 Problem Statement

A web-based system used to identify and analyze the malpractices carried out by students during online examination using artificial intelligence.

1. Input: Live stream video
2. Process: Data preprocessing:

   - Face recognition (using mouth and iris)—Haar cascade method
   - Head pose detection—inclined method
   - Mobile detection—convolutional neural network

3. Output: CNN model classifies the object in the image frame.

3 Scope

The scope of the project will be:

- Our project goal is to design and implement malpractice detection system for online and offline exam. This system can detect whether the students are referring answers online or from the stored resources in desktop or mobile.
- In online/offline exams if the students are carrying mobile, our system will detect and alert the examiner.
- During the exam if the candidate is talking to someone, it can be detected by our system. The student activities during the exam will be captured and monitored by our system.
- So, whenever malpractice is recognized, alert will be sent to the examiner.

4 Review of Literature

The study by Tiong et al. [1] corrects current issues in online assessments, which are especially relevant during the COVID-19 epidemic. Academic dishonesty related to online assessments is the focus, investigates potential e-fraud regimes using case study, and proposes preventive measures. The authors here use the e-cheating intelligence agent as a mechanism to detect online fraudulent practices, which include two main modules: Internet Protocol (IP) Detector and Behavior Detector. Intelligence agents monitor student behavior and can detect any malicious practices. It can be used to assign random MCQs to course tests and combine them with online learning programs to monitor student behavior. Tests on several datasets have confirmed the effectiveness of the proposed method. The results of the accuracy measure are shown:

(i) Deep nervous system (DNN) -----> 68%.
(ii) Long short-term memory (LSTM) -----> 92%.
(iii) Dense LSTM -----> 95%.
(iv) Recurrent neural network -----> 86%.

In the research paper [2], students and instructors are faced with challenges and opportunities when it comes to online education. The important challenge faced is that academic probity for online exams can be artificially compromised by undetectable fraud and malpractices. The development of proctoring software addresses these concerns while preventing academic fraud. The main objective of this study is to inform the results of the online test conducted through the proctored versus unstructured online tests. Linear mixed models were used to compare test performance results for 147 students. Proctoring is not offered to half of the students, and the other half uses an online proctoring system. Students scored an average of 17 points less [95% CI: 14, 20] and significantly less time on online exams using proctoring software and unproctored tests.

According to Atoum et al. [3], the popularity of massive free online courses (MOOCs) and other forms of distance education is on the rise and reaching more students. The ability to proctor remotely online tests is a key limiting factor for this next level of scalability in education. Currently, human procturing is the most common and clinically used assessment method. Test takers can visit the test center or monitor them visually and soundly during the exam through a webcam. However, these types of methods require more labor-intensive work and are relatively expensive. Researchers presented a multimedia analytics system that can proctor online tests automatically. The system consists of six basic components that continually assess key behavioral cues: user verification, text detection, voice detection, active h window detection, gaze estimation, and phone detection. In combination with a temporary sliding window, continuous estimation components can be combined. The authors have designed high-level features to see if the examiner is cheating or is in the act of any misconduct at any moment during the test. To evaluate the proposed system, multimedia (audio and video) data is collected from 24 different types of fraud when taking online testing.

Samba Siva Rao et al. [4] states that thousands of people have problems with testing in the offline process, and this is a very popular field to avoid those problems when the process goes online and has made many security promises. Failing to control fraud, online tests are not widely adopted, but online education is being adopted and prioritized most parts of the world with no major security issues. Its goal is online testing for any subject competitions at any stage of domain studies and examinations in online university courses with students in various districts. Online tests are the easiest solution to the security and fraud/malpractice problem and use the enhanced security control system for online testing (SeCOnE), which deals with group cryptography with an e-monitoring project. By analyzing the relative failures and consequences of the existing system, the researchers identify the proposed processes for dealing with them.

5 System Requirement Specification

In order to develop the system, the necessary information was extracted to create the system requirement specifications. The elaborative conditions are what the system must meet. Additionally, the SRS provides a thorough understanding of the system, enabling users to comprehend the project's intended outcomes without being constrained by any specific methods. This SRS conceals the strategy but withholds the knowledge from outside parties.

Functional requirements

Useful requirements define the internal operations of the product, including the technical details, data monitoring and processing, and other specialized functionalities that demonstrate how to fulfill the use cases. They are supported by non-utilitarian conditions that compel the formulation or realization of requirements. System should process the data.

- System should detect the face.
- System should detect the object.
- System should check the multiple persons.

Nonfunctional requirement

Prerequisites that indicate parameters that may be used to evaluate a framework's performance rather than particular actions are referred to be unnecessary requirements. This has to be separated from practical requirements that reveal explicit behavior or ability. Common non-practical requirements include affordability, flexibility, and dependability. System utilities are frequently used to describe impractical prerequisites. Limitations, quality qualities, and administrative prerequisites are other words for non-practical requirements. It is in the unlikely event that any unique circumstances arise while the product is being used. Incorporating additional modules and features into the design will encourage the creation of new applications. Because programming packages are publicly available, the cost should be low.

- Usability
- Reliability
- Performance
- Supportability.

Design consideration

A brief description of how images are handled by the system in the design consideration briefs and classify the corresponding notation for the given dataset is given. Preprocessing, segmentation, feature extraction, and classification.

1. **Preprocessing**: Dataset will take sample images and convert them to grayscale images and finally to binary images.
2. **Segmentation**: Segmenting the item from the background picture was the next operation following image preprocessing. Thresholding was a great choice for the challenge because there was a significant color difference between the item and the face. To improve segmentation, the contrast of a black-and-white image was altered.
3. **Feature Extraction**: In order to extract the information from an image, feature extraction is crucial. Here, we are analyzing the texture of a picture using GLCM. The spatial dependence between picture pixels is recorded using GLCM. The most prevalent features including contrast, entropy, energy, homogeneity, correlation, ASM, and cluster-shade are captured by GLCM using the gray-level image matrix and contrast by measuring certain variables or features that aid in differentiating one picture from another, feature extraction (glcm). This method suppresses the original dataset of the image.
4. **Classification**: A convolutional neural network (ConvNet/CNN) is a deep learning method that can accept an input picture, give various elements and objects in the image importance (learnable weights and biases), and be able to distinguish between them. Comparatively speaking, a ConvNet requires substantially less preprocessing than other classification techniques. While filters are manually constructed in basic approaches, ConvNets might learn these filters/ characteristics with adequate training.

6 System Architecture

The suggested method's system architecture is shown in Fig. 1. Here, we can see that a camera records the user's action, and the picture input is sent to the recognition system, where it is effectively processed and compared. The user interacts with the machine with the aid of a user interface display, and the virtual item is updated appropriately. The system design mainly consists of:

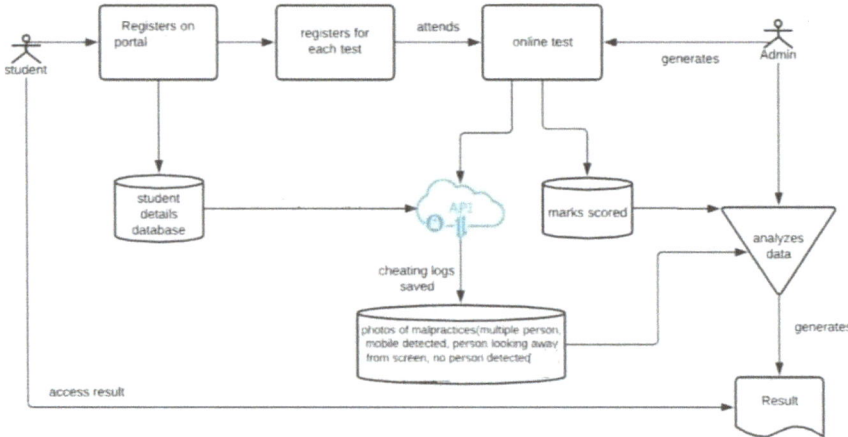

Fig. 1 System architecture diagram

1. Image collection
2. Image preprocessing
3. Image segmentation
4. Feature extraction
5. Training
6. Classification

- Image Collection:

    Real-time video is the input to the proposed system. Users can capture real-time video from their computers or laptops by using a webcam.

- Image Preprocessing:

    Preprocessing aims to improve picture data by minimizing undesirable distortions and enhancing specific image attributes crucial for additional image processing. There are three primary components to image preprocessing. Converting to grayscale and eliminating noise.

- Image Segmentation:

    Segmenting the item from the background picture was the next operation following image preprocessing. Thresholding was a great choice for the challenge because there was a significant color difference between the item and the face. To improve segmentation, the contrast of a black-and-white image was changed.

- Feature Extraction:

    In order to extract the information from an image, feature extraction is crucial. Here, we are analyzing the texture of a picture using GLCM. The spatial dependence between picture pixels is recorded using GLCM. The most prevalent features including contrast, entropy, energy, homogeneity, correlation, ASM, and cluster-shade are captured by GLCM using the gray-level image matrix. By measuring certain variables or features that aid in differentiating one picture from another, feature extraction (glcm) aims to suppress the original image dataset.

- Training:
   From the photographs captured during registration, create a training dataset. On the newly constructed training dataset, train classifiers. Make a test dataset and place it in a folder for now. Forecast the outcomes of the test scenarios. Graph the classifiers. To improve the accuracy of machine learning models, add feature sets to the test case file.

- Classification:
   A convolutional neural network (ConvNet/CNN) is a deep learning algorithm which can take in an input image, give significance (learnable weights and biases) to distinct characteristics and objects in the input image, and be able to distinguish between them. Comparatively speaking, a ConvNet requires substantially less preprocessing than other classification techniques. ConvNets are capable of learning filters, but with manual techniques, filters are hand-engineered.

7 Dataflow Diagram

Local Binary Pattern Histogram (LBPH):
   LBP classifier.

(i)   By passing the image path as an input parameter, we load the input image using the built-in function cv2.imread (img_path).
(ii) It is then converted to grayscale and displayed.
(iii) It is time to load the LBP classifier.

Each pixel's LBP is determined. For each pixel p, the 8 neighbors of the center pixel are compared to the pixel p, and if x is larger than or equal to p, the neighbors are given a value of 1. Equation is defined as the LBP classifier calculation formula, which is provided as (Fig. 2):

$$\text{LBP} = \sum^{7} \; n = 0 \; s(in - ic)2n$$

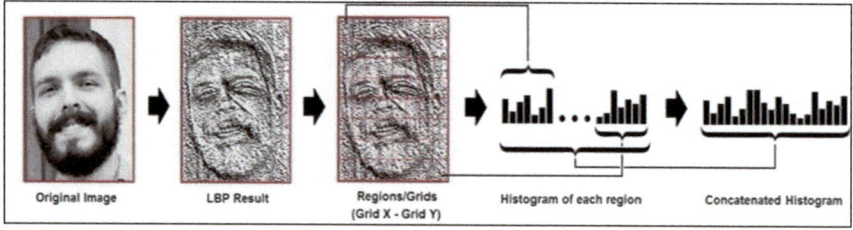

Fig. 2 LBPH process

8 Results

In Figure 3, a student is writing exam, and his head pose is detected by our system detecting it, storing that pose in dataset. Figure 3a shows that students head pose is at center. Figure 3b shows that students head pose at right, and similarly Fig. 3c shows that students head pose at left.

In Figure 4, a student is writing exam, and we are detecting whether his face is at screen level or no.

Figure 4a shows student's face is at screen level, and Fig. 4b shows student's face is not at screen level.

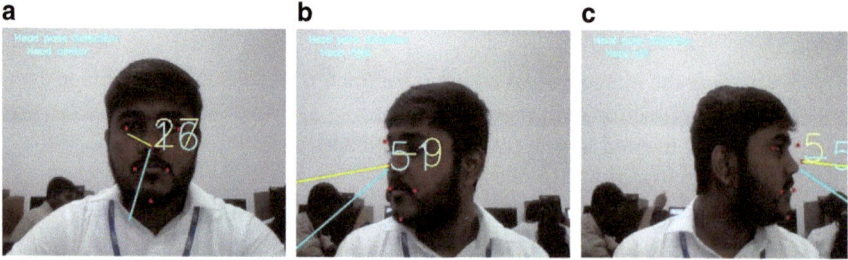

Fig. 3 **a** Students head pose is at center, **b** students head pose at right, **c** students head pose at left

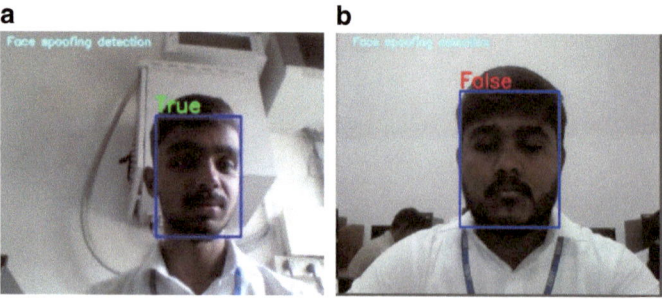

Fig. 4 **a** Student's face is at screen level, **b** student's face is not at screen level

In Figure 5, a student is writing exam, and we are detecting whether he is using mobile phone. Figure 5a, b shows that students are using phone, and they are detected by system.

From Fig. 6, we can see that multiple students are writing exam, and our system detected multiple students writing exam.

In Figure 7, we can detect student looking at left. Figure 7a shows threshold at which eyes are moving, show left. Figure 7b displays message that student eye looking at left.

In Figure 8, we can detect student looking at right. Figure 8a shows threshold at which eyes are moving, show right. Figure 8b displays message that student eye looking at right.

In Figure 9, we can detect student looking at right. Figure 9a shows threshold at which eyes are looking down, show down. Figure 9b displays message that student eye looking down.

Figure 10 shows that all behavior detection features as one system which was our goal to reach.

a   **b**

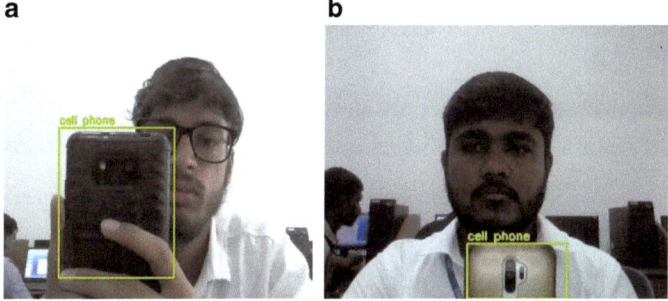

Fig. 5 a, b Students are using phone, and they are detected by system

Fig. 6 Multiple person detection

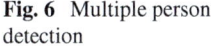

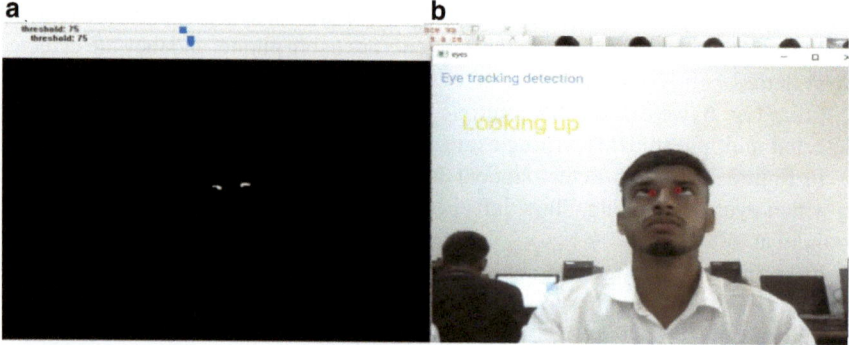

Fig. 7 **a** Threshold at which eyes are moving, show left, **b** displays message that student eye looking at left

Fig. 8 **a** Threshold at which eyes are moving, show right, **b** displays message that student eye looking at right

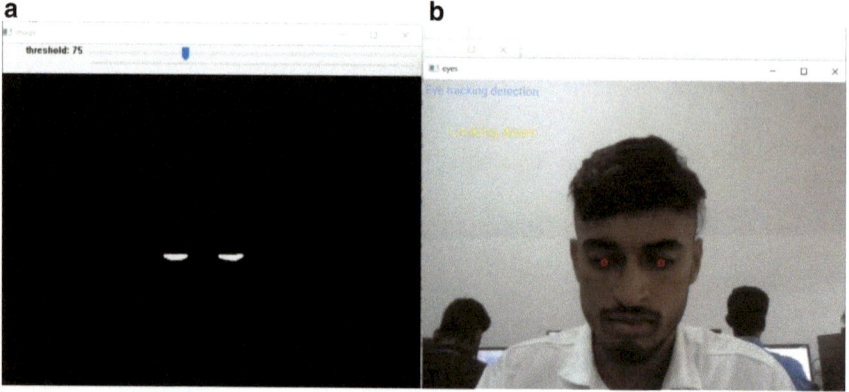

Fig. 9 **a** Threshold at which eyes are looking down, show down, **b** displays message that student eye looking down

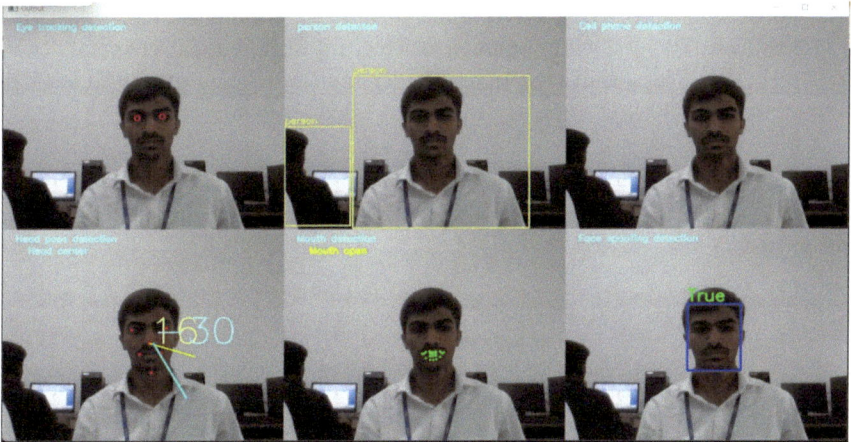

Fig. 10 Multiple detection features

9 Conclusion and Future Work

After implementing all objectives with most cost-effective and reliable algorithms such as Haar cascade, CNN, and LBPH, system has overcome all the disadvantages of previous system and can be used in real time. Combining multiple features as one system leads to library/module-related errors. But it is more reliable than traditionally used "live proctoring" and "record and watch proctoring". As we all know every project gets outdated with time, so following are the features that can help to keep up with time.

10 Future Work

Although we have covered most of major objectives of our project which help in smooth and efficient functioning of online exam mode, there are many other aspects need to be covered as future work which will help in more accurate tool to conduct exams; some of the works for future are:

(i) Voice detection: This feature helps detect any type of noise, which helps reduce the risk of misconduct by using any type of speech/voice.
(ii) Brower lock system: This system does not allow you to open any type of tab in any browser, as testers can take the help of other sources using online browsers.
(iii) Biometric authentication: It is a special feature to authenticate a candidate using any form of government or recognized identity for verification. Fingerprint scanner can also be used as part of authentication in critical situations.

(iv) 360° camera proctoring/surveillance: In normal webcams, the camera captures less than 180°, and it is not enough to detect any malicious chits or notes behind webcams. We need to adopt a 360° camera for surveillance to prevent this kind of fraud.

References

1. Tiong LCO, Lee HJJ (2021) E-cheating prevention measures: detection of cheating at online examinations using deep learning approach—a casestudy. J Latex Class Files
2. Alessio HM, Malay N, Maurer K, Bailer AJ, Rubin B (2017) Examining the effect of proctoring on online test scores. Miami University
3. Atoum Y, Chen L, Liu AX, Hsu SDH, Liu X (2017) Automated online exam proctoring. Accepted with minor revision by IEEE transaction on multimedia. https://doi.org/10.1109/TMM.2017.2656064
4. Samba Siva Rao N, Harshita P, Dedeepya S, Ushashree P (2011) Cryptography—analysis of enhanced approach for secure online exam process plan. Int J Comput Sci Telecommun 2(8)

Squid Game Implementation Using Advanced Open Computer Vision and Music Synchronization

K. R. Nataraj, R. Puneeth Kumar, Rahul J. Nayak, V. Vedashree, L. Kiran, and Rakheeba Taseen

Abstract According to a recent info trends study, in 2021, mobile and camera device users will have taken more than 1.5 trillion images, a sharp increase from the data from 2016. These image data will be used in a variety of real-time applications, including visual video surveillance, object identification, object detection, and classification. Advanced computer vision algorithms that were an upgrade over traditional computer vision techniques were created to manage these enormous volumes of data automatically. One of the most crucial tasks is object detection, which can greatly enhance the functionality of a variety of computer vision-based applications, including object tracking, license plate detection, mask and social distance detection, etc. To create a comprehensive.

Keywords Detection · Computer vision · Image

1 Introduction

Saliency and scalable object detection are two recent examples of object detection methods based on computer vision systems. The conventional object detection procedure can be divided into region detection, feature extraction, and classification. These systems have a number of problems with variations in pose, changing lighting settings, and increased complexity of localization of an object within the given image.

In the world of computer vision, the world is displayed in 3D, but the input to humans and computers is two-dimensional. In addition, computer systems can only process binary bits of data, but some useful programs only work in two dimensions. To take full advantage of computer vision, you need 3D information, not just a collection of 2D views. In these limited opportunities, 3D information is portrayed directly and is clearly not as good as people use. However, designing a robust, feature-rich extract

K. R. Nataraj (✉) · R. P. Kumar · R. J. Nayak · V. Vedashree · L. Kiran · R. Taseen
Department of CSE, Don Bosco Institute of Technology, Bengaluru, India
e-mail: director.research@dbit.co.in

© The Author(s), under exclusive license to Springer Nature Singapore Pte Ltd. 2023
A. Kumar et al. (eds.), *Proceedings of the 4th International Conference on Data Science, Machine Learning and Applications*, Lecture Notes in Electrical Engineering 1038, https://doi.org/10.1007/978-981-99-2058-7_39

for all types of objects is considered a tedious task due to the different number of lighting variations. To perform visual recognition, you need a classification model and library to further distinguish the categories of objects.

The application of these advanced computer vision techniques along with various machine learning algorithms like NumPy and TensorFlow which provide support for advanced mathematical calculations along with vast repositories of libraries and packages available in Python like Tkinter for front-end design and development can solve various real-world problems, and these coupled with deep learning algorithms and accurate object detection using Yolo and various iterations for neural networks prove to be a boon in improving the accuracy of object detection and tracking and application of these into the real world. In context to the current issue concerning expensive games with augmented reality and virtual reality because of their costly sensors and high-end processing power required for their application also not convenient for small-scale manufacturers to develop without sufficient research and development, that is when the application of advanced computer vision comes into picture which has proven to be useful in various application like yoga pose estimation and various complex exercises like posture management; thus, success in this field provides a view to approaching other aspects of applicability.

An approach to gaming implementation of a television series which proved to be a motivation for this implementation of a game based solely just on the camera sensor and advanced computer vision technologies like TensorFlow for artificial intelligence and NumPy for calculations and Tkinter for front-end development, and we intend to replicate the real-world game played in the movie scene into a gaming scenario using the cheapest of resources as possible and proving that immersive games can also be cheap and not needing any complex consoles.

2 Existing System

Many types of immersive games, many with respect to virtual reality, and recently augmented reality applications have increased in the basic format, but they are quite expensive as their need arises for expensive sensors, and customized hardware (consoles) and software are needed.

Previous use cases of computer vision had very complex process of object detecting and tracking which proved to be slower and more process intensive with so many processes needed for detection and its application and unable to compete with their consoles counterpart.

3 Proposed System

Recent advancements in advanced computer vision and popularity of Python as the go to programming language for artificial intelligence and machine learning applications led to many developers getting involved in making custom libraries and packages; also improvement in TensorFlow allowed faster insight discovery from data.

Due to these aspect, the lack of intelligence in current generation gaming consoles meant that we could try to develop immersive games with just basic sensors and cameras. Furthermore, as success and acceptability were seen in AR in the use cases of posture and yoga training as well as trying on of glasses and visualizing a piece of furniture in the living environment.

Thus as a first approach, we intend to replicate implementation of a real-world game (squid game—red light green light) with just the camera backed by artificial intelligence.

4 System Design

The proposed design consists of application of computer vision and artificial intelligence and machine learning algorithm and libraries (Fig. 1).

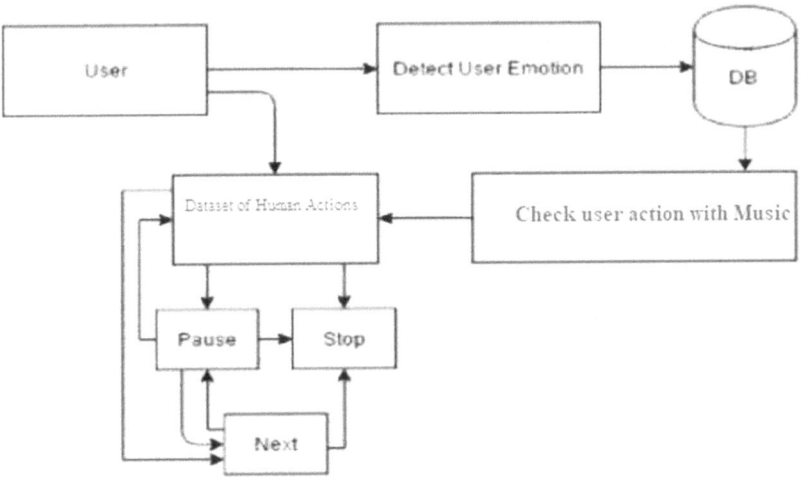

Fig. 1 Squid game system diagram

5 Module Description

Open CV: This research focuses on development of an immersive gaming application with bare minimum hardware requirements as possible, that is where application of advanced computer vision comes into picture which is used to understand the characteristics of the image frames obtained from the video or a live stream, as it is an open-source library for computer vision containing machine learning, and it plays a major role in carrying out tasks that are ubiquitous in today's world (Figs. 2 and 3).

Fig. 2 Open computer vision for squid game

Fig. 3 .

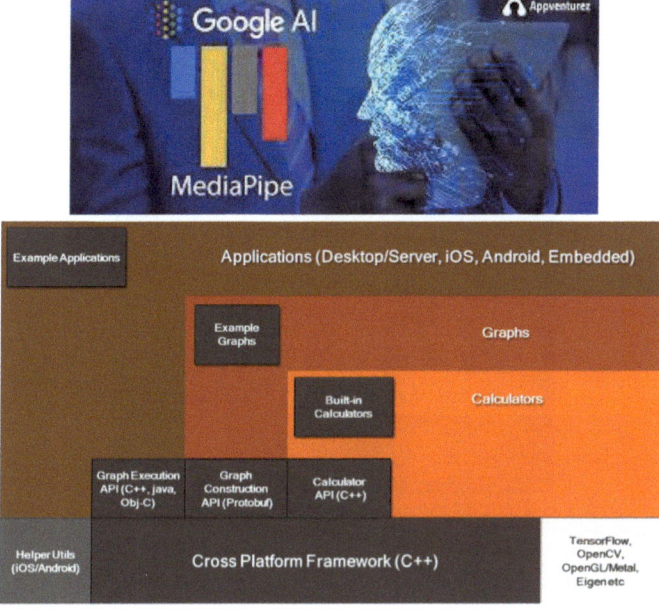

Fig. 4 MediaPipe and its framework

5.1 Media Pipe

MediaPipe is a Python package built by Google for building machine learning pipeline models for processing time series of data like videos, audio, etc., its cross-platform framework works in desktops, servers, Android as well as IOS use case [11, 12].

Like Raspberry Pi, MediaPipe is powered by revolutionary product and services that we use daily [11, 12]. Unlike power-hungry machine learning frameworks, MediaPipe requires minimal resources, and MediaPipe opened up a whole new world of opportunity for researchers and developers following public release [11, 12]. MediaPipe Toolkit comprises the framework and the solutions. The diagram shows how MediaPipe is organized with its features [11, 12] (Fig. 4).

5.2 Tkinter

The Tkinter package is standard Python user interface graphical user interface toolkit. Both Tk and Tkinter are available on many platforms and are cross-platform compatible [10].

Fig. 5 Tkinter

Running Python Tkinter was done as a standard installation procedure from the command line of scripts using the pipe command and later import and used as an object during implementation [10].

In this project, we use Tkinter to represent an image of how the gameplay is going to be along with button operations and playing of instructions along with linking of threading modules [10] (Fig. 5).

5.3 Multi-threading

Multi-threading is a way of multitasking using threads, as the modern processors have the capability to run multiple threads at a time, and it is highly beneficial to run different tasks on different threads of a processor to reduce the delay and improve performance (Fig. 6).

6 Methodology

See Fig. 7.

1. In level 0, the user interacts with the graphical interface developed with Tkinter.
2. A recording of the instructions of the game is played via the play-sound module.
3. The start button is connected to the back-end code via an OS module (latency is reduced, and performance is increased by implementing multi-threading).

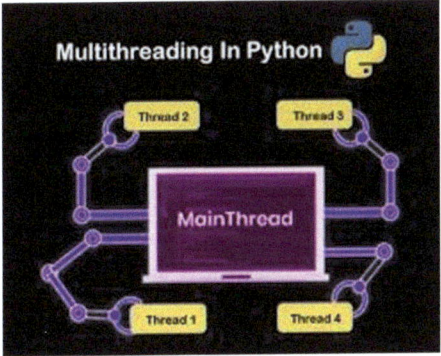

Fig. 6 Multi-threading

DFD-Level 0:

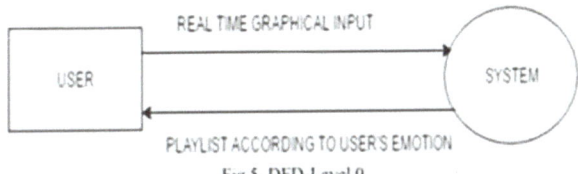

Fig.5. DFD-Level 0

DFD-Level 1:

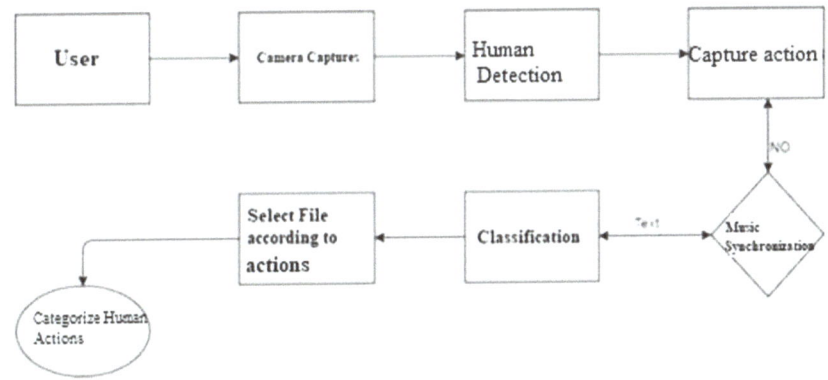

Fig. 7 Data flow diagram

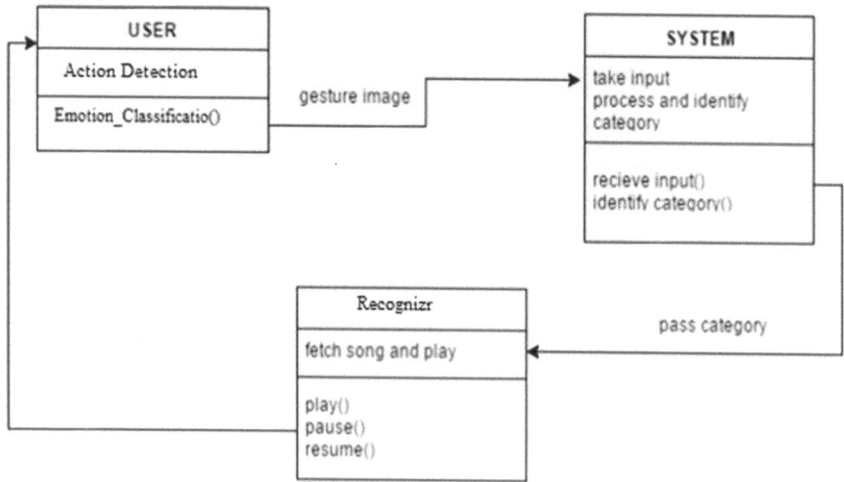

Fig. 8 Class diagram

4. When the start button is pressed, the open computer vision kicks into action and connects to the camera thereby displaying a visual as well as analyzing the frames.

5. Human isolation, detection, and segmentation are done by MediaPipe which is an ML-based Google-developed module.

6. The action of the user is captured via the landmarks drawn according to the necessity by the MediaPipe module.

7. The landmarks are generated as a set of matrices or co-ordinate equations which are then fed into NumPy library and TensorFlow for gathering insights from the data.

8. Later insights are analyzed, and a threshold is given to identify whether the subject is completely visible or not.

9. When the subject (player) is completely visible, the gameplay begins along with music synchronization (Figs. 8 and 9).

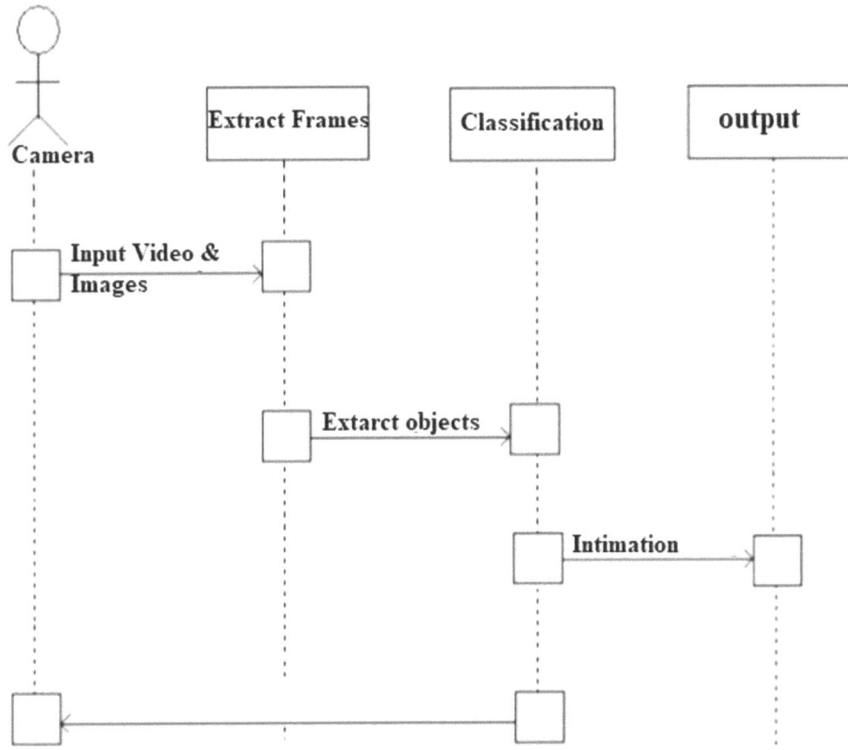

Fig. 9 Sequence diagram

7 Experiment and Result

See Figs. 10, 11, 12, and 13.

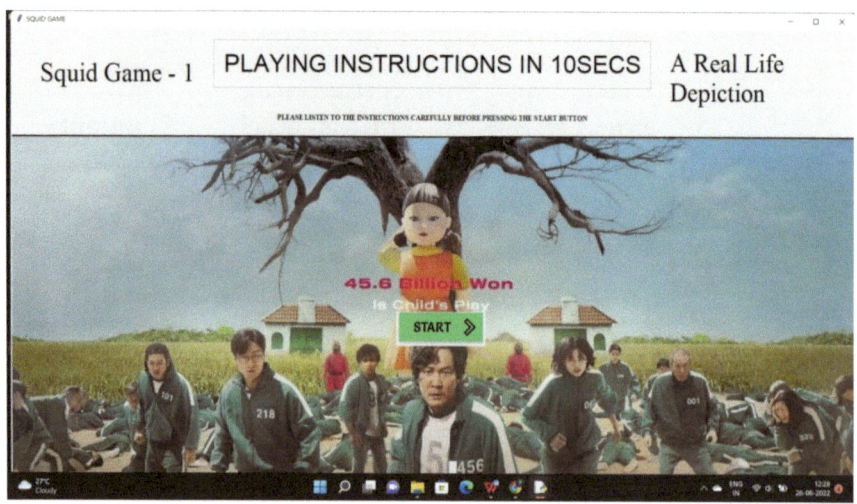

Fig. 10 Front-end using Tkinter

Fig. 11 Out of frame (complete landmarks not visible)

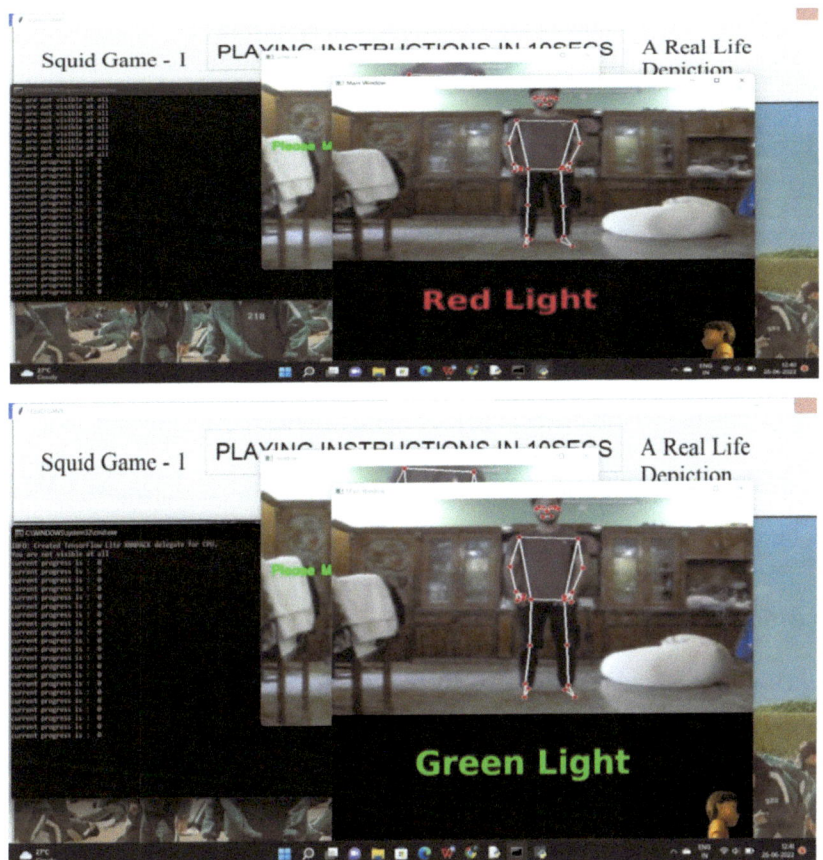

Fig. 12 Red light and green light (music sync)

Fig. 13 Eliminated with gunfire sounds

8 Conclusion

When compared to the conventional immersive gaming technologies and the hardware and processing capabilities needed for its effective functioning, we have proposed implementation of artificial intelligence and machine learning into the immersive gaming technology, which was possible due to the previous success observed in implementation of MediaPipe in exercise posture as well as yoga pose detection which facilitated its application in the field of immersive gaming. With the help of MediaPipe and AI and ML implementation using TensorFlow, the capabilities of the system needed for an immersive gaming experience are considerably reduced.

References

1. Ramana V, Prasanna L (2021) Human activity recognition using opencv. Int J Creative Res Thoughts (IJCRT)
2. Khattar L, Aggarwal G (2021) Analysis of human activity recognition using deep learning. In: 2021 11th international conference on colud computing, data science & engineering
3. Cheng L, Guan Y (2017) Recognition of human activities using machine learning methods with wearable sensors. In: IEEE Members Research and Development Departmenting
4. He R, Sun Z (2019) Adversarial cross spectral face completion for NIR-VIS face recognition. In: IEEE paper received on January 2019
5. Wang J, Cheng J (2016) Walk and learn: facial attribute representation learning. In: 2016 IEEE conference on computer vision and pattern recognition
6. Lin Y, Xie H (2020) Face gender recognition based on face recognition feature vectors. In: International conference on information systems and computer aided education (ICISCAE)

7. Babiker M, Zaharadeen M (2017) Automated daily human activity recognition for video surveillance using neural network. In: International conference on smart instrumentation, measurement and applications (ICSIMA) 28–30 November 2017
8. Ghosh NS, Ghosh A (2020) Detection of human activity by widget. In: 2020 8th international conference on reliability, infocom technologies and optimization (ICRITO) June 4–5, 2020
9. Yang AN, Kan W (2016) Segmentation and recognition of basic and transitional activities for continuous physical human activity. In: IEEE paper on 2016
10. https://docs.python.org/3/library/tkinter.html
11. https://www.assemblyai.com/blog/mediapipe-for-dummies/
12. https://google.github.io/mediapipe/